SOIL MICROBIOLOGY

Robert L. Tate III

Rutgers University
New Brunswick, New Jersey

JOHN WILEY & SONS, INC.

New York · Chichester · Brisbane · Toronto · Singapore

Library of Congress Cataloging-in-Publication Data

Tate, Robert L., 1944–
    Soil Microbiology / Robert L. Tate.
       p.   cm.
    Includes bibliographical references.
    ISBN 0-471-57868-1 (acid-free paper)
    1. Soil microbiology.   I. Title.
QR111.T28   1995
576'.190948—dc20   94-22272

Printed in the United States of America

10 9 8 7 6 5 4 3

To
Ann, Robert, and Geoffrey

PREFACE

In the past, soil microbiology could have been described as a second tier science resulting from the merger of various aspects of soil science and microbiology—two disciplines of somewhat recent birth compared to the longer-lived basic sciences of chemistry, physics, mathematics, and biology. Now soil microbiology has reached sufficient maturity to provide an underpinning of newly emerging disciplines such as the environmental sciences.

The basic information that soil microbiology provides is exemplified by the necessities that arise from managing the long under-valued soil ecosystem. A myriad of environmental complications have been produced that can be solved only by applying the primary principles of soil microbiology. Answering complex questions emerging from the reclamation of polluted or contaminated soil as well as the requirement to manage properly less impacted, perhaps pristine, terrestrial ecosystems requires an appreciation of the delicate balance between the soil microbial community and its environment. This habitat of the soil microbe not only includes the minerals and organic matter of the soil matrix but also encompasses levels of life occurring therein or indeed thereupon.

The growing necessity to understand the diversity of life contained within a gram of soil as well as the intricacies of linking these living entities to terrestrial life in general provided the impetus for writing this treatise on soil microbiology. It is hoped that this book will provide a clear understanding of the principles upon which the discipline of soil microbiology is built as well as illuminate their use in current soil-related environmental problems. A concerted effort was made in assembling the material for this text to avoid the myopic impression that soil microbiology and related disciplines have developed strictly to meet the needs of agricultural scientists. Indeed, the roots of soil microbiology have long been fed by the requirements to provide food and fiber to a growing society, but the principles of the science are essential for answering soil-related questions throughout our society. Hence, a conscious attempt has been made to apply soil microbiology to the grander array of environmental science problems facing our society.

Another basic assumption in writing *Soil Microbiology* is that the science of soil microbiology has grown to the point that it now is self-sustaining. In other words, students of the science should be sufficiently versed in the basic sciences that their repetition within these pages would be redundant. Thus, introductory chapters describing the basic tenants of soil science and microbiology are as inappropriate in a soil microbiology text as introductions to biology, chemistry, and physics would be. Students are expected to have a fundamental understanding of these basic course materials before beginning to study the soil microbe. Discussions of elementary chemical and biological principles in this text are presented only in the context of their applicability to the soil.

A book of this type cannot be prepared without the input of many friends and associates. I especially thank those unnamed colleges and students who in their discussions with me have provided the seeds of inspiration that have matured into the concepts of our study. The capable assistance of Holly S. Kennington in drafting the figures of this treatise is gratefully acknowledged. Lastly, the inputs of the many soil microbiologists who have conducted the experiments upon which the basic tenants of soil microbiology have been constructed are gratefully acknowledged. It is my hope that including the products of their research in this treatise will provide the basis for extending the introductory information contained herein far beyond our current images of our science.

A portion of the time devoted to this project was supported by the New Jersey Agricultural Experiment Station (Publication No. A-07128-1-94).

ROBERT L. TATE III

New Brunswick, New Jersey

CONTENTS

Introduction

Members of the soil microbial community have been appreciated as interesting life-forms and have been exploited as producers of societally important substances, such as a variety of antibiotics. This knowledge alone has been sufficient to produce enough books to occupy a multitude of library shelves. But, such information alone is inadequate to provide the solutions to the microbially associated environmental problems facing society today. Environmental scientists, regulators, as well as the lay public are faced with decisions regarding the proper stewardship of our soil resources and the management or reclamation of damaged systems. We may obtain solutions to these, at times, life-style-altering problems if we think of soil as a closed system where only external manifestations of the processes occurring therein are worthy of consideration; but, such a viewpoint frequently only provides short-sighted, temporary solutions to long-term problems. Indeed, not only is an appreciation of the biological processes occurring in soil essential to achieve societal goals of caring for terrestrial systems but true wisdom in decision making is also predicated upon an understanding of how the individual processes combine to produce a vital, sustainable whole. The parts assembled into a viable soil system clearly produce a whole much more dynamic and much greater than can be predicted by their simple summation.

This thesis is underscored by the magnitude and multitude of anthropogenically impacted soil sites currently demanding some form of reclamation management. For example, refuse from energy fuel or mineral recovery forms unsightly slag piles extending across the countryside. From within these piles, a yellow leachate with a pH of nearly 1 frequently flows. Nearly sterile soil and waters can result from encounters with this deadly by-product of resource recovery. The questions become, How can the production of the substance commonly known as acid mine drainage best be prevented? How can the impacted soils and waters be restored to functional, aesthetic ecosystems? and How can the remains of the mining industry be managed to prevent further environmental degradation?

Perhaps, less dramatic but of similar concern is the problem of evaluating potential difficulties of genetically engineered microorganisms (gems). Common concerns involve unanticipated ecosystem degradation from the introduction of laboratory-created, alien organisms into established soil communities. Will the introduced microbe survive sufficiently long to achieve the objective of its use? Will the unique gene carried by this gem be transferred to indigenous organisms, thereby creating an individual with capacity to wreak havoc to an otherwise stable soil system? These questions represent concerns to which soil microbiologists must respond. Resolution of conflicts in response to these questions requires a clear understanding of the behavior of alien microbes within a functioning soil ecosystem as well as of the dynamics of gene transfer within soil populations.

With these types of concerns in mind, this treatise is presented. The overall objective of this study is to provide the student with a strong basic knowledge of the biological, physical, and chemical properties of soil and the microbes contained therein and of the impact of the combination of these aspects of the soil system on overall ecosystem development and function.

SOIL MICROBIOLOGY

1 The Soil Ecosystem: Physical and Chemical Boundaries

Soil Microbiology Defined: The world of the soil microbe presents a formidable challenge to those seeking to understand and/or manage the processes occurring therein. The soil ecosystem is the product of intricate interactions between a physical and chemical matrix of highly variable composition and living communities composed of essentially all life-forms. Soil biological communities are sustained across the entire range of chemical and physical conditions within which life can exist and function.

Due in large part to the complexity of the soil community, soil microbiological experiments historically have involved the study of individual microbial species purified and cultivated under controlled conditions in laboratory media. This simplification has allowed for a reduction in the complexity of the system of study and of data interpretation. Recently, use of consortia of bacteria or fungi for the study of more complex processes, such as decomposition of xenobiotic compounds, has become reasonably common. These latter cultures exhibit some of the more complex interactions that reflect the soil ecosystem. Use of defined cultures or even simple mixtures of microbes is generally justified by the fact that it is necessary to attain maximum control of experimental variables in order to explain the processes of interest.

As we develop an appreciation for the nature of soils, the inadequacy of experiments with axenic cultures (purified cultures of individual microbial species) to define the world of the soil microbe becomes obvious. Generally, the basic properties of the biological processes measured in laboratory culture defy extrapolation to the more complex soil situation. Not only are stresses of the native soil system unmatched with defined culture conditions, but the microbes themselves also change phenotypically and even genotypically in response to the laboratory growth conditions. Although the microbes studied may have been isolated from soil, a microbial variant with appropriate properties for optimal growth in test tubes or industrial reaction vessels is usually selected either spontaneously (generally, inadvertently as a result of laboratory culture methods) or through genetic manipulation of the microbial isolates. Data collected from experiments with isolated microbial species or strains frequently only explain in part or mimic the processes occurring in soil. Thus, soil microbiologists must expand the purview of experimental design to include the complexities—controllable and otherwise—of the soil ecosystem. The realm of soil microbiology must be defined to include both an understanding of the properties of the microbes themselves and an appreciation of the impact of variability of the soil environment on these traits. Soil microbial ecology must involve an evaluation of the behavior of organisms in their native habitats.

The Necessity of Soil Microbiological Studies: To the purist, elucidation of the principles of soil microbiology is immensely interesting and fully justified strictly because the information generated relates to the basic ecological interactions of soil systems. Yet, due to societal management of terrestrial ecosystems, our concern for soil biological processes reaches far beyond the realm of basic science. For example, our myopic exploitation of natural ecosystems has resulted in situations where the function of resident microbial populations is precluded or severely limited by mismanagement. Spills of toxic organic chemicals result in scarred landscapes. Similarly, metal processing has resulted in sites resembling moonscapes (Fig. 1.1a). Reclamation of such sites has been attempted (e.g., Fig. 1.1b), but for attainment of

(a)

(b)

Fig. 1.1. Metal-impacted site in Palmerton, PA (USA). (a) Site without reclamation management. (b) Site following amendment of soil surface with a mixture of sludge, fly ash, lime, and grass seed.

such objectives, management must develop plans which allow for establishment of essential soil biological processes.

In comparison to these obviously detrimental situations, development of soils for agricultural production, a practice historically considered to be acceptable and beneficial to society, may also reduce overall soil quality. Tillage initiates processes that may eventually reduce soil productivity. Loss of desirable soil structure and reduction of soil organic matter reserves are generally associated with implementation of intensive agricultural practices (see Tate, 1987).

A dramatic example of augmented soil organic matter decomposition due to cultivation of a previously undisturbed soil system is provided by observation of changes in soil properties resulting from cropping of peats. Conversion of the generally anoxic swampy soil to a soil where molecular oxygen is no longer limiting results in oxidation of the peat organic matter. This microbially catalyzed oxidation of the plant debris accumulated under the swampy conditions prior to site drainage results in loss of soil mass, which is manifested by subsidence of the soil surface (Fig. 1.2) (Tate, 1980).

Maintenance and restoration of basic soil quality must be a primary societal objective. This end cannot be achieved without a full appreciation of soil microbiological processes for the total existence of a soil ecosystem relies upon sustenance of a functional soil microbial community. To develop appropriate management plans for restoration of damaged systems or for maintenance of ecosystem productivity, a clear understanding of *in situ* biological processes is necessary. This understanding cannot be fully accomplished by using data generated from the observation of bacterial populations growing at an unnatural rate under the optimized conditions of the laboratory culture flask. Realistic estimates of metabolic rates in the microbe's native environment must be developed. Procedures necessary for maintaining the soil chemical and physical factors controlling expression of microbial activity in a balanced state are needed in order to ensure the long-term survival

Fig. 1.2. Subsidence pole located in the Everglades Agricultural Area of South Florida (USA) showing loss of soil elevation between 1924 and 1979.

of critical populations for ecosystem remediation and sustenance. Such factors as temperature, pH, moisture, surface area, osmotic strength, and presence of predators all interact to select for the development and survival of a stable microbial community necessary for long-term ecosystem function.

The Challenge: Once we understand that reality for the microbe exists in its natural habitat, we encounter the challenge of defining that habitat. For the soil microbiologist, this is not an easy task. Soil properties are highly variable. For example, microbes flourish in temperature-limited, native systems, such as hot spring-impacted soils or the cold desert soils of Antarctica. Highly productive soils may become biologically impoverished due to the impact of anthropogenically produced situations, such as those resulting from acid mine drainage or spills of xenobiotic organic compounds. Thus, it becomes abundantly clear that the soil ecosystem can be spoken of in general terms, but a true encounter with soil ecology must include an assessment of the length and breadth of these highly variable systems. Soil microbiologists must realize that they are studying specific soil types, not simply "soil." The too commonly heard statement, "Soil was used in our experiments," sounds as strange to a soil scientist as the statement, "An animal was studied," sounds to a biologist. The properties of the soil type underlying an ecosystem defined by the aboveground communities must be assessed and understood.

To meet this need, the objectives of this chapter are a) to define the properties delimiting the basic properties of soils and soil ecosystems and b) to examine the impact of the variability of chemical and physical properties of this system on microbial growth and development. Specific introduction to the biological aspects of soil ecosystems is presented in chapter 2.

1.1. THE SOIL ECOSYSTEM

The properties of a soil ecosystem are not only a product of the nature of its physical, biological, and chemical components, but they are also modified by the interactions of these entities. This ecosystem can truly be said to be greater than the sum of its parts. Soil physical constituents are not immutable, but rather they are modified by chemical and biological products of their associated life-forms. Even in soil ecosystems where the needs of the microbial community are only minimally met, colonies of living cells develop slowly to the limit allowed by the physical and chemical parameters stressing the organisms. As is explained in chapter 2, the soil biological system modifies its environment to conditions more conducive for growth and community development. Thus, a soil ecosystem could be said to be evolving toward a steady-state condition dictated by the interactions of each of the mutable components.

Hence, any analysis of the structure of soil systems and the principles describing their function must commence with an evaluation of a) the soil physical components, b) their assembly into the physical whole, and c) the implications of this structure on soil biological processes. The study must then proceed to a determination of the feedback mechanisms involved in the interaction of living populations with their physical environment. This analysis is necessary to develop fully the principles or maxims describing microbial interactions in soil and their importance to that portion of the system most commonly regarded aesthetically, the above-ground plant and animal community.

1.1.1. Soil Defined

A study of soil must be based on an understanding of the basic definition of soil combined with a conceptualization of the properties of soils that make each soil type unique. Simply stated, **soil is an unconsolidated mineral or organic substance on the earth's surface that provides a natural medium for the growth of land plants.** This general picture of material covering terrestrial land masses suitable for plant growth must now be enlarged into a picture of the world in which soil microorganisms live and function.

Definition of Soil Types: Most soils are mineral soils; that is, they are soils that are composed primarily of mineral matter and whose physical properties are controlled by the proportions of the mineral particulates contained therein. These soils generally contain from less than 1 to approximately 4 percent organic matter, but they may consist of as much as 20 percent colloidal soil organic matter. Additionally, an organic layer up to 30 cm deep may be found on the soil surface (e.g., a forest litter layer). Contrast these soils with the much less commonly encountered organic soils (histosols). Organic soils contain more than 20 percent organic matter and are exemplified by peats and mucks.

The most obvious components of mineral soils are sand, silt, or clay, which are defined by their size distribution (Table 1.1). A soil is classified into a specific textural category by the quantities of mineral separates contained (Fig. 1.3). For example, a silt loam is a soil containing 50 percent or more silt and 12 to 27 percent clay, or 50 to 80 percent silt and less than 12 percent clay.

This disposition of soil by textural classification is useful for characterizing a particular soil of interest (either for field or laboratory experiments), but it is in reality the soil structure (i.e., how these individual components are assembled), that best defines the habitat of soil microorganisms. Two soils may have nearly identical textural compositions yet exhibit extremely contrasting microbial activities. For

TABLE 1.1. Size Distrubition of Soil Separates.

Particle Class	Subclass	Mean Diameter (mm)
Sand	Very course sand	2.00–1.0
	Course sand	1.0–0.5
	Medium sand	0.5–0.25
	Fine sand	0.25–0.1
	Very fine sand	0.10–0.05
Silt		0.05–0.002
Clay		<0.002

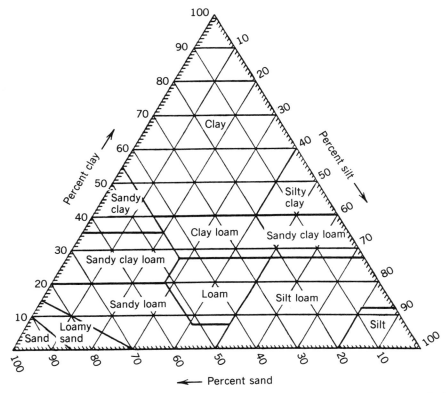

Fig. 1.3. Designation of soil textural classes by principal mineral component analysis (USDA Guidelines).

example, microbial properties of a silt loam existent in an area flooded with acid mine drainage waters reflects the prevailing stresses of oxygen deprivation and extremes of acid, whereas a silt loam from a well-maintained agricultural site is free of such limitations. Clearly, a soil classification system that describes *in situ* ecological influences on biological properties is required. This information is provided by the USDA soil classification system. With this taxonomic system, soils are grouped by field and laboratory properties, among which are soil texture, moisture, pH, temperature regimes, and horizon development. Soils are grouped into orders (e.g., alfisols, inceptisols, oxisols, histosols), which are subdivided into several levels of subgroups that include similarly developed soil types.

Soil Profile Development: An appreciation of soil horizon development and the impact on soil microbial activity is necessary to develop a complete picture of the impact of soil biological processes on the properties of an ecosystem. **A soil horizon is a layer of soil approximately parallel to the land surface and differing from adjacent layers physically, chemically, and biologically or in characteristics such as color, structure, texture, consistency, biotic populations, and pH.** Soil horizonal development is clearly a product of current ecosystem properties as well as of site history. Each soil subgroup description includes an analysis of horizontal structure. Microbiological properties of a specific horizon may be studied in isolation from other portions of the soil profile. In such studies, the designated portion of the soil profile may be defined as an ecosystem in itself.

Most commonly studied soil horizons in microbiological research are the O, A, and B horizons, although considerable research effort is being expended to understand biological processes occurring in deeper-lying aquifer materials as a result of complications of groundwater pollution. **O horizons are dominated by organic material** and are exemplified by forest litter layers. The **A horizons are mineral layers formed on the soil surface or below the O horizon. These regions are characterized by accumulations of organic materials intimately associated with soil mineral matter.** That is, A horizons are usually spoken of as surface soils. The colloidal organic matter concentrations tend to be maximized in this portion of the mineral soil profile. **B horizons are usually formed below an A or O horizon and (a) are dominated by carbonates, gypsum, or silica, alone or in combination; (b) show evidence of carbonate removal; (c) exhibit sesquioxide concentrations; (d) display alterations that form silicate clay; (e) manifest formation of granular, blocky, or prismatic structure; or (f) combinations of these.** In soils without organic surface accumulations, microbial activity tends to be maximized in the A horizon and declines precipitously in the B horizon.

1.1.2. Designation of Soil Ecosystems

Soil microbiologists frequently report results of study of a limited number of soil types and provide minimal descriptions of the soils used. It is not unusual to see statements such as "a garden soil was used" in the literature. From the foregoing discussion, it is apparent that not only the physical and chemical properties of the soil are important for interpretation of the data but also the properties of the site from which the soil was collected and its history must be recorded. Statements simply indicating the ecosystem type from which a particular soil sample was collected are inadequate to allow proper interpretation of the data, extrapolation of the principles revealed by the research, and, frequently, repetition of the study. Although it may seem adequate to report that a grassland soil was evaluated in a particular experiment, application of proper soil classification procedures may reveal that a variety of soils occur within the region classified by an ecologist as a grassland. In other words, to conduct a meaningful examination of a soil ecosystem, the area to be sampled must be appropriately described.

Surface Community Impacts on Subsurface Microbial Activities: The fact that soils are continually changing further complicates the description of soil systems. To the novice, it may seem that soil types and their properties represent a constant in their conceptualization of an ecosystem, but, in reality, the soil lifeforms, climate, and mineral matter are continually altering those properties of the soil matrix that are used to define the soil type Soil pH, salinity, aggregate structure, and cation exchange capacity are but a few examples of soil properties constantly being altered by the actions of the biological community (plant, animal, and microbial). For example, aboveground differences in plant community type and density can cause differences in belowground microbial and enzymological activities that may range over several orders of magnitude. This effect of plant biomass productivity on soil enzymes was shown by analyzing enzyme activity in a reasonably uniform organic soil (i.e., a single soil type) in the Everglades Agricultural Area (South Florida, USA). It was shown that acid phosphatase, invertase, xylanase, cellulase, and amylase activities varied as much as 50-fold between noncultivated and soils cultivate to sugarcane, St. Augustinegrass (*Stenatophrum secundatum* [Waltz] Ktze.), or paragrass (*Brachiaria mutica* [Forsk.] Stapf) (Duxbury and Tate, 1981). A limited understanding of the complexities of these organic-soil-based systems would result if each soil sample (uncultivated or from grass or sugarcane fields) were simply classified as agricultural soils. Yet, inputs from the aboveground plant community necessitates dividing the study area into at least three ecosystem types. Note that in the study cited, all aboveground biomass could have been grouped into a category of grasslands (i.e., both sugarcane and the St. Augustinegrass are by definition grasses). Similarly, as described in section 1.2.2, the interaction of plants and soil microbes with soil particulates is essential for formation of soil aggregates.

Although plant communities are instrumental in soil development, it must be stressed that soil properties can be extremely different in seemingly related ecosystem types. This situation is best exemplified by comparing two large forest groupings, temperate vs. tropical forest soils. Temperate forest soils may be moderately acidic (pH 3.5–4.5) and possess a well-defined horizonal development (Fig. 1.4). They may also have a well-developed surface litter layer plus a subsurface horizon (spodic horizon) with an accumulation of organic matter. Clearing and cultivating such soils in the temperature regions has resulted in moderately productive agricultural systems. In contrast, tropical rain forest soils, which generally lack a surface organic horizon, are usually highly acidic, nutrient poor soils and are poor candidates for agricultural development.

Thus, our operational definition of a soil must include the concept that soils are composed of unconsolidated mineral or organic matter as stated previously, but it must also encompass analysis of the ecosystem properties impacting soil development. The microbiologist must realize that soil is a continuously developing entity with properties that reflect the site in which it has evolved and is still evolving.

State Factor Theory of Soil Development: Interactions of total ecosystem properties on soil development were described succinctly by Jenny (1941) in his State Factory Theory. Factors that moderate soil development include regional climate, biotic factors (properties of the heredity of the organisms), the potential flora on the site, topography, parent material, and time. The theory was recently modified to include anthropogenic influences (Amundson and Jenny, 1991). Therefore, when examining a specific ecosystem, it must be appreciated that the properties of the soil, which control the biological activity and in reality make the soil unique, are a product of a variety of variables interacting over a time frame of

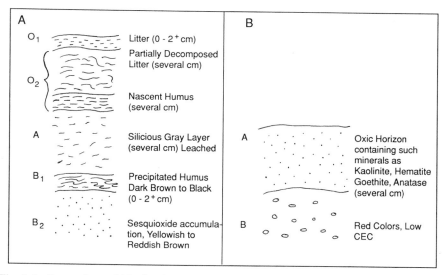

Fig. 1.4. Comparison of idealized soil profile models for temperate and tropical forest soils. A. Spodosol surface horizons with accumulated organic matter, and leached A horizon. Spodosols are exemplified by acidic forest soils. B. Latosilization as would be seen in an oxidol (tropical forest ecosystem). These soils are generally highly leached, acidic soils with no significant humus layer. See Fanning and Fanning (1989) for further discussion of these soil types.

centuries. Thus, the current ecosystem type as well as those preceding it have combined to produce the soil of interest.

Optimization of Study Site Size: Application of this basic understanding of the variability of a soil ecosystem to field studies is solely limited by the viewpoint of the definer. For example, a site for the study of microbial processes could be designated to encompass only a microscopic portion of the soil profile (e.g., the mineral and organic components surrounding a few microbial colonies, i.e., a microsite). Because such colonies frequently consist of a few cells at most, this ecosystem would have boundaries of several microns. Whereas this microsite represents the entire world to the growing microbe—and an appreciation of the situation around the microbe is important for understanding total ecosystem function, practicality dictates that scientists designate their ecosystem of interest on a larger basis.

For soil microbiologists, soils ecosystems are frequently defined by aboveground system properties such as climate (desert vs. swamp), topography (aspect along a hill), plant community type (forest vs. grassland), or even proximity to the growing plant (within row vs. between rows sampling of agricultural systems). In each of these situations, soil microbiologists attempt to define a functional unit of a larger ecosystem (in the extreme, total land mass) that is small enough to allow representative soil sampling and study.

The depth within a soil profile from which samples are collected also affects the types of results gained from study of the soil. Soil scientists are likely to study the A (or surface) horizon of the soil profile since microbial activity is generally maximal therein; but it must be remembered that distinct properties and microbial communities have also developed within all soil horizons. Therefore, for a scientific analysis, a subsurface portion of the soil could be experimentally isolated for study and defined to represent a specific ecosystem. For example, it could be desirable to sample microbial activities within the vadose zone (that portion of the soil profile below the rooting zone, but above the water saturated zone) or the activity of a specific aquifer region deep within the soil profile.

1.1.3. Implications of Definition of the Soil Ecosystem

The intellectual product of delineating a portion of a larger aboveground ecosystem type for soil microbiological analysis (i.e., dividing a whole into a series of subsystems) is seen in the concept of experimental design as well as data interpretation and extrapolation. Indeed, simply deciding to study soil microbial processes clearly separates our terrestrial system into aboveground and belowground processes, but it does not isolate them conceptually because belowground communities do not function independently of aboveground inputs. Carbon fixed in aboveground biomass and nutrients carried in rain or irrigation water necessarily are considered external inputs to our conceptual model of the belowground ecosystem. These inputs must be appreciated as the kinetics of microbial processes are considered therein.

Two complications of selecting operationally defined soil ecosystem types for study are site fragmentation and limitations in capacity to extrapolate the principles derived from the study to new situations. The greater the fragmentation of the overall ecosystem type, the higher the potential for clarifying the details of the microbial processes occurring therein. Unfortunately, the smaller the portion of the overall system studied, the less the potential for valid extrapolation of the data to provide a description of the total ecosystem. Thus, the soil microbiologist faces the dilemma of selecting a soil site small enough that it can be sampled in a representative manner, but sufficiently expansive that the data collected will be at least to some degree representative of the total ecosystem. (See Tate, 1987, for a more detailed discussion of soil-sampling procedural problems associated with the study of larger ecosystems.)

The more limited the dimensions used to define an ecosystem of interest, the more important the considerations of the impact of external processes on the integrity of the system become. For example, shallow aquifer sediments may be examined, but we must remember that energy derived from the oxidation of carbonaceous compounds is dependent upon the rate that the fixed carbon is leached through the soil profile. By drawing boundaries that define the portion of the larger ecosystem to be studied, clearly some aspects of the ecosystem will be self-contained whereas others must be supplied from external sources. The extreme example of this situation is the microsystem introduced previously (i.e., microbial colonies existent upon a soil granule). Assuming that this colony is formed by heterotrophs, the most commonly occurring bacterial grouping within soil, the carbon oxidized for energy must originate from beyond the world of the bacteria themselves. Substrates for carbon-based energy metabolism are ultimately produced photosynthetically by aboveground plant communities. From a short-term view, both in time and space, an endogenous supply of carbonaceous substrate could be considered to exist in the form of native soil organic matter, but this soil organic carbon resource originated as plant biomass and once exhausted it can be replenished only by movement of plant photosynthate into the soil microsite.

1.2. THE MICRO ECOSYSTEM

Existence of aboveground communities literally rests on the stability and function of soil processes occurring at the microscopic level. Due to the fact that microorganisms range in size from less than a micron to a few microns, individual soil particles have a major impact on their function. The challenging aspect of this observation is that the biological processes occurring at the micron level have profound implications on total ecosystem development and longevity. For example, a series of bacterial colonies consisting of as few as a half dozen cells each may determine the fate of a potentially ecosystem-destructive xenobiotic compound. Hence, a clear understanding of the interactions of the soil microbial community and elemental soil components (sand, silt, clay, and colloidal organic matter) is imperative to comprehend total ecosystem function.

1.2.1. Interaction of Individual Soil Components with the Biotic System

The basic components of soil are sand, silt, clay, and humic substances. Of these, the impact of clay and humic substances on soil biological processes has been most studied. Although all soil particulates provide surfaces for microbial colonial development, clay and colloidal organic matter have the smallest diameters and therefore present the largest surface area for interaction with soil microbes and their products. The relationship between estimated particle diameter and surface area is demonstrated in Fig. 1.5. The interactions of clay and humic substances with microbes, enzymes, and nutrients are further accentuated by their ionic associations. The relative contributions of humic substances and clays to microbial processes are detailed later in this chapter.

1.2.1.1. Clay and Ecosystem Function. **Adsorption of Organic Substances to Clay Particles:** Clay, a (positively or negatively) charged soil mineral of less than 0.002 mm in equivalent diameter, provides a surface for adsorption of a variety of individual organic molecules. Not only do whole microbial cells and viruses sorb to soil clay particles, but a diversity of organic compounds also have been demonstrated to interact with clay surfaces. Examples of the types of organic compounds reacting ionically with clay particles includes carbohydrate phosphates (Goring and Bartholomew, 1951), nucleotides (Goring and Bartholomew, 1952), humic and fulvic acids (Greenland, 1971; Kodema and Schnitzer, 1974), and aromatic compounds

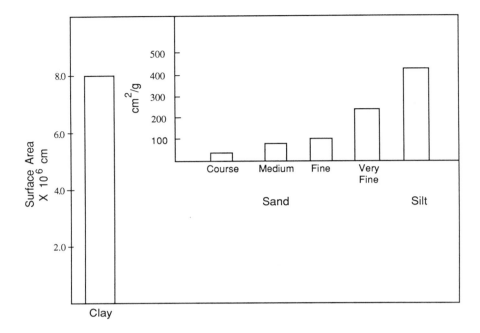

Soil Particle Class

Fig. 1.5. Variation of surface area of soil particles. The particles are assumed to have spherical shapes. Data are selected from Froth and Turk, 1972.

as exemplified by *p*-cresol (Boyd and King, 1984). Most biologically produced compounds interact with clay particles. The environmental effects of this adsorption are variable. Detrimental impacts may result from the potential for increased longevity of clay-sorbed organic molecules. This is particularly significant for materials with negative environmental impacts. Gains from the interaction of carbonaceous substances with clay minerals accrue from those situations where concentration of soil pollutants in or on soil clays increases mineralization rates. Thus, the adsorption potential of any organic compound entering soil should be evaluated in relation to estimation of the effect of this process on its longevity *in situ*.

It must be remembered that both substances entering the soil system from external sources (e.g., pesticides and other soil amendments) as well as those synthesized *in situ* may be sorbed onto clay surfaces. Each of the examples of clay sorption phenomena listed previously were conducted by amending soil or clay fractions with the test material and measuring binding to clay particles. In contrast, Sorensen (1972) found that amino acid metabolites synthesized in montmorillonite-amended soil following cellulose, hemicellulose, or glucose amendment were also bound by soil clays. Clay-amended soil retained two to three times more carbohydrate-derived amino acid carbon than did nonamended soil over a 3-year incubation period. In a related study, ^{14}C-labeled cellulose was added to seven soils with silt plus clay contents ranging from 8 to 75 percent (Sorensen, 1975). As in the previously cited study, the quantities of amino acid carbon synthesized from the cellulose remaining in the soil after 30 days incubation ranged from 6 percent in the lowest silt plus clay-containing soil to 18 percent in the highest silt plus clay-containing soil.

Association of organic compounds with clay particles is not a totally random process. There is an enrichment of specific carbon moieties within clay particles, which is exemplified by evaluating the properties of humic acids associated with clay particles. In native soil samples, aliphatic structures are more common in humic acids derived from the clay fraction than in other soil fractions (Arshad and Lowe, 1966; Schnitzer et al., 1988; Shulten and Schnitzer, 1990). Using chemical, infrared, ^{13}C nuclear magnetic resonance, and gas chromatographic mass spectometric analysis, Schnitzer et al. (1988) found that about half of the clay-associated organic matter consisted of humic materials; the remainder was mainly long-chain aliphatics.

Clay–Organic Matter Adsorption Mechanisms: This list of materials sorbed to clay particles illustrates that both the most commonly occurring clays in temperate region soils and many of the sorbed organic compounds are negatively charged at the predominant soil pH values. Thus, a mechanism must exist to reduce the repulsive forces between the like charged entities. Examples of such mechanisms are reduction of the quantity of negative charges (pH modification or methylation of the carboxyl groups on the organic matter) and cationic bridging of negatively charged functional groups of the humic substances with the clays. The latter process is depicted by a model proposed by Greenland (1971) (Fig. 1.6). The negative charges on the organic and mineral materials are linked in soils by divalent cations.

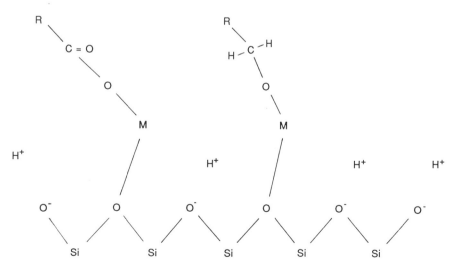

Fig. 1.6. Schematic drawing of divalent metal [M] salt bridge linking the surface of a mica-type clay with a negatively charged organic compound and the distribution of a hydrogen ion cloud around the clay surface.

Organic compounds retained on clay particles through cationic bridges can clearly desorb from the surface with changes in ionic strength and cation content of the soil solution. A more permanent association of organic compounds to clay particles can occur. This association between clay particles and organic materials results from interlayer clay–organic matter complex formation and was demonstrated to occur in two New Zealand soils by Theng et al. (1986). Humic substances with a polymethylene chain structure were found to be regularly interstratified with mica–smectite in highly acidic soils containing low microbial activity.

Any process that alters the negative charges of either the clay or organic matter reduces the repulsive forces. This reduction can be accomplished by altering soil pH or through methylation of the carboxyl groups. Nayak et al. (1990) demonstrated that reduction of the acidic groups of humic substances through methylation increased fixation with montmorillonite. The increased association with the clay particles resulted not only from a reduction of interparticle repulsive forces but also from a reduction in hydration energies.

Biodegradation and Clay Interactions: Association of clay and soil organic matter affects soil microstructure (see section 1.2.2.1 for elucidation of soil structural interactions) as well as the decomposition rate of sorbed organic compounds. Studies of the impact of clay on decomposition of a wide variety of organic compounds has shown that purified clay minerals when added to growing microbial cultures can stimulate, decrease, or have no effect of decomposition of organic materials (see Kunc and Stotzky, 1974). These workers found no clear relationship between the type of organic substance and the type of clay relating to the effect on biodegradation except with aldehydes whose decomposition was accelerated by clay minerals. Lynch and Cotnoir (1956) noted that bentonite inhibited decomposition of a variety of substrates but illite and kaolinite had little effect on decomposition. Similarly, Olness and Clapp (1972) found that the kinetics of the decomposition of the montmorillonite–dextran complex contained periods that were affected by the montmorillonite content more whereas others portions of the degradation curve were impacted more by the dextran concentration.

Clearly, it is difficult to make general statements regarding the impact of clay sorption on biological decomposition kinetics. Some explanation for the seemingly contradictory observations is derived from study of the impact of clay on substrate availability. Substrates existent in concentrations below that easily metabolized by the microbial community may become sufficiently concentrated on the clay surface to induce microbial catabolism. In contrast, for substrates present in high concentration in soils with high clay contents, their effective level in the growth medium can be reduced by clay particle adsorption to levels that preclude decomposition. In this latter situation, strong binding of the carbonaceous substance to the clay particle reduces its contact with the degrader population. The rate of desorption of the growth substances from the clay thus determines the decomposition rate of the carbonaceous energy source.

Thus, the impact of clay sorption is more a function of residual concentrations of organic compound in interstitial waters and in the microsite containing the microbial colony than is the apparent concentration when the bulk soil sample is analyzed (i.e.,

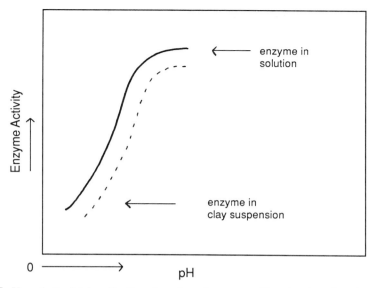

Fig. 1.7. Hypothetical data reflecting alteration of apparent pH optima by microsite variation in pH due to clay micelles.

microsite concentrations of the compound of interest may far exceed macrosite levels). Thus decomposition kinetics of the substance reflect the microsite concentrations rather than those that would be predicted from total ecosystem content of the carbonaceous growth substance.

Alteration of the apparent reaction rate of enzyme catalyzed reactions by soil clays also occurs indirectly. Due to the formation of hydrogen ion cloud around the clay particle, the localized pH is reduced (Fig. 1.6). Thus enzyme-catalyzed reactions in clay soils may appear to have pH optima lower than that detected in the absence of clay particles (Fig. 1.7).

1.2.1.2. Humic Substances and Ecosystem Function. Soil organic matter is commonly considered to be equivalent to soil humus (Stevenson, 1982). The soil humus fraction consists of all organic compounds normally found in soil (with the exception of recognizable plant or animal components), their partial decomposition products, and microbial biomass. Studies of the nature of soil organic matter are complicated by the diversity of materials present and the fact that the definition does not coincide with what can be reasonably separated and analyzed (e.g., clear separation of plant and animal biochemicals from the bulk soil organic matter is not operationally possible). The complexity of soil organic components ranges from simple monomeric components of living cells to the random polymers classified as humic acids. Essentially any organic compound contained in or synthesized by living cells occurs in soil, although some may be present in minuscule amounts due to their being relatively easily decomposed by soil microbes. Microbial biomass function in soils is presented in chapter 2, whereas the role of these materials in nutrient cycling is analyzed in chapter 9.

Soil Humic Substances: Of more interest for the present discussion is that portion of the soil organic matter classified as humic substances. **Humic substances are a series of high molecular-weight, brown to black substances formed by secondary synthesis reactions in soil.** Humic substances are not a functionally described soil component but rather they are differentiated strictly by solubility. Humic substances are separated from soil in alkaline solutions. Acidification of the alkaline extract to a pH between 1 and 2 results in precipitation of the humic acids. Fulvic acids are soluble in both acid and alkali. Humin is that portion of the colloidal soil organic matter fraction that is not solubilized with alkaline solutions. (See Stevenson, 1982, for a description of extraction methods and properties of these soil organic components.)

To understand the impact of humic substances, especially humic acids, on microbial function in the soil ecosystem, an appreciation of the structural components interacting with soil particles and the microbial community is necessary. (For a complete discussion of theories regarding structure and origin of these compounds, see Stevenson, 1982.) Humic and fulvic acids are heterogeneous, random polymers of a variety of aromatic and aliphatic components. No single chemical structure can be proposed to represent these molecules. Indeed, because a wide diversity of aromatic and aliphatic compounds are linked via a variety of covalent linkages that form humic acids with molecular weights in the thousands of daltons, it can reasonably be concluded that few if any of the molecules are identical. In contrast, fulvic acids are more oxidized than humic acids (i.e., they have a higher oxygen content and a lower carbon content and are of lower molecular weight).

Structural Implications of Humic Acids: The nature of the constituents of humic substances impinges upon their importance to total ecosystem function. The heterogeneity and complexity of humic and fulvic acids has limited the detail with which we can understand the specific chemical reactions between humic and nonhumic soil organic components. An appreciation of the types of reactions that occur between microbial cells or individual biochemicals and humic acids has developed from our basic concept of the molecular content of humic substances and the use of related model compounds.

The application of spectroscopic analytical techniques, such as ^{13}C nuclear magnetic resonance (NMR) spectroscopy, has recently expanded our understanding of humic acids structure and their *in situ* interactions in soil. These data suggest that aromatic rings comprise a smaller portion of the humic acid molecule than previously believed (e.g., see Hatcher et al., 1981, and Malcolm, 1980). Hatcher et al. (1981) found that the aromaticities of humic acids from a variety of soils collected from Prince Edward Island ranged from 35 to 92 percent. Significant aliphatic structures were generally noted but were not detected by chemical oxidation procedures. These authors also suggested that phenolic carbons may be minor components of humic acids. They were previously considered to be major substituents. This conclusion was based on the observation that NMR spectra did not contain intense signals corresponding to the presence of these molecules. Schnitzer and

Preston (1986) suggest that, because several independent methods confirm the presence of phenolic hydroxyls, failure to detect them by NMR analysis merely indicates a shift in the NMR signal resulting from these moieties. These studies exemplify the evolving nature of our understanding of humic acids and their structure and the limitations imposed by our inability to develop detailed structural formula for these molecules. (For a more detailed analysis of humic and fulvic acids structure and origin in soil, see Stevenson, 1982.)

The primary conclusion regarding the structure of humic substances and their importance to the soil microbiologist is derived from the complexity of their composition. They contain varying ratios of saccharide, phenolic, methoxyl, aromatic, hydrocarbon, amino acid, and nitrogen moieties. The ratio of these substituents present is dependent on the source of the humic acid preparation. Great variation in composition of humic and fulvic acids is found when preparations of soil, stream, and marine samples are prepared (Malcolm, 1990), but humic acids prepared from a variety of soil types—alfisols and mollisols (Novak and Smeck, 1991)—and soil management histories (Stearman et al., 1989) also show significant variation.

Humic Substances, Organic Compounds, and Biodecomposition: Although a variety of microorganisms have been shown to metabolize humic substances, they are unlikely to be major participants in the energy metabolism of soil microbes. This conclusion results primarily from observations of the complexity of the structure of the molecules—they contain a large variety of substituents that are essentially randomly assembled. (See chapter 4 for further elucidation of microbial energy metabolism in soil.) Humic substances affect soil biological processes to the greatest extent by interacting with soil minerals in soil aggregate formation (see section 1.2.2) and by their role in removing biodegradable substances from the pool of easily metabolized substrates (i.e., humification). Enzymatic interaction with humified substrates is inhibited physically by the imposing humic acid physical structure.

As discussed previously with clays, humic substances sorb organic compounds through temporary associations. For these organic matter interactions, sorption results from π–π interactions, van der Waals forces, hydrogen bonding, as well as ionic associations. More permanent linkage of organic compounds to humic substances also occurs due to the formation of covalent bonds. Examples of organic compounds that may be linked covalently to humic acids include amino acids, peptides, proteins, aromatic compounds, and polysaccharides.

Covalent linkage of phenolic compounds to humic acids may be catalyzed by phenoloxidase (Bollag et al. 1980, Liu et al. 1985). Carbon–carbon and carbon–nitrogen linkages are catalyzed through free radical formation by these laccase-type enzymes (Fig. 1.8). The linkage of aromatic-ring-containing compounds occurs in *ortho–ortho* or *para–para*, or *ortho–para* type associations by this biologically catalyzed processes (Sjoblad and Bollag, 1981). Aside from being a mechanism for the formation of humic substances *in situ*, these oxidative processes provide a means for the stabilization of a variety of xenobiotic compounds, such as pesticides, into soil organic matter.

The impact of association of organic compounds with humic substances on their biodecomposition kinetics is more predictable than was observed with clay–organic

Fig. 1.8. Oxidative coupling of aromatic compounds: Mechanism showing formation of free radicals and examples of some products of *p*-toluidine polymerization. See Sjoblad and Bollag (1981) for further discussion and examples.

matter interactions. Humification of organic compounds results in their stabilization in soil. For example, in a study of the impact on microbial degradation by nucleophilic addition of amino groups to the aromatic nuclei or quinones, Bondietti et al. (1972) found that covalently linking amino sugar units to model humic acids resulted in major reductions in the decomposition rate. Over a 12-week incubation period more than 70 percent of free glucosamine was oxidized to carbon dioxide while only 15 to 23 percent of the humic-acid-model, polymer-bound amino sugar was oxidized. The importance of covalent linkage of the humic acids and sugar was shown by the fact that simply mixing the amino sugar with the model humic acid molecule did not reduce the oxidation rate.

It is important to note that the humic-acid-stabilized organic compounds are still available for biological degradation and that only the rate of oxidation is altered.

Decomposition kinetics of humified organic compounds tend to resemble those of total soil humic acids. This observation has environmental significance. Covalent bonding to humic acids could be a viable mechanism for renovation of xenobiotic-compound-contaminated soils if the substances are permanently bound or if they are released in an inactivated (i.e., partially or totally catabolized) form. Unfortunately, this condition is not always met. Release of humic-acid-bound pesticides intact can occur (Bartha, 1980; Lichtenstein, 1980; Still et al., 1980). Thus, from an agricultural viewpoint, difficulties may arise from release of a pesticide that had been previously applied to an appropriate crop and stabilized in soil humic substances several years after application. Particular concern would be precipitated should this pesticide reappear in a crop where use would be inappropriate either from a crop sensitivity or regulatory view.

It must be stressed that, in most cases, the more probable fate of a biodecomposable substance added to soil is mineralization, either before humification or even subsequent to humification. For example, Haider and Martin (1988) found that *Phanerochaete chrysosporium* could mineralize xenobiotic compounds at a reasonable rate (13 to 56 percent of xenobiotic carbon as carbon dioxide over an 18-day period). Similarly, Saxena and Bartha (1983a) found that humus bound 3,4-dichloroaniline (a biodegradation intermediate) was mineralized at a rate approximately the same as that of the humic acid molecule itself. Adsorbed 3,4-dichloroaniline was mineralized faster than humic acid molecules. That fertilization of agricultural soil with anhydrous ammonia can mobilize humus-bound 3,4-dichloroaniline (Saxena and Bartha, 1983b) is noteworthy.

1.2.2. Soil Aggregate Structure and Biological Systems

Interest in the microstructure of soil is derived from an ecological viewpoint as well as an operational laboratory angle. A complete evaluation of field processes requires an understanding of the physical relationship of biotic and abiotic soil components *in situ* and how this association affects overall ecosystem development. Furthermore, due to the fact that it is difficult to analyze soil microbial processes in the field, techniques are required to fractionate soil into functionally related components in the laboratory so that the nature of the biological processes occurring therein and the factors controlling their rate can be determined.

1.2.2.1. Native Soil Aggregate Structure and Its Impact on Ecosystem Function.
The world of the soil microbe does not consist of individual sand, silt, or clay particles or even of organic matter *in toto*. Microorganisms may appear to be associated with one specific ecosystem component. For example, rhizosphere organisms may be growing essentially on the root surface. Microbes may be physically attached to partially decomposed plant debris. Also, colonies may be found attached to the surface of sand or silt particles. But, even in these situations, the metabolism of soil microbes is affected by associating with the soil particles to which they are attached within the larger groupings called aggregates. The indirect impact of soil structural development on the metabolic rates of individual microbial colonies

results from the fact that the degree of soil aggregate formation controls such biological-process-limiting soil properties as water infiltration and availability, oxygen tension, and nutrient movement.

Soil structure is defined at the level of the microbial cell by the association of soil particulates and colloidal organic matter into soil aggregates. **Soil aggregates are random combinations of soil organic and mineral components assembled into micro-(<50-μm mean diameter) and macroaggregates (>50-μm mean diameter particles).** By their nature, soil aggregates are a product of interactions of the soil microbial community, the soil mineral components, the nature of the aboveground plant community, and ecosystem history.

Soil Aggregate Formation: Assembly of soil particulates into aggregates is an excellent example of the interaction of abiotic and abiontic organic matter, living cells, and mineral components affecting soil physical properties. A conceptual model of this interaction can be described as a coupling of feedback-controlled processes (Fig. 1.9). Biological participants in the interactions are higher plants and

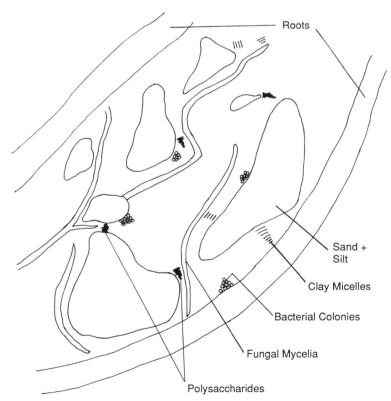

Roots

Sand + Silt

Clay Micelles

Bacterial Colonies

Fungal Mycelia

Polysaccharides

Fig. 1.9. Basic elements of soil substituents contributing to aggregate formation (not drawn to scale).

soil microbes. The physical variable that is altered by the synergistic interactions of the biological entities is the degree of grouping of soil particulates into the at least transiently stable structures called aggregates. Microorganisms metabolize photosynthetically-fixed carbon to produce the compounds responsible for binding soil mineral matter together—polysaccharides and humic substances. Improvement in the size and number of aggregate formed enhances plant community growth, which in turn stimulates even further the microbial development that may result in even greater improvements in soil structure. In contrast, poor soil structure retards air diffusion, water movement, and root growth, thereby resulting in a less active soil microbial community than would be attained under more optimal growth conditions.

Tisdall and Oades (1982) developed a conceptual model that describes the interaction of soil biotic and mineral components to form water-stable aggregates and explains the impact of soil system management and the duration of structural relationships. Organic binding agents were divided into three categories: transient (mainly polysaccharides), temporary (roots and fungal hyphae), and persistent (humic substances associated with polyvalent metal cations). Soil aggregates are formed in a hierarchical arrangement by stabilization of smaller aggregates into larger structures by different binding agents (Waters and Oades, 1991). Microaggregates are stabilized by persistent binding agents whereas polysaccharides and biomass result in the association of the microaggregates into macroaggregates (Elliott, 1986). Microaggregates consist primarily of clay and humified organic materials linked by polyvalent metals (Edwards and Brenner, 1967). Due to the biodegradation resistance of the persistent binding agents, the microaggregate structure is long term and appears to be more of a characteristic of the soil rather than a product of soil management procedures.

This model emphasizes the importance of polysaccharides and humic substances in soil aggregate formation, but it must be understood that any organic compound that stimulates or decreases microbial growth in soil indirectly affects aggregate formation. This results from the production of the polysaccharide binding agents by actively growing soil microbes as well as from the direct association of microaggregates into macroaggregates by fungal mycelia.

Role of Carbohydrates in Soil Aggregate Formation: The importance of polysaccharides as binding agents for adhesion of soil mineral components into aggregates has been demonstrated by a) assessing the correlation of polysaccharide levels with quantities of water-stable aggregates, b) quantifying the impact of periodate oxidation of polysaccharides on aggregation, and c) analysis of the influence of polysaccharide adsorption on the mechanical strength of clay minerals. For example, Chaney and Swift (1984) found that total organic matter, total carbohydrate, and humic material gave highly or very highly significant correlations with aggregate stability in 120 soils. To rule out other soil physical properties as effectors of aggregate formation, 26 agricultural soils were evaluated. Only organic matter and associated properties gave significant correlations with aggregate stability. Sand, silt, clay, and, iron contents, and cation exchange capacity did not correlate with aggregate formation.

Direct evidence for the role of polysaccharides in soil has been gained by assessing the effect of their selective destruction on soil aggregation. Periodate treatment of soil results in oxidation of polysaccharides contained therein. Not all polysaccharides are destroyed by such treatment. The proportion of polysaccharides oxidized depends on the duration of the treatment and the degree of contact of the sugar with the oxidizing agent. The polysaccharide may be occluded by close association within soil particulate structures. Even with these limitations, a major role of polysaccharides in aggregate formation is revealed by this method. Periodate treatment of soil from a long-term grass field destroyed microaggregates greater than 45 μm in diameter along with the carbohydrates (Cheshire et al., 1983). Residual carbohydrate was enriched in sugars commonly found in plant materials. This enrichment suggested preferential destruction of microbial polysaccharide by the periodate. The relationship between soil polysaccharide and aggregate stability was verified in 15 soils and 7 soil series (Cheshire et al., 1984).

Chenu and Guérif (1991) assessed the strength of the association between polysaccharides by assessing the mechanical strength of mixtures of kaolinite or montmorillonite plus or minus a polysaccharide, scleroglucan. The polysaccharide increased the mechanical strength in mixtures of both clays through the formation of polymer bridges between the clay particles and from the progressive coating of the clay surface by the polysaccharide. The primary impact of the polysaccharide on soil aggregation was therefore the result of increased strength of interparticle bonds.

Humic Acids and Aggregate Stability: Microaggregate formation results from interactions between humic substances and clay particles (e.g., Capriel et al., 1990; Dinel et al., 1991; Schnitzer and Kodema, 1992). Aliphatic side chains are particularly instrumental in this process. A high correlation between soil aliphatic fractions (Capriel et al., 1990) and aliphatic-enriched humic fractions (Dinel et al., 1991) with aggregate stability underscores the importance of these fractions in microaggregate formation. As already noted, humic-associated aliphatics can form interlayer associations with clays (Theng et al., 1986), and these materials are enriched in soil clay fractions (Arshad and Lowe, 1966; Schnitzer et al., 1988). The fact that these associations are a property of the soil itself and are not due to management is supported by the observation that, although the quantities of aliphatic fraction extracted with supercritical-hexane varied with management, the chemical properties revealed by ^{13}C nuclear magnetic resonance and Fourier-transform infrared differences were independent of management (Schnitzer et al., 1988).

Soil Management and Aggregate Stability: The stability of microaggregates and macroaggregates reflects the biodegradation susceptibility of the primary organic polymers or biological biomass instrumental in their formation. Microaggregates stabilized through association of humic substances with clay particles are reasonably resistant to management procedures and are more reflective of soil physical properties.

Soil management of ecosystem development typically results in the modification of factors controlling soil microbial activity. For example, the cultivation of soil can cause catabolism of organic matter previously occluded by soil minerals, increases in soil aeration, and modification of soil pH through liming. In contrast to the degradation resistance of humic substances, polysaccharides and microbial biomass are biodegradation-susceptible. Portions of these soil organic fractions could be depleted under appropriate conditions in a matter of days or weeks. Therefore, soil management procedures that result in a depletion of soil organic components reduce the quantity of the water-stable macroaggregates produced by their interaction with soil particulates. The best examples of such declines in soil structural properties are found in studies of the effect of long-term cultivation on soil properties. For example, Tiessen and Stewart (1983) evaluated the effect of cultivation for between 4 and 90 years on soil particle size distribution of three North American grassland soils (Cryoborolls). They observed the continuous depletion of organic matter and the loss of particles >50 μm diameter with time of cultivation, indicating that the more degradable, metabolizable soil organic matter components that support the soil aggregate structure were being depleted with time.

Conversely, soil management procedures chosen to increase microbial activity increase biomass and polysaccharide synthesis and improve soil aggregate structure. This result was reflected in the study of Tisdall and Oades (1982), where the duration and magnitude of the improvement in soil microstructure was shown to be dependent on the nature of the metabolized carbon substrate amended to the soil samples. Maximum aggregation occurred in soil amended with glucose, rapidly metabolized substrate, whereas a more lasting but less dramatic improvement in soil structure was observed with more slowly metabolized cellulose. Although cellulose is much more biodegradation resistant than is glucose (days compared to months for depletion from soil), each of the materials could be considered to be rapidly catabolized when compared with humic substances (degradation times measured in decades or more).

The impact of carbon inputs from crop growth on aggregate formation also can vary with the species grown. For example, Angers (1992) compared the effect of alfalfa (*Medicago sativa* L.) or corn (*Zea mays* L.) growth with the aggregate structure of soils from bare fields and found maximum aggregation under alfalfa with minimal changes in this soil property during the study in the fallow field and in the field cropped to corn. Most likely, the effect on soil aggregation resulted from a combination of variations in the quantity and quality of organic matter available to the soil microbial community plus the resultant alteration of the activity of these populations.

Plant biomass may be incorporated into soils through natural cycling processes, as is observed in grasslands where a distinctive thatch layer can develop (see Tate, 1987) or in agricultural soils when crop residues are returned to the soil as well as when exogenously produced organic materials, such as composts and sludges, are amended to soil. The effect of such an amendment on soil aggregate formation can be substantial. For example, Glauser et al. (1988) found an increase of water-stable aggregates to 85 percent on soils receiving sludge amendments compared to 45 percent in nonamended soils.

1.2.2.2. Separation of Soil Particulates by Density Fractionation. A primary objective of many soil microbiological endeavors is to assess potential for a metabolic process to occur (e.g., nitrogen mineralization) by analyzing specific soil components or reactions occurring in isolated soil fractions. Achieving this objective is most often precluded by the fact that soil fractionation procedures tend to be based on physical or chemical properties, not biologically functional relationships. A procedure that appears to have potential to overcome this limitation is density fractionation. This procedure is still a physical differentiation method—soil is separated by density in sodium iodide solutions (Spycher et al., 1983; Strickland and Sollins, 1987)—but a functional partitioning of soil organic matter is achieved. For example, suspension of whole soil in a high-density sodium iodide solution yields a light and a heavy soil fraction. The light fraction is a reservoir of labile plant carbon and nutrients because it is enriched in partially decomposed plant material including dead root fragments and fungal hyphae. Organic matter contained in the heavy fraction is enriched in humic acids. The combination of water-dispersion and density fractionation demonstrated enrichment of organo-mineral particles in the light clay-size fraction with organic carbon characterized by wide carbon–nitrogen ratios and low amounts of alkali-extractable carbon (Young and Spycher, 1979). Similarly, Sollins et al. (1984) noted enrichment of biodecomposable plant remains and microbial remnants in the light fraction compared with more biodegradation-resistant materials in the heavy fractions. In an examination of cultivation effects of carbohydrate distribution in soils, Dalal and Henry (1988) found that the light fraction was enriched with carbohydrate carbon (27 to 43 percent of the carbohydrate carbon was found in this fraction). Cultivation resulted in a rapid decline in the light fraction and a substantial loss of carbohydrate, especially in clayey soils. These observations are significant ecologically because the primary energy source driving soil microbial processes is labile plant carbon. Thus, assessing reactions that occur in the light soil density fraction and contrasting them to rates found in the heavy fraction may provide a good indication of metabolic potential in soil ecosystems.

1.3. THE MACRO ECOSYSTEM

An appreciation of the variation in soil properties on a macroscale is equally important to understanding the microsite characteristics previously described. Both the vertical and the horizontal variation in soil properties that limit microbial activity must be quantified.

Although a considerable interest is developing in estimating microbial function in deep aquifer materials, most soil microbiology research is directed at developing an understanding of processes that occur at the point of contact between the aboveground and belowground communities. Thus soils from the O, the A, and to some extent the upper portions of the B horizon are most frequently examined. Interactions in the B horizon are especially important in sites where the O horizon is nonexistent and the A horizon is thin or has been lost through erosion.

The importance of these major surface soil horizons is derived from the fact that the primary energy source for the soil microbial community is derived from photosynthetate entering soil primarily as plant biomass or root exudates. Although some water-soluble plant carbon may leach through the soil profile, especially in soils with measurable organic horizons, the primary microbial encounter with organic-carbon-containing materials is in the top few centimeters of the soil profile. Thus a major concern in assessing the microbial community is the distribution of carbon inputs on the soil surface (i.e., plant population growth and dispersion of fixed carbon within the soil profile due to past biomass productivity and its transformation and redistribution within the soil matrix). Secondary considerations relate to the variation of physical and chemical factors that impacts parameters across the soil surface and throughout the profile. Again, many of these variations are a result of the activity of the aboveground community. Factors that must be considered are temperature (which varies diurnally as well as seasonally and is affected by plant density and degree of canopy closure) and moisture (which may also vary seasonally but is also impacted by topography). Closely allied with moisture is soil aeration. Soil system properties affected to a large degree by the soil parent material and its developmental history are pH and salinity. Distribution and range of variation of each of these factors must be considered when extrapolating microbiological data to the ecosystem level.

The effect of energy inputs and distribution across ecosystem types is exemplified by the variations in soil microbial activity in surface soils of forest, grassland, agricultural and swamp soils. Within the forest soil, microbial energy is derived from the oxidation of fixed carbon contained in litter and from decomposition of root exudates, sloughed cells, and fine roots. The contribution of fine roots to the soil nutrient pool is frequently overlooked, but it can constitute a major source of nutrients and energy for soil microbes. (See McClaugherty et al., 1982; McClaugherty et al., 1984; Vogt et al., 1983 for examples.) In two temperate forest stands, fine root production was roughly equivalent to leaf biomass production (McClaugherty et al., 1982).

A similar distribution pattern of plant biomass is observed in a grassland soil. Grass biomass enters the soil as leafy material in the thatch and as root exudates and decaying root biomass in a dense root structure. The highest quantities of water-stable aggregates are generally found in soils collected from grassland ecosystems compared to soils from other major ecosystem types, if all other limiting factors are equivalent. This distribution results from the high input of biomass, the consistent production of this fixed carbon throughout the growing season, and the proportion of readily decomposable components comprising the plant debris as compared to alternate ecosystem types such as forest or agricultural soils. For example, Jenkinson and Ayanaba (1977) noted that 70 percent of ryegrass (*Lolium multiflora*) decomposed with a half life of 0.25 years in agricultural soils of Rothamsted, England. The remainder had a half life of 8 years.

With agricultural soils, the organic matter inputs are reasonably equal to that of the grassland in quality, but their distribution and quantities are more limited than observed with the grassland. Thus soil structural development is reduced. In intensively cultivated sites, reduction of colloidal organic matter levels in the soil result in a loss of soil microstructure. (See Tate, 1987, for further analysis of limitations and soil structure in this ecosystem type.) Analysis of microbial activity in these soils must be conducted with an understanding of crop-planting patterns, cropping history, and tillage practices. In widely spaced grass crops, such as sugarcane (*Saccharum* spp. L.), biomass inputs to the microbial community between rows is minimal.

Organic matter accumulates in swamp ecosystems. Organic matter inputs from the aboveground plant community are limited by the high soil water table, as is the decomposition rate of the decaying plant biomass. The accretion of organic matter indicates that biomass production occurs more rapidly than decomposition. Other organic inputs to the swamp system originate from erosion of soils from adjacent upland areas. Since the A horizon soils contain most of the colloidal soil organic matter, washing these soils down slope results in importing this organic matter into the swamp environment.

1.4. CONCLUDING COMMENTS

Handling a soil sample does not leave us with the impression that a living, respiring community is functioning therein. We could easily conclude that the utmost importance in the various soil types must be attributed to their physical components (i.e., their sand, silt, clay and, perhaps, rock and pebble contents). Although the apparent lack of importance of biological entities is at first disappointing, such an observation is critical to the development of our understanding of soil microbiology. It is too easy for soil microbiologists to concentrate on the extremely diverse community of microbes existing in a soil sample and their vast array of genotypic and phenotypic properties and to forget that these organisms are of no benefit to the total ecosystem if they cannot function within the soil physical matrix. The physical and chemical properties and their variability (macro- and microsite) must always be appreciated if a true understanding of the limitations and capabilities of the soil microbial community is to be fully developed.

The variability of the soil environment and its impact on life processes is both intriguing and frustrating. Because microbiologists are keenly aware of the fact that there is no typical microbe for even the simplest physiological or ecological study, they must also appreciate the fact that there is no standard soil. Soil components and their interactions are at least as variable as the characteristics of their associated microbial populations. Thus a complete understanding of soil microbial behavior and its contribution to total ecosystem function must include an appreciation of their physical environment—both the mix of individual soil components as well as the physical impediments presented to the soil microbes by the assembly of these discrete particles into a defined soil structure, including horizonal development as well as aggregate formation. Assessment of the soil type and its management history is as important to conducting "good science" as is provision of a genus or species name to the microbes studied. That is, a complete soil microbiological study must result not only in an appreciation of the nuances of the biological aspects of the soil ecosystems but also in a conceptualization of the impact of variations in the physical and chemical environment on the expression of the living capacities.

REFERENCES

Angers, D. A. 1992. Changes in soil aggregation and organic carbon under corn and alfalfa. Soil Sci. Soc. Am. J. 56:1244–1249.

Amundson, R., and H. Jenny. 1991. The place of humans in the state factor theory of ecosystems and their soils. Soil Sci. 151:99–109.

Arshad, M. A., and L. E. Lowe. 1966. Fractionation and characterization of naturally occurring organo-clay complexes. Soil Sci. Soc. Am. Proc. 30:731–735.

Bartha, R. 1980. Pesticide residues in humus. ASM News 46:356–360.

Bollag, J.-M., S. Y. Liu, and R. D. Minard. 1980. Cross-coupling of phenolic humus constituents and 2,4-dichlorophenol. Soil Sci. Soc. Am. J. 44:52–56.

Bondietti, E., J. P. Martin, and K. Haider. 1972. Stabilization of amino sugar units in humic-type polymers. Soil Sci. Soc. Am. Proc. 36:597–602.

Boyd, S. A., and R. King. 1984. Adsorption of labile organic compounds by soil. Soil Sci. 137:115–119.

Capriel, P., T. Beck, H. Borchert, and P. Härter. 1990. Relationship between soil aliphatic fraction extracted with supercritical hexane, soil microbial biomass, and soil aggregate stability. Soil Sci. Soc. Am. J. 54:415–420.

Chaney, K., and R. S. Swift. 1984. The influence of organic matter on aggregate stability in some British soils. J. Soil Sci. 35:223–230.

Chenu, C., and J. Guérif. 1991. Mechanical strength of clay minerals as influenced by an adsorbed polysaccharide. Soil Sci. Soc. Am. J. 55:1076–1080.

Cheshire, M. V., G. P. Sparling, and C. M. Mundie. 1983. Effect of periodate treatment of soil on carbohydrate constituents and soil aggregation. J. Soil Sci. 34:105–112.

Cheshire, M. V., G. P. Sparling, and C. M. Mundie. 1984. Influence of soil type, crop and air drying on residual carbohydrate content and aggregate stability after treatment with periodate and tetraborate. Plant Soil 76:339–347.

Dalal, R. C., and R. J. Henry. 1988. Cultivation effects on carbohydrate contents of soil and soil fractions. Soil Sci. Soc. Am. J. 52:1361–1365.

Dinel, H., G. R. Mehuys, and M. Lévesque. 1991. Influence of humic and fibric materials on the aggregation and aggregate stability of a lacustrine silty clay. Soil Sci. 151:146–158.

Duxbury, J. M., and R. L. Tate III. 1981. The effect of soil depth and crop cover on enzymatic activities in Pahokee muck. Soil Sci. Soc. Am. J. 45:322–328.

Edwards, A. P., and J. M. Bremner. 1967. Microaggregates in soil. J. Soil Sci. 18:64–73.

Elliott, E. T. 1986. Aggregate structure and carbon, nitrogen, and phosphorus in native and cultivated soils. Soil Sci. Soc. Am. J. 50:627–633.

Fanning, D. S., and M. C. B. Fanning. 1989. Soil Morphology, Genesis, and Classification. John Wiley & Sons, New York. 395 pp.

Foth, H. D., and L. M. Turk. 1972. Fundamentals of Soil Science. John Wiley & Sons, New York. 459 pp.

Glauser, R., H. E. Doner, and E. A. Paul. 1988. Soil aggregate stability as a function of particle size in sludge-treated soils. Soil Sci. 146:37–43.

Goring, C. A. I., and W. V. Bartholomew. 1951. Microbial products and soil organic matter: III. Adsorption of carbohydrate phosphates by clays. Soil Sci. Soc. Am. Proc. 15:189–194.

Goring, C. A. I., and W. V. Bartholomew. 1952. Adsorption of mononucleotides, nucleic acids, and nucleoproteins by clays. Soil Sci. 74:149–164.

Greenland, D. J. 1971. Interactions between humic and fulvic acids and clays. Soil Sci. 111:34–41.

Haider, K., and J. P. Martin. 1988. Mineralization of ^{14}C-labelled humic acids and of humic-acid bound ^{14}C-xenobiotics by *Phanerochaete chrysosporium*. Soil Biol. Biochem. 20:425–429.

Hatcher, P. G., M. Schnitzer, L. W. Dennis, and G. E. Marciel. 1981. Aromaticity of humic substances in soils. Soil Sci. Soc. Am. J. 45:1089–1094.

Jenkinson, D. S., and A. Ayanaba. 1977. Decomposition of carbon-14 labeled plant material under tropical conditions. Soil Sci. Soc. Am. J. 41:912–915.

Jenny, H. 1941. Factors of Soil Formation; a system of quantitative pedology. McGraw-Hill Book, Co., New York. 281 pp.

Kodema, H., and M. Schnitzer. 1974. Adsorption of fulvic acid by non-expanding clay minerals. Trans. Int. Congr. Soil Sci. 10the. 2:51–56.

Kunc, F., and G. Stotzky. 1974. Effect of clay minerals on heterotrophic microbial activity in soil. Soil Sci. 118:186–195.

Lichtenstein, E. P. 1980. "Bound" residues in soils and transfer of soil residues in crops. Residue Rev. 76:147–153.

Liu, S.-Y., A. J. Greyer, R. D. Minard, and J.-M. Bollag. 1985. Enzyme-catalyzed complex-formation of amino acid esters and phenolic humus constituents. Soil Sci. Soc. Am. J. 49:337–342.

Lynch, D. L., and L. J. Cotnoir, Jr. 1956. The influence of clay minerals on the breakdown of certain organic substrates. Soil Sci. Soc. Am. J. Proc. 20:367–370.

Malcolm, R. L. 1990. The uniqueness of humic substances in each of soil, stream, and marine environments. Anal. Chem. Acta 232:19–30.

McClaugherty, C. A., J. D. Aber, and J. M. Melillo. 1982. The role of fine roots in the organic matter and nitrogen budgets of two forested ecosystems. Ecology. 63:1481–1490.

McClaugherty, C. A., J. D. Aber, and J. M. Melillo. 1984. Decomposition dynamics of fine roots in forested ecosystem. Oikos 42:378–386.

Nayak, D. C., C. Varadachari, and K. Ghosh. 1990. Influence of organic acidic functional groups of humic substances in complexation with clay minerals. Soil Sci. 149:268–272.

Novak, J. M., and N. E. Smeck. 1991. Comparisons of humic substances extracted from contiguous Alfisols and Mollisols in Southwestern Ohio. Soil Sci. Soc. Am. J. 55:96–102.

Olnes, A., and C. E. Clapp. 1972. Microbial degradation of a montmorillonite-dextran complex. Soil Sci. Soc. Am. J. 36:179–181.

Saxena, A., and R. Bartha. 1983a. Microbial mineralization of humic acid-3,4-dichloroaniline complexes. Soil Biol. Biochem. 15:59–62.

Saxena, A., and R. Bartha. 1983b. Binding of 3,4-dichloroaniline by humic acid and soil: Mechanism and exchangeability. Soil Sci. 136:111–116.

Schnitzer, M., and H. Kodema. 1992. Interactions between organic and inorganic components in particle-size fractions separated from four soils. Soil Sci. Soc. Am. J. 56:1099–1105.

Schnitzer, M., and C. M. Preston. 1986. Analysis of humic acids by solution and solid-state carbon-13 nuclear magnetic resonance. Soil Sci. Soc. Am. J. 50:326–331.

Schnitzer, M. J. A. Ripmeester, and H. Kodama. 1988. Characterization of the organic matter associated with a soil clay. Soil Sci. 145:448–454.

Shulten, H.-R., and M. Schnitzer. 1990. Aliphatics in soil organic matter in fine-clay fractions. Soil Sci. Soc. Am. J. 54:98–105.

Sjoblad, R. D., and J.-M. Bollag. 1981. Oxidative coupling of aromatic compounds by enzymes from soil microorganisms. *In* E. A. Paul and J. N. Ladd (eds.), Soil Biochemistry, Vol. 5, pp. 113–152. Marcel Dekker, New York.

Sollins, P., G. Spycher, and C. A. Glassman. 1984. Net nitrogen mineralization from light- and heavy-fraction soil organic matter. Soil Biol. Biochem. 16:31–37.

Sorensen, L. H. 1972. Stabilization of newly formed amino acid metabolites in soil by clay minerals. Soil Sci. 114:5–11.

Sorensen, L. H. 1975. The influence of clay on the rate of decay of amino acid metabolites synthesized in soils during decomposition of cellulose. Soil Biol. Biochem. 7:171–177.

Spycher, G. P. Sollins, and S. Rose. 1983. Carbon and nitrogen in the light fraction of a forest soil: Vertical distribution and seasonal patterns. Soil Sci. 135:79–87.

Stearman, G. K., R. J. Lewis, L. J. Tortorelli, and D. D. Tyler. 1989. Characterization of humic acid from no-tilled and tilled soils using carbon-13 nuclear magnetic resonance. Soil Sci. Soc. Am. J. 53:744–749.

Stevenson, F. J. 1982. Humus Chemistry. Genesis, Composition. Reactions. John Wiley & Sons, New York. 443 pp.

Still, C. C., T.-S. Hsu, and R. Bartha. 1980. Soil-bound 3,4-dichloroaniline: Source of contamination in rice grain. Bull. Environ. Contam. Toxicol. 24:550–554.

Strickland, T. C., and P. Sollins. 1987. Improved method for separating light- and heavy-fraction organic material from soil. Soil Sci. Soc. Am. J. 51:1390–1393.

Tate, R. L., III. 1980. Microbial oxidation of organic matter of histosols. 4:169–201. *In* M. Alexander (ed.) Advances in Microbial Ecology. Plenum Press, New York.

Tate, R. L., III. 1987. Soil Organic Matter. Biological and Ecological Effects. John Wiley & Sons, New York. 291 pp.

Theng, B. K. G., G. J. Churchman, and R. H. Newman. 1986. The occurrence of interlayer clay–organic complexes in two New Zealand soils. Soil Sci. 142:262–266.

Tiessen, H., and J. W. B. Stewart. 1983. Particle-size fractions and their use in studies of soil organic matter: II. Cultivation effects on organic matter composition in size fractions. Soil Sci. Soc. Am. J. 47:509–514.

Tisdall, J. M., and J. M. Oades. 1982. Organic matter and water-stable aggregates in soils. J. Soil Sci. 33:141–163.

Vogt, K. A., C. C. Grier, C. E. Meier, and M. R. Keyes. 1983. Organic matter and nutrient dynamics in forest floors of young and mature *Abies amabilis* stands in Western Washington, as affected by fine-root inputs. Ecol. Monogr. 52:139–157.

Waters, A. G., and J. M. Oades. 1991. Organic matter in water-stable aggregates. pp. 163–174. *In* W. S. Wilson (ed.) Advances in Soil Organic Matter Research: The Impact on Agriculture and the Environment. The Royal Society of Chemistry, Thomas Graham House. Cambridge, England.

Young, J. L., and G. Spycher. 1979. Water-dispersible soil organic mineral particles: I. Carbon and nitrogen distribution. Soil Sci. Soc. Am. J. 43:324–328.

2   The Soil Ecosystem: Biological Participants

Except for an occasional insect or earthworm, once visible traces of plant biomass are removed, any mineral soil appears to be a lifeless mass. Yet, even desert soils are teaming with microscopic life-forms. The seemingly rigid mass of clay, sand, and silt, which easily slips between our fingers, is home for a complex microbial community, including bacteria, fungi, protozoa, and viruses. This intricate association of microbial cells and soil particulates is critical for total ecosystem survival. The integrity of the total ecosystem, above- and belowground, rests on the stability of the soil microbial community. Destruction of the soil microbiota through mismanagement or environmental pollution results in decline or even death of the aboveground plant and animal populations. Thus developing an understanding of soil microbes, their properties, and the nature of the interactions with and within their environment is essential.

Clearly, study of biological processes in soil presents an unique challenge. The properties of the mineral components, the heterogeneous nature of their assembly into soil aggregates, and the minute size of the living organisms minimizes the potential for directly measuring microbial processes in the field. Thus to evaluate native soil processes, representative soil samples must be collected and handled in a manner that preserves the integrity of the biological community and maintains its function at a level representative of *in situ* rates. Hence, the objective of this chapter is to elucidate the nature of the resident living components of soil and to determine their properties in relationship to total ecosystem function. This evaluation of soil life-forms includes a presentation of some commonly used analytical procedures available for quantifying soil microbial populations and the consequences of their use.

2.1. THE LIVING SOIL COMPONENT

Microorganisms constitute less than 0.5 percent (w/w) of the soil mass; yet, they have a major impact on soil properties and processes. Aboveground manifestations of an ecosystem could not develop without extensive outside intervention were the soil microbial biomass to be destroyed. For example, all macro- and micro nutrients, which are essential for plant growth, would need to be supplied from exogenous sources. Appreciation of this fact has historically stimulated interest in soil microbes from three viewpoints: a) identification of the organisms present and elucidation of their relationship with other members of the soil community (i.e., basic ecological studies); b) assessment of the activity of microbes participating in processes of agricultural importance, such as nitrogen fixation, nitrogen mineralization, or pesticide decomposition; and c) selection of organisms capable of producing anthropogenically important products (e.g., antibiotics). More recently, with the observation that human activities have caused a decline in the quality of soil resources, considerable effort has been expended in the study of soil microbial processes associated with maintaining ecosystem stability or renovation of damaged sites.

Since soil bacterial and fungal communities are prime participants in each of these areas, the most commonly studied soil living fraction has been these organisms. Thus, much of what is described in this chapter relates to these primary decomposers; yet, it must be remembered that interactions with protozoa, nematodes, and other higher life-forms help to maintain the vigor and productivity of the bacterial and fungal communities. (See chapter 6 for a more detailed discussions of these interactions.)

Were the transformations of mineral and organic components occurring in soil strictly the product of chemical reactions, soil system processes would be both predictable and limited in extent. The rate and extent of a chemically driven reaction generally can be predicted if we know the conditions existent at any specific point in time, the quantities of the reactants and products, and the properties of the reactants. This situation is contrasted with systems driven by biological processes, wherein expressed and uninduced metabolic capability contained within the genome of living and resting cells and the capacity for rapid cellular replication provide for potential development of biological processes far greater than those predicted by analysis of existent chemical and biochemical reactants and their products.

2.1.1. Biological and Genetic Implications of Occurrence of Living Cells in Soil

To fully appreciate the consequences of biological activities in soil, several unique properties of the living component of soil must be examined in greater detail. The properties of the microbial community that confer special properties to soil processes are that a) living organisms possess a diverse metabolic capacity encoded in the microbial genome that is not necessarily expressed phenotypically, b) they replicate, c) they have a definite cellular structure, d) they form resting structures, e) they catalyze reactions that change the soil, and f) they are a link between the above- and belowground processes.

2.1.1.1. Gene Pool Potential. A unique property of the living component of soil is that the biochemical potential contained therein is significantly greater than that expressed at any particular point in time. Many of the genes encoded in the genome of both active and inactive soil microbes are not expressed at any specific point in time. Furthermore, many microorganisms are quiescent; that is, they exist in a resting state. Thus essentially none of the products of the genes contained in these inactive microbes are produced. Therefore, two major pools of potential metabolic activity are found in soil: a) active microbial cells with noninduced enzymatic capability and b) cells present in an inactive or resting state or with insufficient population densities to produce measurable effects on the expressed or measurable levels of metabolic activity. In either situation, genotypic traits exist within the soil microbial biomass, which, when induced, could result in a soil system with entirely different properties than currently exist.

Induction of Metabolic Activities in Soils: Synthesis of requisite enzymes for previously unexpressed metabolic activities is induced primarily by changes in the physical or chemical properties of the soil. The development of anoxic conditions is an example of the physical modification of soil properties. An oxygen-free status commonly is imposed as the result of soil flooding where the excess soil moisture is retained under reasonably static conditions. The lack of water movement within and overlying the soil limits the influx of molecular oxygen due to mass movement. This situation coupled with an actively respiring aerobic microbial community results in depletion of soil-resident free oxygen. Hence, enzymes associated with anaerobic respiration are induced. Similarly, chemical stimulation of enzyme synthesis is illustrated by the reaction of the cellular enzyme synthesis mechanisms to soil amendment—naturally through invasion by plant or animal tissues and their products or through anthropogenic intervention, which is exemplified by the addition of sludge, compost, or xenobiotic compounds, such as pesticides, into the soil site.

A metabolic activity stimulated by the development of anoxic conditions is denitrification. An examination of the induction kinetics of this activity can serve as a model for changes in other biologically catalyzed processes by this type of extreme modification of life controlling soil properties. Nitrogen oxides are converted to nitrous oxide and dinitrogen (i.e., denitrified) at a low level in most soils. Following a soil-pore-saturating rainfall, should all other requirements for denitrification be available, nitrogen oxides are reduced to dinitrogen and/or nitrous oxide at rates several orders of magnitude greater than occurred prior to soil flooding. This stimulation of denitrification enzyme activity results from induction of denitrification enzymes under the oxygen-limited conditions imposed by water saturation of a soil. Smith and Tiedje (1979) found that, upon oxygen depletion, two distinct denitrification rates were observed. In phase 1 nitrogen oxide reduction proceeded at rates supported by the quantities of previously existing denitrification enzymes. Phase 2 activity, which was attained after 4 to 8 hours of anoxic conditions, was attributed to full derepression of denitrifying enzyme activity in the absence of significant microbial growth. Should conditions conducive for denitrification persist, the soil denitrification capacity eventually would be enhanced by increases in population densities of the responsible bacteria. A variety of studies have shown field responses of denitrification activity reflective of these laboratory observations (Christensen et al., 1991a; Flühler et al., 1976; Focht, 1974; Parsons et al., 1991; Sexton et al., 1985).

Examples of chemical modification (i.e., provision of metabolic inducers) can be associated with pesticide use (as indicated previously), amendment of soil with sulfur, and any other situation where energy-yielding substrates are introduced into the ecosystem. As with denitrification, a lag in enzyme activity of a few hours following soil amendment is observed before a rapid increase in the enzyme activity is detected.

In summary, enzyme induction provides a mechanism for a rapid, short-term response of the microbial community to ecosystem perturbation. New enzymatic activity can be expressed in existing populations in as little as 30 minutes to a few hours. After the stimulus is removed, enzymatic activities could return to preinduced levels as quickly as they were induced. The longevity of any soil enzymatic activity depends upon the stability of the protein involved, its synthesis rate, and the rate it is decomposed by cellular or extracellular proteolytic activity. Therefore, enzyme induction can be viewed as a rapid biochemical response for maintenance of ecosystem homeostasis.

Cellular Replication and Soil Properties: The traits of any soil and its microbial populations impacting cellular proliferation and the dynamics of metabolic function are:

- Microbial biomass in soil is composed of a mixture of actively growing and resting cells.
- A wide range of population densities exist in soil. Some species are represented by a few individuals whereas other populations dominate the community.
- The relative sizes of various population densities are determined by their ability to compete with companion organisms.
- At least the minimal chemical, physical, and biological requirements of the individual microbial species must be met before growth and replication can occur.
- The chemical, physical, and biological properties of any soil site are in a state of flux. As these properties change or are altered, the activities and nature of the microbial community are also modified.

In contrast to the relatively rapid response associated with induction of degradative pathways previously discussed, cellular proliferation is a slow adaptation to ecosystem modification, but its impact on the ecosystem may be much more lasting. Enzymes or enzyme systems may be synthesized at rates that result in detectable levels of activity in a matter of hours. In contrast, changes in a metabolic rate occurring as a result of cellular proliferation may require several days or longer to become detectable. The time limitations involved with cellular replication are shown when a bacterially catalyzed process is considered. These cells reproduce by binary fission with growth rates in the soil of from less than one to about two or three cell divisions per day. Thus, producing a population of several thousand cells that will yield a detectable change in metabolic activity may require several days to develop. Indeed, in situations where only a few cells with the requisite metabolic capacity existed prior to soil amendment, it may take several weeks incubation in the presence of the growth-inducing substance before population densities of a few hundred individuals are produced. Once the microbial populations have developed, they may survive for decades or longer even though the conditions causing their replication no longer exist. Thus, a large population of inactive or resting cells may persist in a soil in which they can no longer grow. Long-term survival is more commonly the exception (i.e., most nonfunctional, nonspore-forming biomass will be decomposed in periods of a few months to a year), but even this level of phenotypic survival in soil exceeds that resulting from enzyme induction and subsequent repression of the activity.

Relationships of Enzyme Induction and Cellular Proliferation to Steady-State Conditions: The potential for replication of microbial cells and induction of nonexpressed activities requires a consideration of the concept of steady-state conditions in soil. For most scientists, an understanding of steady-state conditions is based on experience with chemical reactions. In the hypothetical reaction

$$A + B \rightleftharpoons C + D$$

an equilibrium state, or steady-state situation, is reached. It is dependent upon the reaction constants for the forward and reverse reaction and the concentrations of the reactants and the products; that is, the forward and back reactions are in equilibrium.

Within the reasonably defined situation of the test tube, predictable levels of all chemical species can be reached. Such is not the situation in a soil ecosystem. The concentrations of reactants, products, and levels of the enzymes involved are highly variable. Parameters affecting reaction conditions change with time, temperature, and certainly season. Therefore, concept of steady state that applies to soil biological populations must be developed.

One option is to define steady state by the growth status of the microbial populations (i.e., population densities are unchanging). In axenic culture, this situation can be represented by stationary growth phase (see chapter 3), but this condition is never really achieved in soil by actively respiring microbes. Growth of microbial cells is encouraged by consumption of microbial biomass by predator populations, augmentation of availability of carbon and energy supplies induced by diffusion of fixed carbon substances to the microbial cells, by mass transfer of nutrient-laden water through the soil pores or through root growth and death associated with plant community growth and decline, as well as by variation in soil physical properties, such as temperature and moisture. Furthermore, cell longevity may be reduced by the stress of coping with the harsh conditions of their environment. Thus, the microbial biomass can be seen as being in a constant state of flux.

Since both the microbial populations and their environment are steadily, albeit at times very slowly, changing, a chronological dimension to the definition of steady state must be applied. Because many soil properties that control microbial population densities vary regularly on a diurnal and seasonal level, steady-state conditions may be determined by considering changes in the mean population densities over time. For example, temperature cycles daily as well as seasonally. Nutrient levels available to microbial populations frequently relate to the plant-growing season because plants are the primary carbon and energy source for most active soil microbes. To accommodate this regular cycling or fluctuation of soil properties controlling microbial activity, a steady state can be said to exist in a soil ecosystem when the cycles of microbial activity or proliferation are replicated on a regular (time) basis (e.g., seasonal, plant growth status, or annually).

Based on even this periodic variation of soil microbial properties and activity, the purist might raise a question regarding whether steady-state conditions really ever occur in soils. The validity of this dissension must be acknowledged. Perhaps a more scientifically defensible expression of the status of the soil microbial community would be to modify our concept of steady state by designating the occurrence of a *quasi* steady-state rather than an absolute steady-state condition. A constant level of biological activity is never reached in any soil system, even on an annual basis. It would be rare to observe a field situation where exactly the same physical and chemical conditions impinging on microbial activity occur repeatedly year after year.

We must make similar considerations when evaluating the microbial activity at the field level at any single point in time. Again, as indicated in chapter 1, the physical properties are extremely variable across a field surface, dependent upon such ecosystem-related properties as the distribution and nature of the plant community. Variation in microbial activity becomes even more acute as the level of the microsite is approached. This heterogeneity in the activity levels of a microbial process in soil was recently clearly demonstrated (Christensen et al., 1991b; Parkin, 1987; Parkin et al., 1987). Denitrification requires anoxic conditions plus metabolizable carbon sources and nitrate to occur. These necessities vary spatially and chronologically in soil. Parkin (1987) found "hot spots" of high specific denitrification in soil that were linked to microsites with elevated particulate organic carbon. In a soil core where a high denitrification rate was measured, 85 percent of the denitrification capacity of a 98-g sample was found in an 0.08-g subsample of the core. Thus, it is easy to postulate that activity could vary from essentially zero (no microbial cells present) to extremely high levels (large colonies in the vicinity of active plant roots), depending upon the distribution of the microbial colonies and the substrates necessary for their function. Again, the mean value of the activity in question over the field of study could be defined as the steady-state level of activity for the duration of the study. The "hot" or "cold" spots of activity may vary in location with time, with the overall mean of metabolic activity in the field remaining reasonably constant.

In conclusion, a clear picture of the nature of a steady-state soil ecosystem could be likened to the daily conditions of any large city. When taken as a whole, the activity of the population appears to differ little on a daily basis (i.e., quasi steady state); cars and busses are streaming into the city and the sidewalks are clogged with people rushing to and from their offices. On any given weekday, the numbers of individuals associated with any particular location appears to be reasonably constant. Yet, from the vantage point of any street corner, the situation appears to be nearly chaotic. Such, is the activity of field soil as exemplified by the reactions occurring along a transect of any field site. Biological activities may be extremely variable along the transect (varying over several orders of magnitude), but an average value representative of the overall soil ecosystem along the transect can be calculated. Furthermore, the activity may be highly variable during the growing season, but again a value representative of the process rates may be estimated. In our model of the city, this variability could be likened to that involved with comparing weekdays with weekends. Operationally the highly variable soil systems could be defined as being at a quasi steady-state level. An apparently chaotic system can be reduced to some semblance of reproducible order when integrated along a long transect or over the time frame of a growing or annual season. It is this potential for predictability and representation that makes possible the description of an ecosystem whose whole existence is based on processes occurring in highly variable microsites.

2.1.1.2. Cell Structure and Biochemical Stability in Soil.

Amending the soil with almost any biodecomposable substrate, assuming conditions are appropriate for microbial metabolism to occur, results in a reasonably rapid depletion of the added material. The fact that a wide variety of decomposable organic compounds contained in living cells are readily isolated from soil appears to contradict that observation. That is, we can observe an apparent stability of labile substances. These contrasting observations result from the definite cellular structure of the microbial biomass. Readily decomposable organic compounds are protected within the cellular structure. Hence longevity of the biochemicals is not determined by the capacity of the microbial community to synthesize the enzymes necessary for their catabolism, but rather decomposition kinetics are determined, at least in part, by the capacity of the microbial biomass to breach the protective cellular barriers. In situations where the carbonaceous substrates are encased within protective cell walls, microbial populations may be starving in a soil that contains large reserves of carbon and energy supplies that can support microbial cellular growth locked within protective cell walls.

2.1.1.3. Resting Structures and Soil Respiration.

Another unique property of the soil biological system is that the cells may enter a metabolic state that allows them to survive under adverse conditions. Two levels of nonreplicating state of viable cells exist in soil. In the absence of appropriate energy supplies, microbes may enter a quiescent or reduced metabolic state or they may actually form resistant structures (e.g., spores, endospores, and sclerotia).

Microbes, which are not replicating and have not formed a resting structure, must continue to consume energy to maintain their cellular structure. Thus, these microbes, by oxidizing internal energy sources, continue to alter at least the microsite in which they exist. At least a minimal localized effect of the carbon dioxide generated from respiration associated with cellular maintenance would be expected to occur. Generation of this gas affects the atmosphere of the microsite where the resting microbial cell resides and alters the pH of the interstitial soil water. For example, nitrifiers in the absence of their energy source, ammonium or nitrate, or any heterotroph existing in the absence of fixed carbon substrates may enter this situation where the only metabolic processes occurring in the cell are those associated with maintaining the cell structure. This form of existence can be seen as a type of starvation where ultimate survival depends upon an eventual influx of an energy-supplying nutrient. Interestingly, microbes may exist for extended periods of time (decades to centuries for *Arthrobacter* sp.) under these conditions. For these microbes, cellular metabolic rates are reduced to those levels necessary to maintain a minimal cell structure (i.e., the level of enzymes synthesis machinery and metabolic tools necessary to synthesize the enzymes and cellular products necessary to take advantage of proper growth conditions when they develop). Considering the chronological as well as physical heterogeneity of organic carbon distribution within soil, we can logically conclude that the majority of soil microbes are found either in this reduced metabolic state or in resting structures.

2.1.1.4. Soil Mineral Transformation by Microbial Cells. Microbes are also unique in soil in that they can alter the solubility of soil mineral components, reduce organic compounds to essentially undetectable levels, modify soil structure, oxidize inorganic compounds, and use a variety of soil components as electron acceptors. Merely through growth and metabolism, soil microbes alter their environment. For example, soil structure is enhanced by production of cellular biomass, which may link soil particulates or production of polysaccharide adherents (chapter 1). In contrast, soil structure may be lost through oxidation of the colloidal organic matter supporting soil aggregate structure by the same microbes. A less obvious but nonetheless important aspect of the impact of microbial metabolism on soil particulates is the dissolution of soil minerals resulting from organic or inorganic acid production. A variety of organic acids, including acetic acid, citric acid, and lactic acid, carbonic acid (carbon dioxide), and mineral acids (sulfuric acid and nitric acid) are produced by the soil microbial biomass. These acids solubilize soil minerals.

Soil organic and inorganic components may also be oxidized or reduced as energy sources or electron acceptors for the microbial community. All nonphotosynthetic microorganisms must oxidize their growth substrates for energy. Commonly considered reactions are the oxidation of ammonium to nitrite or nitrate, organic carbon (fixed by photosynthesis) oxidation to carbon dioxide, and elemental sulfur conversion to sulfuric acid (see Chapter 3). The impact of these oxidative processes on soil is magnified by the nature of the final acceptor of the electrons produced by oxidation of the growth supporting substances. A basic principle of biochemistry is that all oxidative reactions must be balanced by comparable reductive processes. Thus, some environmental substances must be reduced as a result of microbial growth. Denitrifiers oxidize organic carbon while reducing nitrate to nitrous oxide and dinitrogen. Aerobic catabolism of organic carbon most commonly involves reduction of oxygen to water. Furthermore, carbon oxidation may result also in the reduction of sulfur oxides to sulfide or elemental sulfur.

The practical implications of these basic biological principles become apparent when it is considered that both oxidative and reductive processes have major implications in restoring contaminated soils. Whereas aerobic, heterotrophic metabolism is used to purify soils contaminated by organic-carbon-based compounds, reductive processes may be exploited for renovation of metal-contaminated and acid-impacted sites. For example, sulfate reduction to hydrogen sulfide may be facilitated in renovation of acid mine drainage-impacted sites (Mills, 1985). Similarly, hydrogen sulfide production may be used to encourage removal of heavy metals from waters or immobilize these metals in soil systems through precipitation as metal sulfides (see Wildung and Garland, 1985). Frequently, a wetland-type system is developed to accomplish these tasks. Passage of metal-bearing water via overland flow through swampy ecosystems where hydrogen sulfide is generated results in precipitation of the metal sulfides.

2.1.1.5. Microbial Link to Aboveground Communities. It is relatively easy to forget that the processes occurring in soil cannot occur at maximum rates in the absence of fixed carbon inputs from the aboveground plant community. Catabolism of colloidal soil organic matter and more slowly decomposed biological components (complex polysaccharides, lignin, and lignin products) supports some microbial activity, but maximal biomass productivity is derived from catabolism of more easily oxidized substrates, such as simple polysaccharides and proteins. Because the primary, natural source of such compounds in soil is plant biomass, a nearly essential dependency develops between the autotrophic aboveground community and soil microscopic life.

Similarly, the aboveground community is dependent upon the decomposer population in soil to mineralize macro- and micronutrients retained within dead plant biomass. For example, the organic nitrogen contained in the plant biomass must be converted to ammonium and nitrate before it can again be incorporated into new plant biomass. The facilitators of this cyclic process are the soil microbial community. (See chapter 10 for further discussion of these processes.)

2.1.2. Implications on Microbial Properties of Soil Samples Handling

An appreciation of the unique properties of living organisms in soil is mandatory for study and characterization of the versatile and rather fragile material called soil. Soil samples used to characterize native processes must not be treated in a manner that destroys the living systems contained therein. Of first consideration when collecting soil samples is the necessity of remembering that the prime mover in the system is biological. Thus, the sample should be maintained so that biotic components are preserved and induction of major changes in their composition is avoided.

Evaluating the problems encountered by air-drying soil samples provides a primary example of a commonly used procedure in soil sample collection which necessarily results in alteration of soil biological activity. Historically, soil scientists have preserved soil samples by air-drying. With this procedure, moist soil samples are incubated in the laboratory under conditions that allow soil moisture to approach ambient atmospheric levels. The final level of moisture retained in the soil may be considerably below that achieved under field conditions in that periodic influxes of water from rain fall or dew formation is precluded. This procedure is appropriate for preserving many chemical properties of soil, but it is disastrous for evaluating most soil microbial processes.

Air-drying soil creates a new ecosystem that may be quite different from the soil in the field where the sample was collected. For soil microbes to function there must be a coating of water on the soil particles (see chapter 3). When soil is dried, sensitive microbes may die or be induced to enter into a resting stage. The fact that major changes in soil properties occur during soil desiccation is suggested by the observation that a burst of microbial respiration occurs when air-dried soils are remoistened. This increased respiration results from the oxidation of organic carbon liberated from soil aggregates by their destruction during the drying process and from catabolism of the dead microbial biomass. It also results in changes in the levels and nature of soil enzymatic activities, modification of soil organic components, and changes in the composition of the microbial community.

Thus, modification of the physical condition of soil may have a dramatic effect on the nature of the active microbial populations. This situation is particularly acute if these are the microbial communities of interest. For example, anaerobic organisms are killed by contact with molecular oxygen. A sensitivity gradient in such populations occurs (see chapter 3), but in studies where quantification of such populations or their activities is desired (especially with anoxic soil samples such as swampy material), oxygenation of the soil must be avoided. Modification of the aeration status of soils from that existing *in situ* under the oxygen-free or oxygen-limited conditions may also cause increased population densities of aerobic microbes. A pool of partially decomposed fixed carbon substrates accumulates under oxygen-limited conditions. These substances provide carbon and energy for the aerobic microbes once free oxygen is introduced into the system, thereby resulting in an augmentation of their population densities.

2.2. MEASUREMENT OF SOIL MICROBIAL BIOMASS

Soil microbial biomass is a primary catalyst of biogeochemical processes as well as an energy and nutrient reservoir. For example, its significance is exemplified not only by the infinite array of biochemical transformations catalyzed in soil but also by the quantities of fixed nitrogen it contains. Anderson and Domsch (1980) found that in 26 agricultural soils, total soil nitrogen contained in the microbial biomass ranged from 0.5 to 15.3 percent with an average of approximately 5 percent. This nitrogen becomes available to the aboveground community upon death and decay of the microbial cells. (See chapter 10 for further discussion of this cycling of microbial nitrogen.) In soils not receiving exogenously supplied fixed nitrogen (e.g., nonagriculture soils), this nitrogen pool is sufficiently large that its stability is a prime factor in controlling the flux of fixed nitrogen through the ecosystem. In disturbed soils, such as when grassland soils are initially cultivated, a decline in soil microbial biomass results in a release of large quantities of fixed nitrogen. The fixed nitrogen not incorporated into newly synthesized microbial cells or aboveground biomass is lost from the ecosystem through leaching to groundwater or through runoff. Hence, estimating the size of the soil microbial biomass and its stability is important to the understanding ecosystem dynamics.

Microbial biomass is readily assessed in simple ecosystems, such as the growth flask, but its measurement in soil is complicated by the complexity of the system and the fact that these cells comprise a small portion of the total soil mass. Because it is essentially impossible to isolate all the microbial cells from soil to quantify them, we must estimate the microbial biomass by measuring a reasonably easily quantified cellular component that can be extracted from the soil. An ideal indicator of microbial biomass should

- be present in all cells
- occur in all microbial cells at the same concentration, regardless of species designation
- be present in the cells at the same concentration, independent of growth status
- be rapidly decomposed upon the death of the cell
- be quantitatively extractable from soil
- be easily assayed

The latter trait is necessitated by the large number of samples that are generally processed and the fact that the biomass may change during storage. A variety of compounds for quantifying soil microbial biomass have been used, but none meets all these criteria as an ideal indictor of microbial biomass.

Soil microbial biomass has been commonly estimated through

- direct counting of the microbes (Sonderstrom, 1977; Paul and Johnson, 1977; Rosser, 1980, and Lundgren, 1981)
- analysis of specific cellular components, such as adenosine triphosphate (ATP), phospholipids, or muramic acid (Ausmus, 1971, Findlay et al., 1989; King and White, 1977; Paul and Johnson, 1977; Fazio et al., 1979; Jenkinson et al., 1979; Verstraete et al., 1983)
- measurement of specific microbial processes, such as nitrogen mineralization (Alef et al., 1988) and reduction of dimethylsulfoxide to dimethylsulfide (Alef and Kleiner, 1989)
- measurement of respiration rates (Anderson and Domsch, 1973)
- direct analysis of cellular components solubilized by chloroform fumigation (Brookes et al. 1982; Sparling and West, 1988; Tate et al., 1988; Vance et al., 1987a) or the carbon dioxide produced by respiration of these products (chloroform fumigation–incubation method) (Jenkinson, 1976; Jenkinson and Powlson, 1980)

Of these methods, the most commonly encountered procedures are direct counts, ATP analysis, respiration methods, and variations on the chloroform fumigation procedure. Thus, these methods are examined in greater detail to show the strengths and problems associated with soil microbial biomass assessment. (A detailed presentation of the methods can be found in Parkinson and Paul, 1982.)

ATP Measure of Soil Microbial Biomass: For these estimates, soil is homogenized in a buffer solution to extract the ATP (Fig. 2.1). Generally an acidic buffer is used to minimize solubilization of soil humic substances that interfere with measurement of the ATP. The extractant is removed from the soil suspension by centrifugation or filtration, and the ATP is quantified by measuring light production with the firefly luciferin–luciferase system. Soil microbial biomass carbon is calculated

SOIL EXTRACTION

①

```
5.0 g   Soil Sample
50 ml   0.5 N NaH₂CO₃, pH 8.5
15 ml   CHCl₃
```
Blend high
speed, 1 minute

②

```
After first blending, add
75 ml 0.5 N NaH₂CO₃, pH 8.5.
Blend high speed for
second 1 minute period.
```

Remove Soil by centrifugation $\Big\}$
CHCl₃ removed by vacuum $\Big\}$ $\longrightarrow$ extract

ATP ANALYSIS

Soil Extract
+
buffer, pH 7.8   + Luciferase $\longrightarrow$ Light

CALCULATION

1 μg ATP = 250 μg Microbial C

Fig. 2.1. Outline of a typical ATP analysis procedure for estimating soil microbial biomass (Parkinson and Paul, 1982).

by multiplying the measured ATP concentrations by a constant indicative of the amounts of ATP contained per unit of microbial biomass. The procedure is dependent upon the extraction efficiency of ATP from the soil sample and the selection of an appropriate constant for final calculations. An assumption underlying this procedure is that there is negligible variation in ATP contents of microbial cells. The validity of this assumption is questionable because cellular ATP contents are highly dependent upon the metabolic status of the cell. Actively metabolizing microbial cells contain considerably more ATP than is found in resting cells, two types of cells found in varying proportions in soil microbial biomass. (For further information on the use of ATP for microbial biomass measurements, see Ahmed and Oades, 1984; Ausmus, 1971; Jenkinson et al., 1979; Paul and Johnson, 1977; and Verstraete et al., 1983; Verstraeten et al., 1983.)

Soil Aerobic Respiration Measurements: The rate at which fixed carbon substrates are oxidized to carbon dioxide in a soil sample is proportional to the quantities of organisms mediating the reaction (Anderson and Domsch, 1978) (Fig. 2.2).

1. Determination of Optimal Glucose Concentration

Select lowest glucose concentration producing optimal CO_2 flux

2. Assessment of Respiration Rates of Test Samples

3. Calculation

Microbial Biomass = 40.04 (CO_2 flux) + 0.37

Fig. 2.2. Outline of aerobic respiration method for estimating soil microbial biomass (Parkinson and Paul, 1982).

With this procedure, soil samples must be collected and any plant roots or macroscopic biomass must be removed to negate carbon dioxide contributions from these sources. With the soil respiration procedure, a metabolizable carbon source is added to the soil sample, and carbon dioxide evolution can be measured with infrared analysis or gas chromatography. Should sufficiently large quantities of carbon dioxide be produced, it can be quantified by titration of the gas collected in alkaline solutions. Alternatively, ^{14}C-labeled substances, such as glucose, may be added to the soil sample and the ^{14}C-labeled carbon dioxide produced quantified.

This procedure may be used to differentiate fungal and bacterial respiration in soil by amending the samples with population specific antibiotics (Anderson and Domsch, 1973; Anderson and Domsch, 1975). Streptomycin is generally used to inhibit bacterial respiration and cycloheximide for eucaryotic respiration. Each antibiotic should be used singly and in combination, because a proportion of the microbial population is resistant to the antibiotics. Furthermore, several concentrations of antibiotic must be tested with each soil studied to ensure use of optimal antibiotic levels. Inhibition of microbial activity is proportional to the quantity of antibiotic dissolved in the soil interstitial water. Antibiotics sorbed onto soil organic matter or clays is inactivated. Thus, the effective concentration of the antibiotic in soil may be significantly lower than the total quantity added. Higher antibiotic concentrations are required for inhibition of microbial populations in soils containing high concentrations of clay or colloidal organic matter than would be necessitated in a sandy soil, for example. Similarly, too much antibiotic must not be added to the soil, because excessive antibiotic concentrations may inhibit the activity of nontarget microbes.

A major consideration when using any technique involving measurement of carbon dioxide from soil is the soil pH (e.g., see Martens, 1987). Microbial biomass may be underestimated by failure to recover all the carbon dioxide produced during the incubation period. Under neutral or alkaline conditions, carbon dioxide is retained in soil as bicarbonate or carbonate due to the following equilibrium:

$$CO_2 + H_2O \rightleftharpoons H_2CO_3 \rightleftharpoons HCO_3^- + H^+ \rightleftharpoons CO_3^{2-} + 2H^+$$

Chloroform Fumigation (Extraction and Incubation) Technique: Because of the simplicity of this technique and the fact that it is relatively inexpensive, chloroform fumigation procedures have become the most commonly employed method for estimating soil microbial biomass. Thus, the principles, applicability, and problems associated with the technique must be examined in detail.

Chloroform fumigation procedures (Fig. 2.3) are based on a) the disruption of cellular membranes by chloroform and b) quantification of the carbon contained in the biomass of the cells killed by the chloroform vapors. Generally, chloroform

Fig. 2.3. Outline of the chloroform fumigation–incubation procedure for estimating soil microbial biomass (Parkinson and Paul, 1982).

vapors are used. Following fumigation, soils may be extracted directly, and soluble carbon (Sparling and West, 1988; Tate et al., 1988; Vance et al., 1987a), nitrogen compounds (Azam et al., 1988), or phosphate (Brookes et al., 1982; Brookes et al., 1984; Hedley and Stewart, 1982) measured (estimates of microbial biomass carbon, nitrogen, or phosphorus, respectively). Alternatively, for determination of microbial biomass carbon, the soils may be incubated to allow biological conversion of the dead microbial biomass to carbon dioxide to occur. The amount of excess carbon dioxide produced in the fumigated vs. nonfumigated control samples is quantified. This procedure is based upon the assumption that the biomass carbon of the cells killed by the chloroform vapors is used by the surviving soil microbial biomass as carbon and energy sources. (Note that aerobic incubation conditions are required. Anaerobic conditions result in incomplete oxidation of the cellular carbon of the dead cells.) To ensure the existence of adequate microbial populations during the incubation period, the fumigated and control soil samples are usually inoculated with small quantities of fresh soil (see Chapman, 1987). For assessment of carbon dioxide production, soils are sealed in respiration chambers. Carbon dioxide is collected either in alkaline solutions and titrated or assayed with gas chromatographic procedures.

For each of the procedures—incubation or direct extraction methods—assay of nonfumigated soil serves as a control. The total biomass present is calculated by subtracting quantities of carbon dioxide produced in nonfumigated soil samples (controls) from that yielded in the fumigated soil samples and dividing by a constant. The constant is a value representative of the killing efficiency of the fumigation procedure. Most frequently, for the assessment of microbial biomass carbon, killing efficiency is considered to be about 40 percent (i.e., a constant of 0.4).

Some variation in the incubation period of the soil samples following chloroform fumigation has been used. After the treatment and control samples are incubated concurrently for 10 days, the total carbon dioxide yielded can be assayed, or the control sample can be incubated for 20 days so that the carbon dioxide that evolves between days 10 and 20 can be used as the correction factor in the calculations. In this situation, for the calculations, the carbon dioxide produced in control samples during the last 10 days of a 20-day incubation period is subtracted from that yielded in the fumigated samples during the first 10 days of incubation. This procedure avoids inclusion of the carbon dioxide resulting from the commonly observed burst of microbial activity in disturbed soils samples that occurs with control soils in the final calculation. Failure to compensate for this soil-disturbance-induced carbon dioxide production may result in an underestimation of total soil microbial biomass.

Aside from providing a reasonable indication of the quantities of cellular components (carbon, nitrogen, and phosphorus) contained in microbial biomass, chloroform fumigation of soils has proved to be a valuable tool for demonstrating the assimilation of soil amendments by the soil microbial community. For example, Kassim et al. (1982) quantified the incorporation of glucose, acetate, pyruvate, uracil, uridine, amino acids, and a variety of polysaccharides into microbial biomass. To accomplish this, ^{14}C-labeled substrates were added to soil samples that were then incubated for varying time periods to allow their incorporation into various soil organic matter fractions, including microbial biomass. Following incubation, the nonreacted ^{14}C-labeled amendment was washed from the soil. For analysis of the microbial biomass-incorporated materials, the soil samples were fumigated with chloroform. Assessment of the excess ^{14}C-labeled carbon dioxide produced during the incubation period in the fumigated samples provided a measure of the quantity of the metabolized carbon incorporated into microbial cellular components.

The chloroform fumigation procedure is an effective means for estimating microbial biomass in most soils, but difficulties have been encountered with application of the technique to acidic soils (Coûteaux et al., 1989; Vance et al., 1987a; Vance et al., 1987b), wet soils (Ross, 1987; Ross, 1988), and in plant-biomass-amended soils (Martens, 1985). A common trait among these soil situations is that the capacity of the microbial community to recover following fumigation and subsequent catabolism of the dead microbial cell substance has been affected. Large quantities of available carbon preexisting in the soil samples may result in minimal differences between quantities of carbon dioxide produced by control and fumigated soils (i.e., the carbon dioxide production from the killed biomass oxidation is not significantly greater than the large quantities produced from catabolism of native soil organic matter). In extremely acidic soils, data suggest that the nature of the population (fungal vs. bacterial contributions to carbon metabolism) may differ between fumigated and control soils (Tate, 1991). In either case, modification of the incubation procedure and/or method of calculation of the microbial biomass may be necessary (Vance et al., 1987c).

Direct Estimation of Microbial Biomass: Soil bacterial and fungal biomass may be viewed directly in preparations of soil samples. The refraction index of microbial cells is not sufficiently different from that of some soil particulate organic components to allow clear differentiation. Thus, it is necessary to stain the preparation. A variety of methods have been developed to enhance microscopic resolution of microbial cells in soil. Nonspecific staining (e.g., acridine orange (see Mills and Bell, 19860), fluorescent staining of respiring bacteria (e.g., Rodriquez et al., 1992), and ethidium bromide (Rosser, 1980) may be used for general measurements of soil bacteria or fungi (e.g., see Fig. 2.4). Fluorescent-labeled antibodies are useful for quantifying individual species populations.

An advantage of direct observation procedures is the capacity to detect and identify organisms at the species level and to differentiate active from inactive microbial populations. Metabolically active bacteria may be detected by treating soil samples with tetrazolium salts [accumulation of water-insoluble formazan accumulates in the cells], labeling cells with radioactive substrates (autoradiography) (e.g., see Rozak and Colwell, 1987) and counting synthetically active bacteria. With the latter procedure, soil samples are incubated with naladixic acid, an inhibitor of DNA synthesis. This procedure results in an artificial elongation of metabolically active cells. The active cells grow, but do not divide. These microbial cells may be detected in the treated soil preparations with acridine orange stain and epifluorescence microscopy.

Once the microbial cells have been quantified in several microscopic fields, the

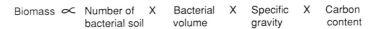

Fig. 2.4. Outline of procedure for estimation of microbial biomass using direct observation of fluorescein isothiocyanate (FITC) label. See Schmidt and Paul (1982) for details of procedure.

values are converted to biomass by multiplying the number of cells per unit of soil by their volume times the cell density. Calculating the number of cells per unit of soil requires a knowledge of the area of the microscope field and dilution of soil used in preparing the slide. Biovolumes may be calculated from cellular dimensions estimated during their enumeration. Difficulties are encountered in converting the cell volume to cell mass (Bakken and Olsen, 1983; Bratbak, 1985; Bratbak and Dundas, 1984; Van Veen and Paul, 1979) because this requires estimation of the impact of moisture on changes in cell dimension during sample preparation.

Once the number of cells per unit of soil on the prepared slides is determined, problems arise regarding the validity of extrapolating data collected from the limited area of the microscope slide to field-scale dimensions. For these studies, small quantities of soil (perhaps less than a gram of soil) are suspended and spread in a thin layer on a microscope slide. Alternatively, the liquid used to suspend the soil is extracted, and the microbes contained therein are collected on an appropriate filter medium. The slide or filter is dried, the microbial cells stained, and a portion of the preparation examined under the microscope. Careful consideration of the extrapolation limits of data from each method is necessary, but it is reasonable to conclude that extrapolation capacities for the direct count procedure is more limited.

Limitations of Microbial Biomass Measurements: A variety of inaccuracies are intrinsically associated with estimation of soil microbial biomass. In selecting procedures, determining constants for calculating results, and interpreting the data, we must consider the basic properties of the soil. These properties include the soil moisture level, pH, and organic matter contents referred to previously as well as soil texture. Extraction efficiency of charged materials from soils with high clay contents may be reduced. Furthermore, high levels of colloidal soil organic matter may sorb products of interest or interfere with analysis (ATP–luciferin–luciferase assay).

Note that the calculation procedure for each of the preceding procedures requires the use of a constant. Selection of the control value assumes a killing efficiency for chloroform fumigation, a concentration of ATP in cells of soil microbes, or a specific effect of soil drying on biovolume for direct counting procedures. All such assumptions have a leveling effect on the data for they may be true for some samples or cells and not for others.

In data analysis the investigator must consider the foregoing difficulties, which are, to a certain degree, beyond experimental control. Other procedures, which are within design control, involve laboratory and field procedural variables. Nearly all procedures require some degree of sample storage prior to analysis. Once soil samples are collected, creation of a new ecosystem commences. Soil properties limiting microbial activity in the field site (e.g., perhaps temperature, soil structural occlusion of organic matter, and moisture variation) are removed, and new delimiters of microbial population densities and activities are created. Such changes must be minimized. The most reasonable means of reducing artifacts due to sample preparation procedures is to assay the microbial biomass as quickly as possible after collection (preferably within minutes but no more than a few hours). Should this assay not be possible, storage at 4°C and avoidance of direct sunlight are mandatory. Good laboratory procedure requires documentation of all sample preparation procedures so that any compromising of the results can be determined.

Finally, the experimenter must ask the question, "What is being measured by the procedure used?" Some methods provide an estimation of the total microbial populations present (e.g., chloroform fumigation procedures or ATP analysis), whereas other methods only quantify the microbes capable of a particular reaction. For example, utilization of glucose in soil aerobic respiration measurements provides a good estimate of the biomass of organisms capable of oxidizing glucose as an energy source. In the former situation (total biomass quantification) both active and inactive microbial populations are quantified whereas in the latter (soil aerobic respiration assessment) only those populations capable of catabolizing a particular biochemical process are enumerated.

2.3. THE NATURE OF THE SOIL INHABITANTS

All six major groups of microorganisms plus higher animals interact in soil to catalyze the biogeochemical processes occurring therein. Soil bacteria, fungi, actinomycetes, protozoa, algae, viruses, nematodes, and mite populations contribute to development of the total community. The arrangement of these classes of organisms may be evaluated in a hierarchical manner. This association provides the basis for fundamental viewpoints in examining native ecosystems. The bottom of this arrangement is the individual organism. The summation of the individuals is a population. All populations of different organisms in soil constitute the community, whereas the interaction of all biological components with the abiotic portions of the environment constitutes the ecosystem.

Soil ecological research can be divided by the portion of this hierarchy stressed in the study. Such research may have an autecological or synecological basis. **An examination of the behavior of a single species in an ecosystem, including a study of the effect of chemical, biological, and physical aspects of the ecosystem on the population is an autecological study.** In contrast, soil may be examined as a whole. In this situation, soil is being treated as if it were a single "tissue." Thus, **the relationships between the environment and community of organisms (synecology)** is being stressed. For synecological research, processes are of more concern than the identity of the individual microbes involved. For example, denitrification kinetics and their variation between soil types may be studied (synecology), or an experimenter may concentrate on the identify of the individual denitrifying species existent in a soil site (autecology).

2.4. AUTECOLOGICAL SOIL MICROBIOLOGY

Quantification of microbial biomass is useful for expressing ecosystem potential (e.g., nutrient or energy cycling or even general biological capacity), but most evaluations of soil community activity require elucidation of specific biological or biochemical processes and their kinetics. The processes themselves may be the center of the experiments (synecology) (see section 2.5), or elucidation of the specific microbes and the system properties controlling their activity may be the pivotal point of the research (autecology). The latter aspect of ecosystem evaluation is the topic of this section.

Before entering into autecological research, we must determine what constitutes a bacterial species. An autecological study by strict definition involves assessing the presence or activity of individual species. This assessment can be accomplished with higher organisms where species designations are more easily determined, but the fluidity of bacterial species complicates such research for the soil microbiologist. This difficulty arises from the reasonably easy transfer of genetic material between bacteria (e.g., transduction, conjugation, and transformation). The definition of a bacterial species resides in the genetic material contained therein and its expression as it impacts cellular morphology, colonial structure, and metabolic capacity. Should the genetic material resulting in distinctive properties that define one of the bacterial species be readily transferred into another species, the distinction between the two groups of bacteria becomes questionable at best. Such problems with speciation of common soil bacteria are reasonably common.

These problems with designation of bacterial species has led to the use of other more easily derived groupings of bacteria. Thus, autecological research may involve the study of bacterial groupings separated by phenotypic traits (functional or even colonial groupings such as guilds) rather than by species designation (Mills and Bell, 1986). Thus, the following discussion of autecological research is based on the expanded definition of autecology, which allows for the study of guilds as well as scientifically accepted species designations.

Limitations to Autecological Research: Note that the compromise involved in delineation of bacterial species incorporates a limitation into the value of the data derived from such studies. A primary objective of any ecological study is not only to understand the particular system of study but also to develop principles applicable to a variety of situations. Taxonomic designation of the living organisms studied has provided a common link between research projects. The phrase "a lion is a lion no matter where it is encountered" can be readily accepted. To some degree, such a statement could be made for many of the historically studied bacterial species, but a generalization of this type regarding bacteria is questionable. Thus, use of the less restrictively derived guilds limits intersystem and interlaboratory comparisons.

A second difficulty associated with autecological research results form the minute dimensions of soil microbes. Autecological research provides meaningful data for understanding biological processes, but extrapolation of such research on an ecosystem basis is limited. Data describing the interactions of microbial colonies and their environment are collected, but the area impacted by an individual microbial colony in soil encompasses a few cubic microns at best. This inherently small size of the microsite impacted by the individual microbe necessitates collection of large quantities of data before a clear picture of all soil microbial colonies emerges (Tate, 1986). Development of automated microbiological procedures for species identification, expanded use of fluorescent antibodies (including monoclonal antibodies), and application of polymerase chain reaction analysis to detect individual microbial species and to quantify their density in soils results in an expansion of autecological soil microbiological research.

2.4.1. Viable Counts/Enrichment Cultures

Prior to the recent development of procedures for detecting nucleic acid sequences specific for selected microbes and their application to physically and chemically complex systems like soil, the initial step in any autecological study of soil involved growth of the microbes of interest in axenic culture. Indeed, this step remains the initiation point—and in many cases the primary limitation—for most such research today. Studies in which the experimenter's interest resides simply with detecting selected microbes and estimating their population density are examples of exceptions to this observation. In that situation, indicators such as species-specific activities or DNA–DNA hybridization techniques eliminate the necessity of some cultural studies. Purification of individual microbial species allows evaluation of their genotypic potential as well as consideration of the factors controlling phenotypic expression.

Microbes may be isolated from soil using nonselective techniques, such as growth on soil extract agar (this technique might better be referred to as "less selective" because growth in any laboratory medium allows growth of only a portion of the microbial population), elective growth, and enrichment cultures. With the latter two procedures, microbes with certain desired physiological properties are directly sought. With enrichment cultures, increases in microbes existing in low population densities are favored prior to isolation of the axenic cultures, whereas with the former, an extract of soil may be cultured directly on selective media.

The first concern in enumerating specific soil microbes, especially bacteria, results from their population density. Typically bacteria exist in soil at densities of 10^8 to 10^9 colony-forming units per gram of dry soil. These bacteria are associated with 10^7 to 10^9 actinomycete propagules per gram of dry soil and about 10^6 fungal and protozoan propagules each per gram of dry soil. Direct plating on a nonselective agar medium would clearly be unfruitful. Soil samples must be diluted to reduce colonial growth on agar plats to 20 to 300 colonies per plate (Fig. 2.5). Consideration of the dilution factor plus the quantity of diluent added to the growth medium allows calculation of viable microbes in the original soil sample. A result of this necessity to dilute soil samples is that the data from such studies are highly variable. A slight variation in dilution precision—especially at the lower dilutions—causes large differences in estimated propagule densities. Variations as large as plus or minus 100 percent are not uncommon.

With viable counts of microbes from any environmental sample, a qualification must be made regarding the meaning of "viable." Those bacteria classified as viable for any specific experiment in reality include only the propagules capable of growth on the media used and only under the particular incubation conditions employed. Comparison of data from direct microscopic analysis of soil samples with results from viable plate counts reveals that in most cases only 0.1 to 10 percent of the observed bacteria grow in culture.

A further complication associated with microbiological analyses of individual species in soil samples results from the variation in microbial growth rates. Typically, those organisms that form visible colonies during the most commonly used incubation times are those with the shortest generation times. Thus, many organisms that are important *in situ* are frequently overgrown. This difficulty may be overcome by using selective media (e.g., media containing antibiotics to which the desired population is resistant but most soil bacteria or fungi are sensitive or by using growth selective substrates) or enrichment cultures, which are, again, based on a unique property of the microbe(s) of interest.

Viable plate counts are also used for counting soil fungal populations. Estimates of fungal population densities in soil are confounded by their prolific spore production. Immense quantities of spores may exist in soil with few or none of the

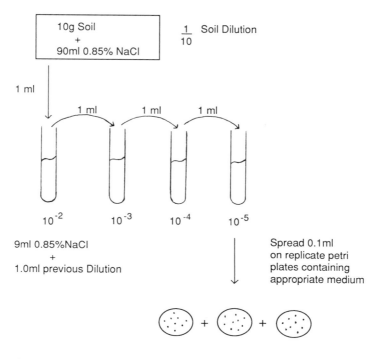

Calculation:

$$\text{Viable cells/g dry soil} = \frac{\text{Mean Plate Count X Dilution Factor}}{\text{Dry weight of soil}}$$

Fig. 2.5. Outline of viable plate count method for estimating numbers of organisms per gram of soil.

organisms existent as hyphae. Generally, fungal population densities are best estimated by direct microscopic observation.

Most-Probable-Number Procedures: A variation of the viable count procedure is the most-probable-number estimations of microbial populations. Many microbial populations (e.g., a variety of protozoa, nitrifiers, and sulfur-reducing bacteria) are difficult to grow in defined media. Also, when all potential contributors to a given metabolic function, perhaps nitrification, are to be quantified, it must be realized that some individual species contributing to the activity may have been grown in axenic culture, but the probability frequently exists that other nonculturable strains occur in soil. Most-probable-number procedures may be used to provide a measure of the density of such populations (Fig. 2.6).

Most-probable-number procedures are based on determining the dilution of the soil sample beyond which no propagule of the population to be quantified can be detected (i.e., the extinction point). Therefore, with this method, soil must be diluted

Fig. 2.6. Outline of most-probable-number method for estimating number of organisms per gram of soil.

until no further propagules of the organisms of interest are present in the highest dilution prepared, and multiple tubes of culture media (e.g., five per dilution) are inoculated from each dilution. After an appropriate incubation period, a trait common to all the organisms to be quantified is measured (visual observation of protozoan cells, nitrate or nitrite production for nitrifiers, or black color produced by precipitation of iron sulfide, for the preceding examples). Population densities are determined using prepared tables relating to the statistical probability of the presence of the organisms of interest to the number of positive samples for each dilution (e.g., Alexander, 1982). This procedure retains the problems of selectivity of media and incubation conditions. Furthermore, due to the nature of the calculation procedure and the range of dilutions used, confidence limits are frequently as great as plus and minus 300 percent.

2.4.2. Intrinsic Limitations of Viable Count Procedures

Each step in any viable count procedure (soil sample collection and storage, preparation of dilutions, colony development) contains major impediments for production of an accurate population census. The most fundamental obstacle is the necessity of collecting site-representative soil samples. Once attained, the soil must be stored in a manner that precludes major changes in the population density. Fortunately, considering the intrinsic variation of the dilution procedure itself, some change in the microbial populations can occur before statistically significant effects are detected. That is, with reasonable sample storage procedures, changes in microbial population densities are usually less than the inherent variation of plate count or most-probable-number data. For most studies, storage of the soil samples for limited times at 4°C results in minimal changes in the data.

Two sources of data variation incurred during preparation of soil dilutions may be controlled by careful attention to proper laboratory procedures and experimental protocol. These sources of error are variability due to dilution procedure precision and the potential for microbial growth during dilution preparation. A third source of error is basic to the procedure—association of microbial cells with soil particles—and likely must be considered during data interpretation, but little can be done to prevent the problem. Association of microbial propagules with soil particulates may result in an underestimation of their number. Typically, microbial colonies in soil consist of a few cells adsorbed to a soil particle or linked through production of a polysaccharide slime. Incomplete mixing of the soil in the diluent could result in data that is an estimate of soil-particle-bound colonies rather than separated microbial propagules. This data compromise can be minimized by thoroughly suspending and mixing the soil sample in the diluent, but realistically it cannot be eliminated.

A problem frequently encountered even when extreme caution is used in preparing the dilutions is microbial growth in the dilution media. Microorganisms generally have generation times of several hours in native soil samples. Thus, that time lapse from preparation of the initial dilution to plating of the sample might seem to have little effect on the colonies developing on the plate. Unfortunately, this is not true. Mixing the soil sample in the dilution medium liberates metabolizable organic material from the soil. The solubolized organic carbon may precipitate a few rapid divisions of the soil heterotrophic microbes under the less-restrictive conditions of the dilution tube. This problem is particularly acute when the diluent contains mineralizable organic carbon. For example, it is not unusual to prepare dilutions in 0.1 percent (w/v) amino acids or 0.1 strength of a growth medium such as trypticase soy broth to minimize the impact of dilution on fragile microbial cells. Hence, it is desirable to minimize the time lapse between preparation of the initial dilution and inoculation of the growth medium.

Interpretation of viable count data must be based on the knowledge that not all soil microbes can grow on the medium and that those organisms detected are a mixture of organisms, which are both active and inactive *in situ*. As indicated previously, the number of colonies developing on the growth media represents only the microbes capable of growth on the selected medium under the prevailing incubation conditions. This is particularly true when selective media are used. These media are necessarily more stringent than those selected for estimating total microbial populations (e.g., bacterial or fungal population densities). For example, it is not unusual to attempt to isolate organisms capable of using a particular pesticide as a carbon and energy source. A medium is prepared containing nitrogen, phosphorus, sulfur, and trace minerals necessary for growth and buffered at an appropriate pH for population development. A concentration of pesticide, which is sufficient for colonial development, but not toxic, is added to the salts medium. This medium appears to be ideal for the study, yet no microbes grow or only a few species are isolated from a soil sample in which it is known that a diverse population capable of catabolizing the pesticide exists. A limiting factor associated with the medium is that all microbes that require vitamins or amino acids are precluded from growth.

A further caveat regarding interpretation of viable count data relates to the nature of the state of the microbial populations in soil. Production of a colony on the test medium indicates that the particular organism was viable in the soil of interest, but nothing is indicated regarding the activity of the microbe in the native soil sample. Actively growing cells, nongrowing cells, and resting cells will develop colonies under appropriate conditions. Thus, isolation of a particular organism in axenic culture means that the organism was present in the soil sample, not that it contributed to the metabolic activity expressed in the soil at the time of sampling.

2.4.3. DNA Hybridization/PCR Procedures

DNA hybridization is a valuable procedure for detection of specific genetic sequences. This technique is useful in screening microbial isolates from soil for specific genotypic strains as well as for detecting specific gene sequences in DNA samples extracted directly from the soil (e.g., Sayler and Layton, 1990). Combining the method with the polymerase chain reaction (PCR) provides a sensitive means of detecting genes present in a few microbes in a soil sample (e.g., Steffan and Atlas, 1991). Evaluation of DNA sequences unique to individual microbial species, strains, or even guilds defined by a trait such as nitrogen fixation allows estimation of their population densities.

The polymerase chain reaction provides for amplification of individual nucleic acid sequences without the necessity of cloning them (Fig. 2.7). Chemically synthesized primers are hybridized to specific homologous DNA sequences adjacent to the target DNA regions. Thermally stable DNA polymerase is used to duplicate the target sequences. The thermal stability of the polymerase preserves the enzyme activity during the repetitive melting of the DNA, which is necessary for yielding the exponential increase in the target DNA.

To apply this procedure to soil systems, the bacteria whose populations are to be monitored may be separated from the soil prior to analysis, or the cells may be lysed *in situ* and the liberated DNA extracted. Humic acids must be removed from the sample since they inhibit the PCR reaction (e.g., Tsai and Olson, 1992a). For detection of microbes in low numbers, the extracted DNA may be concentrated by

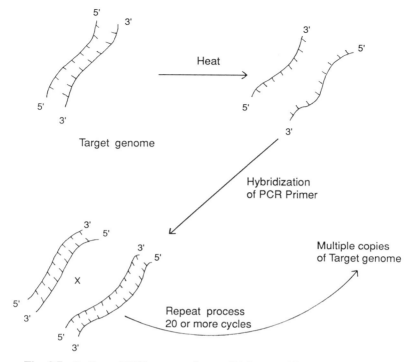

Fig. 2.7. Outline of PCR process for amplifying specific DNA sequences.

filtration (Bej et al., 1991a). Jacobson and Rasmussen (1992) separated bacteria from soil for PCR analysis using cation exchange resin.

The detection limits of hybridization procedures are low. For example, DNA–DNA hybridization limits of 4.3×10^4 cells per gram of dry soil (approximately 0.2 pg of hydrolyzable DNA in a 1-μg DNA sample) were measured for *Bradyrhizobium japonicum* populations in soil (Holben et al., 1988). Polymerase chain reaction increases the assay sensitivity so that as few as one colony-forming unit can be detected per gram of soil (Josephson et al., 1991; Pillai et al., 1991; Steffan and Atlas, 1988). Steffan and Atlas (1988) reported a one-thousand fold increase in sensitivity with PCR compared to analyzing nonamplified DNA. In their study of *Pseudomonas cepacia* AC1100, as few as 100 colony-forming units per 100 grams of sediment were detected in a background population of 10^{11} cells of diverse nontarget organisms. Tsai and Olson (1992b) were able to detect low copy numbers of 16S ribosomal gene fragments of *Escherichia coli* seeded in soils and sediments using PCR amplification. As few as 3 cells per gram of soil were detected. Picard et al. (1992) detected 0.2×10^5 genomes of *Frankia* sp. per gram of soil using PCR techniques. (For further discussion of these procedures, see reviews by Arnheim and Erlich, 1992; Bej et al., 1991b; Gibbs, 1990; Saylor and Layton, 1990; Steffan and Atlas, 1991).

2.4.4. Expression of Population Density per Unit of Soil

The ultimate goal for most ecological studies is to compare data from divergent soil ecosystems and to evaluate the impact of differences in physical, chemical, or biological properties between the sites of the activities. Traditionally, soil population densities, enzymatic activities, and chemical concentrations have been expressed on a per gram of dry soil basis. This method of data presentation may be reasonable with some systems, but with the diversity of soils encountered in field studies, consideration of the bulk density and the volumes of soil represented is important. This conclusion is based on the observation that, in reality, soil processes occur in a three-dimensional system (i.e., a volume of soil). The quantity of soil contained within that volume can be extremely variable. The bulk density of most mineral soils range from about 0.9 to 1.2. Thus, activity per gram of dry soil is approximately the activity per cubic centimeter of soil. In contrast, bulk densities of soils containing high levels of organic matter or organic soils may be quite low. For example, organic soils frequently have bulk densities of 0.1 grams of soil per cubic centimeter or less. Assuming a bulk density of 0.1, any activity expressed per gram of dry soil would represent 10 cubic centimeters of soil within the native environment. Thus, if the results of the low bulk density organic soil were compared on a dry weight basis to those similarly expressed for an average mineral soil, a tenfold error in conceptualization of the field situation would result from the different field volumes represented. Hence, when evaluating published data or preparing one's own data for publication, it is important to include a study of the site soil bulk density so that activities or populations in comparable quantities of soil can be evaluated.

2.4.5. Products of Soil Autecological Research

To the individual interested in studying unusual microbes or even in evaluating the diversity of microbial life forms, soil is a virtual treasure chest. Each gram of soil contains a mixture of both anaerobes and aerobes, heterotrophs and autotrophs, protozoa, nematodes, and viruses. Seemingly independent of current soil situation, we find organisms with temperature optima from cold-loving psychotrophic bacteria to those best adapted to high temperatures, thermophiles. Similar adaptability of populations to soil pH and salinity are found within the population of a single soil.

A list of most frequently encountered bacterial or fungal species, historically the most commonly studied soil populations, can only be criticized by its omissions rather than by its inclusiveness. Not only is there a vast array of species naturally present and functioning within any soil ecosystem, but propagules also gain entry by transport through air, water, or sediment. Furthermore, a variety of anthropogenic activities (such as those associated with waste disposal) augment the range of microbial species present. Commonly encountered soil bacterial species include representatives from the genera *Acinetobacter, Agrobacterium, Alcaligenes, Arthrobacter, Bacillus, Brevibacterium, Caulobacter, Cellulomonas, Clostridium, Corynebacterium, Flavobacterium, Hyphomycrobium, Metallogenium, Micrococcus, My-*

cobacterium, Pseudomonas, Sarcina, Streptococcus, and *Xanthomonas.* These heterotrophic bacterial genera are augmented in soil by autotrophic and mixotrophic representatives, including a variety of nitrifiers, *Thiobacillus* species, and iron bacteria. Actinomycetes populations include the commonly encountered *Nocardia* sp. and *Streptomyces* species. Fungal populations are comprised of a variety of slow- and fast-growing species. Perhaps the most commonly encountered are *Penicillium* and *Aspergilus* species plus representatives of the *Zygomycetes* and the mycorrhizae-associated ascomycetes and basidiomycetes.

The objectives of experiments requiring identification of soil populations range from isolation of individuals with traits of interest to analysis of the diversity of soil microbes by measuring ten to hundreds of different microbial species. The latter studies include evaluation of the interactions between the diverse populations present as well as a determination of the effect of external stimuli on microbial diversity, including cultural impacts such as of clear-cutting forests or of industrial pollution.

Diversity studies of soil microbes have been limited in the past due to the difficulty of determining function *in situ* and the vast quantities of data necessary to identify the species present or to group the organisms into guilds based on structural or physiological similarity. Development of automated techniques for cultural characterization and data analysis has made such studies more practical (see Russek-Cohen and Colwell, 1986; Mills and Bell, 1986; Holder-Franklin, 1986; Holder-Franklin and Tate, 1986). These experiments have provided an understanding of the genetic potential within a given ecosystem and an indication of the resilience of the system to external perturbation (i.e., the capacity for homeostatic stability). Some systems with low species diversity, such as soils receiving thermal hot spring outflows, appear to be reasonably stable, but the classic maxim has been that the greater the species diversity, the higher the probability that the community has the capacity to overcome intrusions such as influx of contaminants or pH or temperature alterations. Much of this stability is derived from latent genotypic capability.

2.5. PRINCIPLES AND PRODUCTS OF SYNECOLOGICAL RESEARCH

Although the sciences of microbiology and ecology have traditionally involved matching a genus and species name with the living creatures studies, as indicated previously, such activities have more limited value in soil ecosystem characterization. This is due in part to taxonomic problems of defining a bacterial species and the capacity of soil organisms to exist in active, inactive, or resting states. Thus, the presence of an organism does not necessarily prove its current participation in ecosystem driving or defining activities. Thus, synecological studies are of growing importance in defining soil activity. This process microbiology research involves assessment of metabolic and enzymatic activities, as well as determination of the properties of the soil ecosystem resulting from and controlling these activities.

As with the evaluation of specific microbial populations discussed previously, a prime concern in examining soil as a "tissue" is collection of a representative soil sample. Because soil enzyme analysis or respiration measurements generally require larger samples than are necessary to detect individual microbial species (grams of soil vs. perhaps milligrams or less), sample heterogeneity difficulties are reduced. Composited soil samples collected in such a way as to average system variability become more practical; for example, a composited soil sample from a corn field may contain equal quantities of soil from between rows and within rows. Soil collection considerations may reside more with concern for aboveground plant distribution affect or soil horizon influences than is possible with autecological studies.

2.6. INTERPHASE BETWEEN STUDY OF THE INDIVIDUAL AND COMMUNITY MICROBIOLOGY

Neither the study of individual species present in soil and their metabolic capability nor consideration of soil metabolic activates provides the sole conceptual basis upon which to construct the science of soil microbiology. Axenic cultures of soil microorganisms provide a detailed understanding of the nuances of the metabolic intricacy of the "players" in the soil ecosystem. Evaluation of enzymatic and respiratory activities of isolated soil samples demonstrate site-specific and perhaps even chronologically limited system traits; however, without an understanding of the organisms present, little is revealed regarding the resilience of the community. That is, both the test tube and the soil sample are requisites for the soil microbiologist.

This conclusion is demonstrated quite clearly when the impact of soil heterogeneity on growth factor limitations and microbial interactions are considered. Microorganisms require growth factors (i.e., vitamins and amino acids) for growth. As discussed in chapter 3, a greater proportion of soil bacteria requires that these factors preexist in their environment than are capable of synthesizing them *de novo*. The effect of this observation can be seen by examining a hypothetical reaction catalyzed by a microorganism requiring a vitamin. Assume that the following reaction sequence in the test tube is catalyzed by a single microbe:

$$A \rightarrow B \rightarrow C \rightarrow D$$

Let us hypothesize that to complete this reaction a growth factor is required to convert product B to product C. This may not be considered in laboratory experimentation because the growth media typically used are supplemented with yeast extract or comparable vitamin and amino acid sources to provide all nondefined growth needs. Thus, experimental results lead us to the conclusion that the bacterial isolate, perhaps one commonly occurring in soil, is perfectly capable of mineralizing the compound of concern, yielding carbon dioxide and water. That is, the compound of interest is reported to be biodegradable. Unfortunately, the requisite growth factor may be missing from the soil microsite. Therefore, instead of observing complete mineralization as the laboratory data would have predicted, product B accumulates. Thus, a substance that laboratory data indicated is biodegradable yields intermediates that accumulate in the ecosystem.

Laboratory data could also predict environmental stability of a product when complete decomposition is observed in native soils. This situation occurs in ecosystems where no single microbial species posses the enzymatic capacity to convert the substance to carbon dioxide and water, but a consortia that can catalyze the conversion exists. Again, consider the following generic reaction sequence:

$$A \rightarrow B \rightarrow C \rightarrow D$$

In this situation, substance A may be converted to product B by a microbial species. A second or even a third microorganism could be postulated to be required for transformation of products B to C and C to D. Thus, data from axenic culture indicate environmental stability of substance A, whereas amendment of soil with substance A results in complete conversion of substance A to product D.

2.7. CONCLUDING COMMENTS

This analysis of the soil biological community justifies reaching a seemingly heretical conclusion that the total potential of a soil ecosystem exceeds the sum of its parts. Conceptually, soil is readily separated into a mass of mineral and chemical components, as discussed in chapter 1, plus a variety of living organisms. It is the living organisms that transform the soil mineral matrix into the ever-changing, life-supporting substance essential for our terrestrial existence. Although a vast array of biological processes occurs in a soil sample, a much greater potential is contained within the unexpressed microbial genome. Thus, understanding the intricacies of the interaction of the biological dimension of soil with its physical and chemical aspects is essential for proper stewardship of our ecosystem. This conceptual understanding is based on the results of a variety of autecological and synecological studies of soil organisms and processes. Validity and interpretation of the data collected in these ecological studies must be predicated on a firm appreciation of the methodological problems associated with soil microbiological research. Even with their inherent limitations, such studies provide the foundation for further incursions into evaluation of the ever-changing world of the soil microbe.

REFERENCES

Ahmed, M., and J. M. Oades. 1984. Distribution of organic matter and adenosine triphosphate after fractionation of soils by physical procedures. Soil Biol. Biochem. 16:465–470.

Alef, K., T. Beck, L. Zelles, and D. Kleiner. 1988. A comparison of methods to estimate microbial biomass and N-mineralization in agricultural and grassland soils. Soil Biol. Biochem. 20:561–565.

Alef, K., and D. Kleiner. 1989. Rapid and sensitive determination of microbial activity in soils and in soil aggregates by dimethylsulfoxide reduction. Biol. Fert. Soils 8:349–355.

Alexander, M. 1982. Most Probable Number Method for Microbial Populations. *In* A. L. Page (ed.), Methods of Soil Analysis, Part 2. pp. 815–820. Agronomy Society of America, Inc. Madison, WI.

Anderson, J. P. E., and K. H. Domsch. 1973. Quantification of bacterial and fungal contributions to soil respiration. Arch. Mikrobiol. 93:113–127.

Anderson, J. P. E., and K. H. Domsch. 1975. Measurement of bacterial and fungal contributions to respiration of selected agricultural and forest soils. Can J. Microbiol. 21:314–322.

Anderson, J. P. E., and K. H. Domsch. 1978. A physiological method for the quantitative measurement of microbial biomass in soils. Soil Biol. Biochem. 10:215–221.

Anderson, J. P. E., K. H. Domsch. 1980. Quantities of plant nutrients in the microbial biomass of selected soils. Soil Sci. 130:211–216.

Arnheim, N., and H. Erlich. 1992. Polymerase chain reaction strategy. Annu. Rev. Biochem. 61:131–156.

Ausmus, B. S. 1971. Adenosine triphosphate—A measure of active microbial biomass. Biologie du Sol 14:8–9.

Azam, F., R. L. Mulvaney, and F. J. Stevenson. 1988. Determination of *in situ* K_N by the chloroform fumigation–incubation method and mineralization of biomass N under anaerobic conditions. Plant Soil 111:87–93.

Bakken, L. R., and R. A. Olsen. 1983. Bouyant densities and dry-matter contents of microorganisms: Conversion of a measured biovolume into biomass. Appl. Environ. Microbiol. 45:1188–1195.

Bej, A. K., M. H. Mahbubani, J. L. Dicersare, and R. M. Atlas. 1991a. Polymerase chain reaction–gene probe detection of microorganisms by using filter-concentrated samples. Appl. Environ. Microbiol. 57:3529–3534.

Bej, A. K., M. H. Mahbubani, and R. M. Atlas. 1991b. Amplification of nucleic acids by polymerase chain reaction (PCR) and other methods and their applications. CRC Crit. Rev. Biochem. Mol. Biol. 26:301–334.

Bratbak, G. 1985. Bacterial biovolume and biomass estimations. Appl. Environ. Microbiol. 49:1488–1493.

Bratbak, G., and I. Dundas. 1984. Bacterial dry matter content and biomass estimations. Appl. Environ. Microbiol. 48:755–757.

Brookes, P. C., D. S. Powlson, and D. S. Jenkinson. 1982. Measurement of microbial biomass phosphorus in soil. Soil Biol. Biochem. 14:319–329.

Brookes, P. C., D. S. Powlson, and D. S. Jenkinson. 1984. Phosphorus in the soil microbial biomass. Soil Biol. Biochem. 16:169–175.

Chapman, S. J. 1987. Inoculum in the fumigation method for soil biomass determination. Soil Biol. Biochem. 19:83–87.

Christensen, S., S. Simkins, and J. M. Tiedje. 1991a. Temporal patterns of soil denitrification: Their stability and causes. Soil Sci. Soc. Am. J. 54:1614–1618.

Christensen, S., S. Simkins, and J. M. Tiedje. 1991b. Spatial variation in denitrification: Dependency of active centers on the soil environment. Soil Sci. Soc. Am. J. 54:1608–1613.

Coûteaux, M. M., R. Henkinet, P. Pitta, P. Bottner, G. Billès, L. Palka, and G. Vannier. 1989. Native carbon mineralization of an acid organic soil after use of the chloroform-fumigation method to estimate microbial biomass. Biol. Fert. Soils 8:172–177.

Fazio, S. D., W. R. Mayberry, and D. C. White. 1979. Muramic acid assay in sediments. Appl. Environ. Microbiol. 38:349–350.

Findlay, R. H., G. M. King, and L. Watling. 1989. Efficacy of phospholipid analysis in determining microbial biomass in sediments. Appl. Environ. Microbiol. 55:2888–2893.

Flühler, H., M. S. Ardakani, T. E. Szuszkiewicz, and L. H. Stolzy. 1976. Field-measured nitrous oxide concentrations, redox potentials, oxygen diffusion rates, and oxygen partial pressures in relation to denitrification. Soil Sci. 122:107–114.

Focht, D. D. 1974. The effect of temperature, pH, and aeration on the production of nitrous oxide and gaseous nitrogen—A zero-order kinetic model. Soil Sci. 118:173–179.

Gibbs, R. A. 1990. DNA amplification by the polymerase chain reaction. Anal. Chem. 62:1202–1214.

Hedley, M. J., and J. W. B. Stewart. 1982. Method to measure microbial biomass phosphorus in soils. Soil Biol. Biochem. 14:337–385.

Holben, W. E., J. K. Jansson, B. K. Chelm, and J. M. Tiedje. 1988. DNA probe method for the detection of specific microorganisms in the soil bacterial community. Appl. Environ. Microbiol. 54:703–711.

Holder-Franklin, M. A. 1986. Ecological relationships of microbiota in water and soil as revealed by diversity measurements. pp. 93–132. In R. L. Tate III (ed.) Microbioal Autecology: A Method for Environmental Studies. John Wiley & Sons. New York.

Holder-Franklin, M. A., and R. L. Tate III. 1986. Introduction of the computer into autecological studies. pp. 76–91. In R. L. Tate III (ed.), Microbial Autecology: A Method for Environmental Studies. John Wiley & Sons. New York.

Jacobsen, C. S., and O. F. Rasmussen. 1992. Development and application of a new method to extract bacterial DNA from soil based on separation of bacteria from soil with cation-exchange resin. Appl. Environ. Microbiol. 58:2458–2462.

Jenkinson, D. S. 1976. The effects of biocidal treatments on metabolism in soil. IV. The decomposition of fumigated organisms in soil. Soil Biol. Biochem. 8:203–208.

Jenkinson, D. S., S. A. Davidson, and D. S. Powlson. 1979. Adenosine triphosphate and microbial biomass in soil. Soil Biol. Biochem. 11:521–527.

Jenkinson, D. S., and D. S. Powlson. 1980. Measurement of microbial biomass in intact soil cores and sieved soil. Soil Biol. Biochem. 12:579–581.

Josephson, K. L., S. D. Pillai, J. Way, C. P. Gerba, and I. L. Pepper. 1991. Fecal coliforms in soil detected by polymerase chain reaction and DNA–DNA hybridizations. Soil Sci. Soc. Am. J. 55:1326–1332.

Kassim, G., J. P. Martin, and K. Haider. 1982. Incorporation of a wide variety of organic substrate carbons into soil biomass as estimated by the fumigation procedure. Soil Sci. Soc. Am. J. 45:1106–1112.

King, J. D., and D. C. White. 1977. Muramic acid as a measure of microbial biomass in estuarine and marine samples. Appl. Environ. Microbiol. 33:777–783.

Lundgren, B. 1981. Fluorescein diacetate as a stain of metabolically active bacteria in soil. Oikos 36:17–22.

Martens, R. 1985. Limitations in the applications of the fumigation technique for biomass estimations in amended soils. Soil Biol. Biochem. 17:57–63.

Martens, R. 1987. Estimation of microbial biomass in soil by the respiration method: Importance of soil pH and flushing methods for the measurement of respired CO_2. Soil Biol. Biochem. 19:77–81.

Mills, A. L. 1985. Acid mine drainage: Microbial impact on the recovery of soil and water ecosystems. pp. 35–81. *In* R. L. Tate III and D. A. Klein (eds.), Soil Reclamation Processes: Microbiological Analyses and Applications. Marcel Dekker, Inc. New York.

Mills, A. L., and P. E. Bell. 1986. Determination of individual organisms and their activities in situ. pp. 27–60. *In* R. L. Tate III (ed.), Microbial Autecology: A Method for Environmental Studies. John Wiley & Sons. New York.

Parkin, T. B. 1987. Soil microsites as a source of denitrification variability. Soil Sci. Soc. Am. J. 51:1194–1199.

Parkin, T. B., J. L. Starr, and J. J. Meisinger. 1987. Influence of sample size on measurement of soil denitrification. Soil Sci. Soc. Am. J. 51:1492–1507.

Parkinson, D., and E. A. Paul. 1982. Microbial biomass. pp. 821–830. *In* A. L. Page (ed.), Methods of Soil Analysis, Part 2. Chemical and Microbiological Properties. American Society of Agronomy. Madison, WI.

Parsons, L. L., R. E. Murray, and M. S. Smith. 1991. Soil denitrification dynamics: Spatial and temporal variations of enzyme activity, populations, and nitrogen gas loss. Soil Sci. Soc. Am. J. 55:90–95.

Paul, E. A., and R. L. Johnson. 1977. Microscopic counting and adenosine 5'-triphosphate measurement in determining microbial growth in soils. Appl. Environ. Microbiol. 34:263–269.

Picard, C., C. Ponsonnet, E. Paget, X. Nesme, and P. Simonet. 1992. Detection and enumeration of bacteria in soil by direct DNA extraction and polymerase chain reaction. Appl. Environ. Microbiol. 58:2717–2722.

Pillai, S. D., K. L. Josephson, R. L. Bailey, C. P. Gerba, and I. L. Pepper. 1991. Rapid method for processing soil samples for polymerase chain reaction amplification of specific gene sequences. Appl. Environ. Microbiol. 57:2183–2186.

Rodriguez, G. G., D. Phipps, K. Ishiguro, and H. F. Ridgway. 1992. Use of a fluorescent redox probe for direct visualization of actively respiring bacteria. Appl. Environ. Microbiol. 58:1801–1808.

Ross, R. J. 1987. Soil microbial biomass estimated by the fumigation–incubation procedure: Seasonal fluctuations and influences of soil moisture content. Soil Biol. Biochem. 19:397–404.

Ross, D. J. 1988. Modifications to the fumigation procedure to measure microbial biomass C in wet soils under pasture: Influence on estimates on seasonal fluctuations in the soil biomass. Soil Biol. Biochem. 20:377–383.

Rosser, D. J. 1980. Ethidium bromide a general purpose fluorescent stain for nucleic acid in bacteria and eukaryotes and its use in microbial ecology studies. Soil Biol. Biochem. 12:329–336.

Roszak, D. B., and R. R. Colwell. 1987. Metabolic activity of bacterial cells enumerated by direct viable count. Appl. Environ. Microbiol. 53:2889–2983.

Russek-Cohen, R., and R. R. Colwell. 1986. Application of numerical taxonomy procedures in microbial ecology. pp. 133–146. *In* R. L. Tate, III (ed.), Microbial Autecology: A Method for Environmental Studies. John Wiley & Sons. New York.

Sayler, G. S., and A. C. Layton. 1990. Environmental application of nucleic acid hybridization. Annu. Rev. Microbiol. 44:625–648.

Sexton, A. J., T. B. Parkin, and J. M. Tiedje. 1985. Temporal response of soil denitrification rates to rainfall and irrigation. Soil Sci. Soc. Am. J. 49:99–103.

Schmidt, E. L., and E. A. Paul. 1982. Microscopic methods for soil microorganisms. *In* A. L.

Page (ed.), Methods of Soil Analysis, Part 2, pp. 803–814. Agronomy Society of America, Inc. Madison, WI.

Smith, M. S., and J. M. Tiedje. 1979. Phases of denitrification following oxygen depletion in soil. Soil Biol. Biochem. 11:261–267.

Sonderstrom, B. E. 1977. Vital Staining of fungi in pure cultures and in soil with fluorescein diacetate. Soil Biol. Biochem. 9:59–63.

Sparling, G. P., and A. W. West. 1988. A direct extraction method to estimate soil microbial C: Calibration *in situ* using microbial respiration and ^{14}C labelled cells. Soil Biol. Biochem. 20:337–343.

Steffan, R. J., and R. M. Atlas. 1988. DNA amplification to enhance detection of genetically engineered bacteria in environmental samples. Appl. Environ. Microbiol. 54:2185–2191.

Steffan, R. J., and R. M. Atlas. 1991. Polymerase chain reaction: Application in environmental microbiology. Annu. Rev. Microbiol. 45:137–167.

Tate, K. R., D. J. Ross, and C. W. Feltham. 1988. A direct extraction method to estimate soil microbial C: Effects of experimental variables and some different calibration procedures. Soil Biol. Biochem. 20:329–335.

Tate, R. L., III. 1986. Importance of autecology in microbial ecology. pp. 1–26. *In* R. L. Tate III (ed.), Microbial Autecology: A Method for Environmental Studies. John Wiley & Sons. New York.

Tate, R. L., III. 1991. Microbial biomass measurement in acidic soils: Effect of fungal: bacterial activity ratios and soil amendment. Soil Sci. 152:220–225.

Tsia, Y.-L., and B. H. Olson. 1992a. Rapid method for separation of bacterial DNA from humic substances in sediments for polymerase chain reaction. Appl. Environ. Microbiol. 58:2292–2295.

Tsai, Y.-L., and B. H. Olson. 1992b. Detection of low numbers of bacterial cells in soils and sediments by polymerase chain reaction. Appl. Environ. Microbiol. 58:754–757.

Vance, E. D., P. C. Brookes, and D. S. Jenkinson. 1987a. An extraction method for measuring soil microbial biomass C. Soil Biol. Biochem. 19:703–707.

Vance, E. D., P. C. Brookes, and D. S. Jenkinson. 1987b. Microbial biomass measurements in forest soils: The use of the chloroform fumigation–incubation method in strongly acid soils. Soil Biol. Biochem. 19:697–702.

Vance, E. D., P. C. Brookes, and D. S. Jenkinson. 1987c. Microbial biomass measurements in forest soils: Determination of K_C values and tests of hypotheses to explain the failure of the chloroform fumigation-incubation method in acid soils. Soil Biol. Biochem. 19:689–696.

Van Veen, J. A., and E. A. Paul. 1979. Conversion of biovolume measurements of soil microorganisms, grown under various moisture tensions, to biomass and their nutrient content. Appl. Environ. Microbiol. 37:686–692.

Verstraete, W., H. Van De Werf, K. Kucnirowicz, and M. Ilaiwi. 1983. Specific measurement of soil microbial ATP. Soil Biol. Biochem. 15:391–396.

Verstraeten, L. M. J., K. De Conick, K. Vlassak, W. Verstraete, H. Van De Werf, and M. Ilaiwi. 1983. ATP content of soils estimated by two contrasting extraction methods. Soil Biol. Biochem. 15:397–402.

Wildung, R. E., and T. R. Garland. 1985. Microbial development on oil shale waste: Influence on geochemistry. pp. 107–139. *In* R. L. Tate III and D. A. Klein (eds.), Soil Reclamation Processes: Microbiological Analyses and Applications. Marcel Dekker, Inc. New York.

3 Energy Transformations and Metabolic Activities of Soil Microbes

A microbe's habitat must not only have physical and chemical properties conducive to the sustenance of cellular viability (chapter 4) but must also contain the requisite energy source(s) to drive cellular metabolic processes as well as essential macro- and micronutrients for biomass synthesis. The assumption here is that the microbes themselves possess the genotypic traits necessary to function in their habitat and that prevailing conditions are sufficient to allow the expression of these traits. Prerequisites for population development must also include a means to cope with a variety of biological stresses, such as competition and predation (see chapter 6). These survival mechanisms may involve a capacity to exploit soil structural components to escape predation (e.g., refuges) or result from cell physiological traits (e.g., efficient energy recovery processes).

An appreciation of the many primary requirements for microbial existence in soil provides a portion of the foundation necessary for evaluating microbial metabolism and energy transformations in soil. Therefore, the objectives of this chapter are to determine the applicability and implications of laboratory derived principles of microbial growth kinetics to soil ecosystem function and to elucidate the diversity of energy-yielding processes existent in soil microbial communities. It is assumed in this quest that the student is familiar with basic microbiological and biochemical principles. A clear understanding of the fundamentals of intermediary metabolism, energy-yielding processes in cellular metabolism, regulation of gene expression, and microbial growth parameters is also required. Elementary microbiology and biochemistry textbooks should be consulted if deficiencies are noted. The following study is prepared with an emphasis on the elucidation of the nuances of the repercussion of soil physical, chemical, and biological realities on microbial metabolic function.

3.1. SOIL MICROBIAL GROWTH KINETICS

Since the primary benefits to microbes in soil are growth and/or maintenance of population densities, consideration of general microbial growth kinetics and specific examples of these replication and survival patterns in native soil ecosystems is a necessary starting point for analysis of soil physiological processes. To achieve this

Fig. 3.1. Examples of laboratory incubated open and closed systems for study of microbial processes.

goal, the applicability of laboratory-derived relationships to field situations will be examined.

Open vs. Closed Biological Systems: In the laboratory, microbial growth kinetics are assessed as closed systems (e.g., the culture tube or batch soil incubation chambers) or as open systems (e.g., a chemostat or in some primary aspects, soil columns) (Fig. 3.1). A closed ecosystem is self-contained in that external contributions to microbial function are excluded, whereas open systems receive inputs from processes occurring external to the defined boundaries of the site. As a consequence, in the closed system nutrient supplies are limited and metabolic waste materials or by-products accumulate, whereas in the open system microbial cells are constantly bathed in growth nutrients, and wastes do not necessarily accumulate in the vicinity of the colony or individual cell.

Although it is essentially impossible to exclude external influences from any portion of a soil site, conditions may exist locally where microbial population dynamics mimic those of a culture vessel incubated in the laboratory. Microbial community development in soil surrounding a portion of decaying insect biomass or plant debris exemplifies this situation. Growth of the microbial population is initially controlled locally by the quantities of readily metabolized substrates exuded from the animal or plant tissues. As the microbes grow and replicate, these easily metabolized substituents are depleted or are reduced to growth-limiting concentrations. Then, the rate of solubilization or release of nutrients from within the chitinous cell walls of the insect or the cellulosic walls of the plant material controls cellular function. In other words, nutrients for microbial growth are produced by solubilization of simple sugars by chitinase or by release of nutrients protected within cell structures through breaching of protective barriers. Upon depletion of all energy and nutrient supply pools in the vicinity of the microbial colony (i.e., total catabolism of the fixed carbon contained in the plant or animal debris), microbial growth ceases, and the population become quiescent. This state could be achieved prematurely if growth-limiting cellular waste products accumulate to inhibitory levels prior to nutrient depletion or if another essential nutrient supply is exhausted (e.g., free oxygen for aerobic microbes). At this point, continued activity of the microbial population depends on an influx of nutrients from outside its boundaries as previously described.

The artificiality of this comparison of the native soil condition with closed cultural systems emerges when all the soil properties that influence microbial growth are considered. The primary delimiters of microbial replication in the previously described site were energy and nutrients, but imposition of moisture limitations, deficiencies in fixed-nitrogen resources, or even difficulties with temperature modulation can easily by envisioned. Even with this small list of soil properties affecting microbial function, it is clear that both internal and external forces are involved in sustaining the soil biological community. In a soil site contiguous with a variety of different soil habitats, stresses and the means of their relief for microbial populations may easily originate in regions beyond the artificial boundaries of an arbitrarily defined closed system. Hence, in soil, closed ecosystem properties may be invoked to simplify data interpretation, but the experimenter must always bear in mind that the microbes of interest are functioning in a system of interchelated habitats.

Not only does a closed system imperfectly describe soil microbial ecosystems, but the system also is poorly described by the kinetics of an open system as represented by the model of the chemostat. True, many delimiters of biological activity are imposed from outside sources. This external control of microbial growth could result from a variety of conditions, such as from fixed carbon contributions of root exudates in rhizosphere soils, leaching of soluble fixed carbon sources from soil surface organic matter accumulations (e.g., forest litter layers, thatch accumulated in grassland soils), or anthropogenically created situations such as where industrial or domestic waste systems enter into a soil ecosystem (e.g., sewage or factory effluents).

These similarities to nutrient control of microbial growth in a chemostat are matched by an equally significant array of dissimilarities. A good example of these differences is the fate of microbial biomass. The influx of nutrients into the system can control biomass production in a manner reminiscent of those existent in a chemostat until space limitations are imposed. In open systems described by a chemostat in the laboratory, microbial biomass remains constant although the cells are in an active growth state. New cells are washed from the system as rapidly as they are produced. This condition is not reproduced in soils. Microbial cells tend to be trapped within the soil particulate structure. In soil, nutrients may flow to the cells, and waste products may be removed by the dynamic fluid system, but microbial growth eventually approaches the volume limitations imposed by the microsite structure.

3.1.1. Microbial Growth Phases: Laboratory-Observed Microbial Growth Compared to Soil Population Dynamics

The intricacies of microbial growth and the impact of soil properties on this process are best illustrated by contrasting microbial growth principles derived from laboratory study to those occurring in the soil ecosystem. The generalized microbial growth curve (Fig. 3.2) is usually divided into four phases; lag, exponential growth, stationary, and decline. Each of these portions of the growth curve can be observed in soil ecosystems. Yet, it must be noted that large changes in soil microbial populations are unusual. A major change in population density generally occurs only with an input of easily metabolized organic matter or release of another impediment to microbial proliferation. This pattern of population development is most commonly associated with soil sites where the native microbial growth rates are stimulated by alternating soil properties, such as introducing plant debris into a carbon- or energy-limited community, oxygenating an anaerobic system, or increasing the availability of native soil organic matter through disruption of the soil structure, which pre-

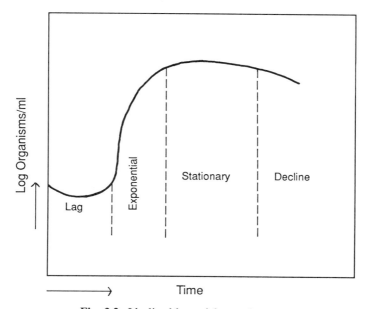

Fig. 3.2. Idealized bacterial growth curve.

cludes its catabolism. Once the previously existing impediments to microbial growth are removed, the cells replicate until a new limitation to community development is encountered.

Lag Phase: A lag phase occurs when conditions are appropriate for microbial growth, yet there is a delay before a measurable change in population density is detectable. This cellular division could be due to site amendment with metabolizable substances that were previously not present (e.g., pesticides, xenobiotic compounds) or alteration of the physical conditions (e.g., imposition of anaerobic conditions, which would stimulate growth of obligate anaerobes, or sufficient reduction of soil pH to favor acidophilic populations).

In axenic cultures, a sensitive measure of replication is usually employed. Hence, an accurate assessment of the lag time is gained. This assay sensitivity is difficult to achieve in soil. Most methods for quantifying soil microbial population density do not detect small changes in microbial population densities. Thus, several cell division cycles must occur before a change in the microbial populations can be recorded. Therefore, at least some delays in response of soil populations due to system modification are the product of methodological limitations.

A variety of methods are available for assessing increases in microbial biomass and biomass of individual microbial species in soil, including techniques for assessing the synthesis rates for various biomass-specific compounds, such as lipids (^{14}C-acetate incorporation into microbial lipids, Barnhart and Vestal, 1983; Federle et al., 1983; Tate, 1985) and nucleic acids (^{3}-H-thymidine incorporation, Christensen et al., 1989). (See chapter 2 for an analysis of the methods for detecting microbial populations.) Accuracy of the data derived from these procedures is limited by the variability of biomass within soil samples themselves, the extraction efficiency for recovery of the population of cellular component assayed from the soil sample, and the minimal populations that can be detected with each procedure.

A variety of implications of the occurrence of a lag phase for microbial growth in soils on interpretation of data describing soil process rates can be conceived. A positive aspect of the occurrence of a lag phase in soil microbial growth kinetics can be illustrated by the use of biodegradable soil amendment (e.g., pesticides). The additive must survive long enough in the soil to allow its interaction with the target species. The most time is allowed if the population density of the microbes catalyzing destruction of the soil amendment is minimal. With populations of microbes capable of mineralizing the soil amendment of a few hundred or even thousands of organisms per gram of soil, several cell divisions may be required before detectable declines in the pesticide concentration occurs. Thus, lag periods of days to weeks can be anticipated with pristine systems. In such systems, the chemical amendment must have an adequate longevity to accomplish the desired ends before its equally desirable biodegradation occurs. The efficacy of repeated uses of the soil amendment in the system is dependent on the longevity of the microbial populations involved in its decomposition.

A negative environmental implication of extended delays in biological decomposition of soil amendments is the potential for subsoil or groundwater contamination. Again, biodegradable compounds may be added to soil advertently or accidentally through mismanagement or through spills. To minimize the environmental impact in these situations, rapid biodegradation of the contaminant is necessary. The longer the lag in development of requisite biochemical activity, the greater the potential for leaching of the substance to a site where its presence is inappropriate and its decomposition is less likely to occur (e.g., the A horizon) to subsoil regions where decomposition is limited or groundwater contamination can occur.

This lag in the development of new biological activity may be exploited during laboratory estimation of *in situ* biochemical activities. Proper collection and storage of soil samples is essential for assessing soil processes. Unfortunately, disturbance of the soil matrix induces growth of many microbial populations. Fortunately, the delay in growth (lag phase) and use of methods such as storage at 4°C to reduce the microbial growth rate may afford the experimenter several hours to days to complete the laboratory assessments before statistically significant changes in population densities occur. The length of storage allowed before data are compromised by microbial growth depends upon the properties of the species of interest and the sensitivity of the assay method. The more sensitive the assay (i.e., the capacity to detect changes in population density of a few cells per gram soil) or the more prolific the microbes of interest (i.e., growth rates of a few hours under the storage conditions), the greater will be the problems associated with estimating *in situ* soil microbial population densities. Less storage stability is anticipated for samples to be assayed using PCR procedures or fluorescent antibody detection of microbial cells, because these methods are sufficiently sensitive to allow detection of increases of population densities of a few cells per gram of soil. In contrast, considerable increases in populations quantified by most probable number procedures is possible before significant changes in data can be shown. Confidence limits of ±300 percent are not unusual with these types of procedures. In either case, a control study must be conducted to determine the acceptability of the chosen storage procedure for the soils and populations in question.

Exponential Growth Phase: Conceptually, occurrence of a period of exponential growth, a blooming per se, of specific microbial populations in soil is easily envisioned. Amendment of soil with essentially any biodegradable substance at a nontoxic, growth-supporting concentration creates a new niche for the resident microbes.The genetic diversity of the indigenous community as well as the commonly occurring immigration of propagules from adjacent as well as distant ecosystems (e.g., via wind and water erosion) virtually ensures inoculation of the incorporated materials with organisms capable of their mineralization.

Other less appreciated, but perhaps more commonly occurring, processes inducing microbes to enter an exponential growth phase are associated with indigenous ecosystem processes. Intrusion of predators into a microsite where bacterial growth is controlled by space limitations may reduce the population densities sufficiently to allow several rounds of cell division. Indeed, balanced feeding of the predator could maintain the prey population in an active growth phase for extended time periods. When this equilibrium occurs, the microbial population could be dividing at a relatively rapid rate, but ingestion of the newly produced cells by the predators would prevent development of an observable increase in prey population density.

Similarly, a burst of cell division occurs after freeze–thaw cycles in soil. Freezing and thawing of soil results in the death of susceptible microbial populations. For example, Biederbeck and Campbell (1971) found that one freeze–thaw cycle in recently cropped soil resulted in the death of 92 percent of the bacterial population, 55 percent of the fungi, and 33 percent of the actinomycetes. The organic substances comprising the cell mass of the deceased organisms provided carbon and energy as well as a variety of cofactors for replication of the surviving microbial populations. This provision of nutrients and cofactors as well as the relief of any existent space limitations results in a rapid increase in microbial biomass.

Natural soil disturbance, which is initiated by soil animal activity or anthropogenic intervention such as cultivation, may also induce increases in microbial biomass. The disturbance may release occluded organic matter (i.e., substances trapped in soil aggregates) and/or relieve space or free oxygen limitations to growth. For example, plowing a meadow soil resulted in a decrease in native organic matter and an increase in microbial biomass carbon in the top 6 cm of coil by 40 to 50 percent (Angers et al., 1992). Although initially increases in microbial populations result from cultivation of virgin soils, ultimately the microbial biomass level in the soil declines due to depletion of easily metabolized organic matter resources. (See Tate, 1987, for a review of this topic.)

Stationary Phase: For many microbial species in soil, the stationary phase is their primary mode of existence. Growth is limited to the degree that new cell production is dependent upon accompanying cell death. For example, most of the surface area in soil is not occupied by microbial colonies; colonial development is controlled by the space available for expansion. Organisms lacking the capacity to migrate to new habitats soon become constrained by the accessible volume of the soil pore in which they are developing. Hence, division of the parent cells depends upon production of space by death and lysis of companion cells. Similar situations may result from the immobilization of a growth-limiting, essential nutrient into microbial biomass. Again, replication is controlled by releasing the limiting nutrient from biomass via cellular death and decomposition.

Decline (Death Phase): This portion of the bacterial growth curve is essential for the maintenance of soil community homeostasis. Under relatively constant conditions, system survival depends upon maintaining the status quo, which includes a balance in microbial population size. Population densities may increase in response to ecosystem perturbation, but to maintain community stability, the augmented population must return to its preexisting density. This conclusion is predicted on the assumption that a permanent change or relatively long-lived modification of system dynamics has not occurred. In that situation, all microbial population densities are modulated to the new levels optimal for the newly established system parameters. In a climax community (one with inputs and outputs in equilibrium), population increases are dampened by death and decay of the newly synthesized cell mass, predation by protozoa or other predators and subsequent decay of their populations, or any other comparable mechanism for returning the microbial populations to the previously existing equilibrium conditions.

Diauxic Growth: In laboratory culture, microbes are generally grown in a medium containing a single carbon and energy source plus a variety of cofactors. The growth curve depicted in Fig. 3.2 is characteristic of such a situation. Because soil microbes rarely, if ever, encounter chemically pure growth substrates, this growth model must be modified to reflect the more complex carbon and energy supplies that exist in soil and the microbial response to them. If the microorganism lacks the genetic capacity to metabolize more than one of the fixed carbon substrates present in its habitat, the presence of multiple organic compounds is of little consequence. This situation is rare. Microbes with the means of catabolizing a variety of carbon sources, many of which are concurrently present in the growth environment, are more common.

In a uniform system with no other limitations, a principle of microbial physiology which applies to microbial growth kinetics in the presence of multiple carbon and energy sources is catabolite repression. Catabolite repression, also called the glucose effect, occurs when one growth substrate represses synthesis of enzymes necessary for the metabolism of alternate compounds. Thus, use of a second substrate for growth cannot occur until the initial substance is exhausted or reduced in concentration below growth-supportive levels. This minimization results in diauxic growth kinetics (Fig. 3.3). The growth pattern reflects successive exhaustion of the available substrates. Once the first metabolized substance is depleted, a lag in growth is observed before the microbes commence to use the second substrate for energy and carbon. This pattern can be repeated several times if the microbes exist in a complex mixture of metabolizable organic compounds.

Demonstrating diauxic growth in the heterogenous world of soil microbes is difficult. A preferred carbon and energy source for a particular soil population may

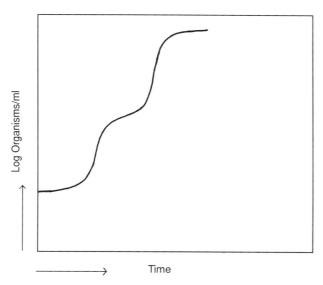

Fig. 3.3. Bacterial growth curve resulting from sequential catabolism of two energy substrates (diauxic growth).

become exhausted in one microsite while an adjacent site retains a reasonable supply of the substance. Thus, in a cross section of a soil community, a variety of microsites occur wherein many different organic substances are concurrently metabolized. Studies in pure culture may predict occurrence of diauxic growth kinetics, but the true field data may not reflect its occurrence because of the nonhomogeneity of the field soil. A more specialized growth curve as exemplified by diauxic growth kinetics may be more significant in a soil contaminated with a mixture of organic substances where sequential mineralization of the substituents could occur.

3.1.2. Mathematical Representation of Soil Microbial Growth

A variety of complex mathematical models have been developed to describe microbial metabolism in soil ecosystems (e.g., see Tate, 1986, for a review of this topic). These relationships are commonly based on combinations of simple microbial growth relationships derived from population behavior assessment in laboratory culture. One set of such equations, which may be used to elucidate the nuances of microbial growth in soil, is the classical Monod equations (Fig. 3.4).

Growth Rate Relationships to Microbial Biomass: Equation 1 ($dM/dt = \mu M$) depicts the fundamental association of growth rate with total population biomass present. The specific growth rate μ is not only intrinsic to the microbe itself (i.e., there is a maximal growth rate characteristic of the organism functioning under optimal conditions) but is also controlled by the physical and chemical conditions of the environment in which the organism exists. In soil the chemical and physical limitations of the habitat result in microbial population replication at a rate far below the maximum.

Classical Monod Equations for Microbial Growth

1. Equation for Logarithmic Growth:

$$dM/dt = \mu M$$

2. Growth Substrate Impact on Specific Growth Rate

$$\mu = \frac{\mu_{max} [S]}{K_s + [S]}$$

3. Growth Yield

$$y = -\frac{d[S]}{dM}$$

Where μ = specific growth rate, M = microbial biomass, t = time, μ_{max} = maximum growth rate, K_s = saturation coefficient, y = growth yield

Fig. 3.4. Classical Monod equations describing microbial growth, cell mass yield, and growth substrate interactions.

Because bacteria divide by binary fission, the growth rate is related to the generation time as follows:

$$T_d = \text{generation time} = \frac{\ln 2}{\mu}$$

Under the harsh conditions of the soil environment, generation times of several hours to days are common. These extended generation times compare to division cycles of as short a duration as 20 minutes in laboratory culture conditions.

While individual microbial populations in a soil community may be actively replicating, changes in cell mass as well as total microbial biomass may remain unchanged. Individual populations frequently increase or decline, but it is not uncommon to detect a relatively constant level of microbial biomass in an undisturbed soil site. This phenomenon is explained by the fact that it is not unusual to find that most of the microbes in a soil sample are in a resting (or nondividing) state. Thus, a small portion of the microbial community could be actively dividing, but their increase in numbers results in an insignificant alteration of the total soil biomass due to the quasi buffering effect of the inactive portion of the community. Similarly, microbial growth can be shown to be occurring; yet, population densities appear to be unchanging. In this type of situation, fate of the microbial biomass carbon must also be evaluated. Recall that a variety of predictors and parasites feed upon the microbes that decompose fixed carbon in soil. Thus, the microbes could be consumed by predators at a rate approximating their growth rate. Thus, quantification of growth rate indicates an actively multiplying community whereas total cell mass remains constant.

Growth Substrate Effects on Microbial Growth Rate: Although it is reasonable to assume that the microbial growth rate in soil will be proportional to the concentration of growth-limiting carbon and energy substrates as indicated by Monod equation 2 (Fig. 3.4), the relationship is not applicable to soil communities because it is essentially impossible to determine the concentration of growth substrate occurring in the soil microsites where microbial growth occurs. Assumptions underlying the use of concentrations of total substrate in a soil sample as a predictor of microbial growth rate are a) the substrates and microbes are evenly distributed throughout the soil sample and b) all the carbon and energy sources present in the culture (or vicinity) of the microbial cell are equally available for catabolism. These assumptions clearly do not apply in a heterogenous system such as soil.

Along with the variability in microbial oxidation of their energy sources resulting from the admixture of living cells, soil particulate mineral matter, and colloidal organic matter, heterogeneity of fixed carbon distribution within the soil matrix also results from the exhaustion of nutrients in microsites due to the activity of resident microbial populations. Once the local nutrient supply is depleted, the nutrient input rate controls the microbial growth rates. Growth kinetics then are described by diffusion equations or flow dynamics (where energy resources are contained in water flowing through soil pores).

Similarly, adequate energy or carbon sources may be available to the microbes in soil microsites, but they may occur in a form that precludes microbial attack. That is, they may be water-insoluble or retained within water-insoluble structures. Decomposition of these materials and hence the growth rate of microbial populations is controlled by the surface area of the particulate material, the rate of breaching of the protective barriers, or even the dissolution rate of the particulate substance.

Surface-area-determinated microbial growth is exemplified by the oxidation rate of elemental sulfur particles. Sulfur is minimally soluble in water, but its oxidation to sulfate provides an excellent energy source to a variety of sulfur-oxidizing bacteria. Physical attachment of the bacteria to the sulfur particle is not only required for oxidation of this substance but it is also frequently the rate-controlling factor. Early studies by Vogler and Umbriet (1941) revealed the necessity for this physical interaction. Recently, in an examination of the kinetics of oxidation of sulfur by *Thiobacillus ferrooxidans,* Espejo and Romero (1987) noted that only the bacteria attached to the sulfur particles were capable of growth. Therefore, because the elemental sulfur particles are not soluble in water and direct contact of the microbial cells is necessary for its oxidation, we can conclude that the surface area of the particles controls not only substrate oxidation rate but also subsequent microbial biomass production. (See chapter 14 for a more detailed examination of the role of surface area of sulfur particles in determining their oxidation rates.) Thus in this situation microbial growth rate is proportional to substrate surface area, not its concentration.

In contrast, Stucki and Alexander (1987) found that dissolution rate, not surface area of the particles, controlled the decomposition of biphenyl by *Moraxella* sp., *Pseudomonas* sp., and *Flavobacterium* sp. Once the water-soluble substrates were depleted, dissolution rate of the hydrocarbon apparently controlled the microbial growth rates.

Many biodecomposable xenobiotic compounds are detected in soils at micro- and nanogram concentrations per gram soil. Due in part to the common practice of evaluating biodegradation potential in culture media containing at least a thousandfold these concentrations (greater than 0.05 percent [wt/wt]), it has been assumed that microorganisms could not mineralize the minute quantities of organic substrates. Several laboratory studies have shown the mineralization of nanogram concentrations of a variety of organic substrates in culture (Simkins and Alexander, 1984), in sewage (Simkins and Alexander, 1984; Simkins and Alexander, 1985) and in soil (Scow et al., 1986). In the latter study, soil was amended with phenol or aniline and incubated in biometer flasks at 70 percent of -0.3 bar soil moisture. The soil moisture level resembled one that would be commonly encountered in a well-drained field soil. The phenol and aniline were mineralized, but not at rates resembling Monod kinetics. A two-compartment model provided the best fit of the data. These data suggest that biological oxidation can occur in soil at concentrations that would not be anticipated to support synthesis of cell mass.

Growth Yield as a Measure of Biomass Production in Soil: In culture the quantity of biomass yielded is proportional to the rate of substrate oxidation (Monod equation 3, Fig. 3.4). This relationship is applicable to microbial growth under well-defined cultural conditions where biomass production is totally determined by oxidation of a known energy source, such as occurs in a chemostat. Because of the simplicity of the relationship and the general ability to quantify microbial biomass and growth substances, application of this relationship to soil ecosystems is tempting. Unfortunately, the idealized conditions of the chemostat are rarely, if ever, replicated in soil. Using this relationship with soil systems is precluded a) by the variable contributions of energy sources to biomass production in soil, b) by difficulties in attribution of biomass yields with oxidation of a specific growth substance, and c) by complexities in linking biomass synthesis and growth substrate oxidation chronologically in complex soil ecosystems. In soil a variable proportion of the energy yielded from oxidation of the growth substrate by the slowly replicating cells is used for cellular maintenance (maintenance energy). Thus, a clear relationship between energy production and biomass yields is not always evident in soil ecosystems. Furthermore, with the number of fixed carbon and other energy sources available in the soil microbe's habitat, at least a portion of the growth assessed in an experiment could result from oxidation of energy sources not quantified or even considered in the experimental design. Over the duration of the study, the microbes may catalyze a number of different energy sources.

A further complication associated with complex soil systems relates to the capability to attribute changes in concentration of energy substrates directly with microbial growth yields. The concurrent observation of products of microbial metabolism in soil is not sufficient evidence to assume their linkage in a cause and effect relationship. For example, nitrate is commonly detected in soil samples. This compound is the product of energy production by nitrifiers that oxidize nitrite to nitrate. It could be easily assumed that the nitrite oxidizers in a soil sample arose from the production of nitrate associated with them in their habitat, but this nitrate could just as easily have entered the soil from external sources (carried with percolating water) or have been produced by previous generations of nitrite oxidizers. Direct linkage of the two variables—nitrifier biomass production and nitrite oxidation—is appropriate only if their augmentation occurs concurrently. Because this prerequisite generally cannot be assumed to be met, avoid applying the third Monod equation to analysis of soil systems.

3.1.3. Uncoupling of Energy Production from Biomass Synthesis

Cometabolism: Organic carbon compounds may also be oxidized in soils in a manner in which energy production, substrate oxidation, and microbial biomass yields are uncoupled. That is, a microbe produces enzymes that function in the oxidation of a compound without providing a metabolic benefit to the microbe. Energy to maintain the microbial cell mass is gained by oxidation of a second substrate, which may or may not be chemically related to the cometabolized material. A variety of aromatic-ring-containing compounds have been shown to be catabolized in soil through this mechanism. For example, catabolism of a variety of polychlorinated biphenyls (PCBs) comprising the mixture Arochlor was stimulated by the addition of biphenyl to soil samples (Focht and Brunner, 1985). In the presence of the biphenyl, between 48 and 49 percent of the PCBs were converted to carbon dioxide compared to less than 2 percent in the absence of the biphenyl.

Similar cometabolic responses have been noted for aniline, phenol, and their mono-chlorinated derivatives (Janke and Ihn, 1989), catabolism of trichloroethylene in the presence of methane (Henry and Grbic-Galic, 1991), and chloroparaffinic hydrocarbons (Beam and Perry, 1973; Beam and Perry, 1974). The role of cometabolism in the oxidation of chlorinated aromatic-ring-containing compounds has recently been reviewed by Boyle (1989).

A fortuitous observation from the view of environmental reclamation is the fact that the energy providing substance and the chemical cometabolized need not be present in equivalent amounts. Growth-supporting concentrations of an energy-producing substrate can facilitate the catabolism of nanogram concentrations of the cometabolized substrate (e.g., see Schmidt and Alexander, 1985; Schmidt et al., 1987; Wang et al., 1984; Wiggins and Alexander, 1988). Hence, amendment of soil with metabolizable substrates may encourage decomposition of contaminating organic compounds present in trace quantities.

Maintenance Energy: A proportion of the energy derived from oxidation of growth substrates is used for cell maintenance. This is termed maintenance energy (the portion of the energy yielded by substrate metabolism that is not used for biosynthetic purposes). Maintenance energy is used to provide for such basic cellular functions as motility and control of internal osmotic pressure. In a culture limited by the quantity of energy source, the growth yield is reduced by the amount of energy source expended to meet these needs. The slower the generation time of a microbe, the larger the proportion of energy required to maintain cellular integrity. Because of the long generation times of soil microbial populations and the observation that these organisms are generally energy limited, significant reductions in growth yield results from maintenance energy expenditures by these populations.

An example of the impact of maintenance energy expenditures on growth yield of microbes growing at reduced growth rates was provided by Traore et al. (1982). Reduction of the apparent growth rate in chemostat culture from 0.206 to 0.0125/hr reduced the growth yield of *Desulfovibrio vulgaris* from 6.65 to 2.06 g/mol. That is, at the lowest growth rate, 68 percent of the energy was used for cellular maintenance. This phenomenon combined with the low free-energy efficiencies explain how relatively high respiration rates may be detected in soil with a minimal increase in microbial biomass. The microbes are replicating, but they are doing so rather inefficiently.

3.2. IMPLICATIONS OF MICROBIAL ENERGY AND CARBON TRANSFORMATION CAPACITIES ON SOIL BIOLOGICAL PROCESSES

Although elucidation of biochemical pathways used by soil microbes and the physiological control of the expression of these capacities generally is studied in axenic culture, the real value of microbial metabolic processes resides in their expression by the organisms in their native environment. The metabolic diversity of the soil microbial community has long been appreciated.

A manifestation of this understanding is the long-held concept of microbial infallibility. It was commonly believed in the mid-twentieth century that microbes that can decompose any organic compound could be anticipated to occur in soil. This opinion justified the optimistic treatment of soil as a vast repository for processing human organic wastes. With the advent of the chemical era and initiation of what could be termed a time of excessive exploitation of soil resources, this concept of microbial infallibility has frequently been transformed into a feeling of extreme microbial fallibility. Many difficulties have arisen because organic compounds, which contain chemical linkages rarely if ever encountered in biologically synthesized compounds, are produced industrially and added to soil systems on a routine basis. These same compounds may have a negative or positive effect on the function of the soil microbial community, but the primary environmental impact is their extended persistence. Accordingly, the concept of microbial infallibility needs to be modified to include the concept of natural synthesis; that is, the soil microbial community is considered to be capable of decomposing all naturally synthesized organic compounds, assuming all other physical and chemical requirements for microbial function are met.

To develop soil management techniques that optimize native biological processes as well as to produce beneficial xenobiotic compounds that are biologically decomposed in all target ecosystems, an appreciation of the diversity of metabolic potential and its control in soil ecosystems must be possessed. Thus, this discussion explains the impact of microbial metabolic and energy transformation capacities on ecosystem function and resilience.

3.2.1. Energy Acquisition in Soil Ecosystems

Soil microbes, as all living entities, derive the energy necessary for cellular function through oxidation reactions. A basic principle of biochemistry is that all oxidations must be balanced by reductions (i.e., electrons generated in the oxidation processes must be transported to an electron acceptor). Terminal acceptors include oxygen, nitrate, nitrite, sulfate, and a variety of organic compounds, including acetate and pyruvate.

The quantity of energy provided to the growing cell in the form of adenosine triphosphate (ATP) is dependent upon the intermediate steps between the initial oxidative reaction and the final electron acceptor. A direct transfer of the electron from the oxidized substrate to an organic recipient such as occurs in glycolysis (Fig. 3.5), yields a single ATP molecule per electron transfer, whereas passage of the electron through a series of intermediate reduction and oxidation reactions (Fig. 3.6), as seen in the oxidative metabolism of glucose through the citric acid cycle, yields as many as three ATP molecules per electron transferred. Thus, the incomplete oxidation of glucose to organic intermediates by glycolysis produces a net gain of two ATP molecules, whereas a maximum of 38 ATP molecules is produced by the complete conversion of glucose to carbon dioxide (Fig. 3.7).

Aerobic vs. Anaerobic Processes: Environmental metabolic processes can conveniently be grouped into two rather broad categories: aerobic and anaerobic conversions. Maximal biological energy is made available for cellular function when

Embden - Meyerhof Pathway

Fig. 3.5 Glycolytic conversion of glucose to pyruvate. Note the net production of 2 ATP and 2 NADH molecules. Under fermentative conditions, organic intermediates such as acetaldehyde or pyruvate serve as the terminal electron acceptors.

free oxygen is the terminal electron acceptor. Inversely, incomplete oxidations of fixed carbon substrates in the absence of free oxygen results in reduced energy recovery and reduced biomass production. Based on the foregoing discussion of substrate level phosphorylation and the recovery of energy through an electron transport chain, such conclusions are reasonable and are applicable to the majority of metabolic processes.

Fortunately, the underlying assumption of this generalization—cytochrome-based respiration (the basic process when molecular oxygen serves as the terminal electron acceptor) occurs only under aerobic conditions—is *not* true. Denitrification (e.g., Koike and Hattori, 1975; Kristjansson et al., 1978) and sulfate reduction (e.g., see Kim and Akagi, 1985; Postgate, 1984), two conversions that require anoxic conditions, yield energy to the growing cell through cytochrome-based electron transfer chains similar to those depicted in Fig. 3.5. Thus, for these two anoxic processes, cell biomass production and total mineralization of organic substances

Fig. 3.6. Production of ATP via transport of electrons through an electron transport chain with molecular oxygen as the terminal electron acceptor.

approaching those achieved when free oxygen is the terminal acceptor are found to occur.

A basic principle to be considered in these dissimilatory processes (oxidation of carbon substrates coupled to cytochrome oxidation under anoxic conditions) is that the use of the specific electron acceptors is mutually exclusive. In other words, the dissimilatory reduction of nitrate or sulfate does not occur until the oxygen supply

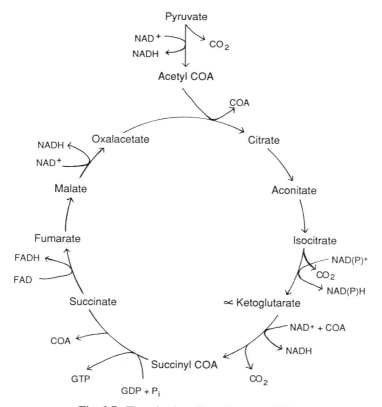

Fig. 3.7. The tricarboxylic acid cycle (TCA).

TABLE 3.1. Reduction Potentials of Some Common Electron Acceptors Functioning in Soil Biological Systems.

Redox Couple	E_0 (volts)
O_2/H_2O	0.82
Fe^{3+}/Fe^{2+}	0.77
NO_3^-/NO_2^-	0.42
Cytochrome c Oxidized/Reduced	0.25
Acetaldehyde/Ethanol	0.20
Cytochrome b Oxidized/Reduced	0.07
SO_4^{2-}/S^{2-}	−0.22
$NAD^+/NADH$	−0.32
CO_2/CH_4	−0.35
H^+/H_2	−0.42

around the microbial cell is exhausted. This principle results in a reduction in the local reduction–oxidation (redox) potential sufficiently to allow use of the alternate electron acceptors. Each of these substances function at specific redox potentials (Table 3.1). That is, the redox potential of the environment of the microbial cells is perched at the specific value associated with a given electron acceptor until it is exhausted, then the potential drops. For example, in a system containing nitrogen oxides (but no oxygen), the reduction potential remains at the level dictated by the nitrogen oxides until all of them have been reduced through denitrification, then organic substances or other acceptors (e.g., sulfate), if present, become the primary terminal electron acceptors for the active biological community. A more complex example of this phenomenon involves the sequential oxidation and reduction of growth substrates and terminal electron acceptors in a flooded soil containing, nitrate, ammonium, manganese(II), and ferrous iron (Patrick and Jugsujinda, 1992). No overlap in the oxidation or reduction of nitrate and manganese was detected, and little overlap in the transformation of manganese and ferrous ions occurred.

The environmental implications of the potential for dissimilatory oxidation of fixed carbon with nitrate or sulfate as the final electron acceptor include the fact that greater biomass is supported by the processes than occurs with classical fermentation reactions and that the organic substances providing the energy are completely oxidized to carbon dioxide and water. The latter trait is particularly significant in bioremediation processes. (See Tiedje et al., 1984, for a review of the capacity to exploit anaerobic processes for soil remediation.) This trait is especially important in situations where the oxygen supply is exhausted due to the rapid catabolism of the energy-supplying substrate. Should nitrogen oxides or sulfate be present, the contaminant could still be reduced rapidly to carbon dioxide and water even under the anoxic conditions.

The potential for exploiting dissimilatory processes under anoxic conditions for decomposition of soil contaminants is underscored by the variety of substrates catabolized by denitrifiers and sulfate reducers. Denitrifiers have been shown to oxidize a variety of aromatic-ring-containing compounds, including phthalic acid (Aftring et al., 1981; Nozawa and Maruyama, 1988), toluene (Altenschmidt and Fuchs, 1991; Evans et al., 1991a), xylene (Evans et al., 1991b), benzene and a variety of alkyl benzenes (Hutchins, 1991), resorcyclic acids and resorcinol (Kludge et al., 1990), plus the more complex acenaphthene and naphthalene (Mihelcic and Luthy, 1988a; Mihelcic and Luthy, 1988b). An added advantage of oxidation of these substances via denitrification is that a second soil pollutant, nitrate, is also removed from the system through its conversion to nitrous oxide and dinitrogen. Similarly, sulfate can serve as a terminal electron acceptor for oxidation of such compounds as chlorophenol (Häggblom and Young, 1990), aniline and dihydroxy-benzenes (Schnell et al., 1989), benzoate (Tasaki et al., 1991), catechol (Szewzyk and Pfennig, 1987), and m-cresol (Ramanand and Suflita, 1991).

3.2.2. Microbial Contributions to Soil Energy and Carbon Transformations

The diversity of habitats available to soil microbes is reflected in their versatility in carbon and energy metabolic capacities. Essentially all microbial groupings (Table 3.2) are found at least to a limited extent in soil ecosystems. Some are important contributors to total microbial biomass and are versatile in their capacity to oxidize fixed carbon substrates (e.g., the heterotrophs) whereas other, although limited contributors to total soil microbial biomass, are important because of their pivotal role in soil biogeochemical processes (e.g., nitrifiers). Others pique the interest of soil microbiologists not because of the imposition of their presence on soil system function but more due to their unusual or precarious existence in the system (e.g., soil algae or photosynthetic bacteria).

To gain a more complete understanding of metabolic properties of soil microbes, the function and biological limitations of autotrophs and heterotrophs (aerobic and anaerobic) are examined. Although these organisms are found in a variety of habitats, their function in soil acquires unique characteristics imposed by the soil physical and chemical environment.

Chemoautotrophic Existence in Soils: Chemoautotrophic bacteria gain their energy through the oxidation of inorganic compounds. Primary electron sources are ammonium, nitrite, hydrogen, and a variety of sulfur-based anions. A characteristic of these microbes is their specificity for specific energy sources. For example, autotrophs that oxidize nitrogenous compounds use only ammonium or nitrite as electron sources, whereas sulfur autotrophs are restricted to oxidation of sulfur compounds. Furthermore, the specificity generally extends to individual compounds within each grouping, for example nitrifiers are divided into ammonium and nitrite oxidizers. Chemoautotrophs may assimilate simple organic compounds, but their primary cellular structure is provided through the reduction of carbon dioxide. Many of the autotrophic bacteria are obligate autotrophs; that is, they can use only carbon dioxide as a carbon source. Facultative autotrophs are capable of growing either autotrophically or heterotrophically.

Chemoautotrophic organisms are of pivotal importance to soil biotic function. Although their contribution of fixed carbon to the ecosystem is minor compared to

TABLE 3.2. Some Groupings of Soil Microbes Based on Their Primary Energy Source.

Microbial Group	Energy Source	Carbon Source	Terminal Electron Acceptors	Examples
Heterotrophs:				
Aerobes	Fixed carbon	Fixed carbon	O_2, NO_x, SO_4^{2-}	*Arthrobacter* sp. *Pseudomonas* sp. *Bacillus* sp. Denitrifiers Sulfate reducers
Anaerobes	Fixed carbon	Fixed carbon	Fixed carbon	*Clostridium* sp.
Fermentaters	Fixed carbon	Fixed carbon or O_2	Fixed carbon	Enterics Lactic acid bacteria
Chemoautotrophs:				
	Inorganic chemicals (aerobes)	CO_2	O_2, NO_3^-	Nitrifiers *Thiobacillus*
	H_2 (anaerobes)	CO_2	CO_2	Methanogens
Photoautotrophs:				
	Light (aerobes)	CO_2		Algae
	Light	CO_2		Photosynthetic bacteria

that incorporated by higher plants, oxidation of their inorganic substrates is frequently central to the completion of various soil biogeochemical cycles. For example, ammonium oxidation to nitrate is primarily catalyzed by autotrophic nitrifiers in soil.

Although neither heterotrophs or chemoautotrophs capture all the energy released from oxidation of their substrate, chemoautotrophic bacteria are especially inefficient in energy recovery (free-energy efficiency). (Free-energy efficiency equals the free energy captured by the microbe divided by the free energy available in the substrate oxidized for energy.) The free-energy efficiency is generally measured in culture by quantifying the heat of combustion of the cells produced by metabolism of a specific substrate and the heat of combustion of the quantity of substrate oxidized in this process. This is not measured directly in soil because determining the specific substrate catabolized among the vast array available to the cell is difficult. The free-energy efficiency can be as low as 5 to 10 percent for oxidation of nitrite to nitrate by *Nitrobacter* species (Alexander, 1977) to 50 percent for the oxidation of elemental sulfur to sulfate by *Thiobacillus thiooxidans.*

For the obligate autotrophs, all the cellular substituents must be synthesized from the reduction of carbon dioxide. Considering that the conversion of ammonium to nitrate yields 66 kcal of energy per mole of substrate oxidized, the oxidation of nitrate to nitrate produces only 20 kcal of energy per mole of substrate, and the cellular substituents of both the ammonium and nitrite oxidizers are the same, approximately threefold more nitrite must be oxidized to produce the same biomass of nitrite oxidizers as for the ammonium oxidizers. Approximately 35 nitrogen atoms are oxidized per carbon fixed for the ammonium oxidizers as compared to about 100 nitrogens for the nitrite oxidizers. Thus, the microorganisms using these low-energy substrates must oxidize large quantities of their substrates to produce the same biomass as those growing by more efficient metabolic processes.

In summary chemoautotrophs are of minor importance as a contributor to biomass in soil ecosystems. Their primary contribution to ecosystem development resides in their capacity to oxidize soil mineral components. They are the primary if not sole contributor to nitrite formation, catalyze some sulfur oxidations, participate in oxidation of ferrous iron to ferric ion, are instrumental in the genesis of acid mine wastes, contribute to corrosion and weathering of rocks and statues, and prevent the accumulation of molecular hydrogen.

Anaerobic Heterotrophs and the Soil Ecosystem: Anaerobes are those microorganisms that grow and reproduce in the absence of free oxygen. Organisms using cytochrome-based electron transport systems, even when functioning under anoxic conditions are, by definition, excluded from this class of organisms. Although there is in reality a gradient of sensitivity to free oxygen, oxygen is lethal to anaerobes.

Anaerobic microbes may be divided into two classes, strict or obligate anaerobes and facultative anaerobes. In the presence of molecular oxygen, strict anaerobes do not grow or survive in the vegetative phase for extended periods. Examples of this group of organisms include the clostridia (which are ubiquitous in soil), methanogens, and some protozoa (which occur in the rumen and sediments). No anaerobic fungi have been reported to occur in soil although anaerobic fungi may be found in the rumen (Billon-Grand et al., 1991; Dore and Stahl, 1991; Gordon and Phillips, 1989; Lowe et al., 1987; Teunissen et al., 1991; Webb and Theodorou, 1991). Facultative anaerobes can grow in the presence or absence of free oxygen. Examples of these organisms include the commonly encountered enteric bacteria.

In spite of their sensitivity to molecular oxygen and the abundance of this molecule in our terrestrial system, anaerobic microbes are nearly ubiquitous in soil systems. In fact, in a well-aggregated soil, anaerobic and aerobic metabolism occur concurrently. Aerobes frequently function on the surface of a soil aggregate at a rate sufficient to exhaust free-oxygen supplies such that the internal portions of the granule are anaerobic. Because it is not uncommon for the requisite nutrients to support growth of anaerobic microbes to be contained within the aggregate structure, these populations not only survive but also contribute significantly to overall soil metabolic activity.

Flooding of the soil, which can result from a heavy rainfall, may result in an expansion of the anaerobic microbial biomass. A flooded soil is not necessarily an anaerobic soil. The oxygen tension of a particular soil system depends on a) the oxygen diffusion rate in water, b) the rate of consumption of this molecule by the living community plus any depletion due to chemical processes, as well as c) the potential for oxygenation of the sites through influxes of oxygen-bearing water. In other words, a flooded system in which the water level is maintained by inputs of oxygenated waters may not become anoxic. Soil sites commonly associated with oxygen-depleted conditions are swamps or marshes, rice paddies, sediments, and any other sites receiving inputs of easily decomposed organic matter. The latter soils include a variety of environmentally stressed situations such as buried organic matter in land fills or soils impacted by spills of biodecomposable societal products.

The products of complete mineralization of an organic substance containing carbon, hydrogen, oxygen, and sulfur in the absence of oxygen are carbon dioxide, methane, ammonium ion, hydrogen sulfide, and water. This list of products contrasts to the carbon dioxide, ammonium ion, water, and hydrogen sulfide yielded by aerobic decomposition of the same compound. Of more general importance to soil systems are the products of incomplete mineralization of organic substances by facultative and obligate anaerobic organisms. In these situations, alcohols and organic acids accumulate. These fermentations leave a major portion of the energy contained in the substrates in the fermentation products. Thus, the cell yield is minimal compared to that which would be produced through complete, aerobic oxidation of the fixed carbon sources.

Methanogenesis: Methane generation provides an excellent example of complete mineralization of organic compounds under anaerobic conditions and the interaction of a variety of anaerobic bacteria to achieve these ends. Complex substrates are decomposed by anaerobic and fermentative bacteria to the simple substrates that can be converted to methane. Methane synthesis is continuous in marshes and bogs where the extremely reducing environment necessary for the process (-200 to -1000 mV) commonly exists.

Methane production is catalyzed by a highly specialized group of obligately anaerobic microbes, the methanogenic bacteria. This unique group of anaerobic bacteria includes the following species: *Methanobacterium arbophilicum, M. for-*

micum, M. ruminatum, M. mobile, M. thermoautotrophicum, Methanococcus vanielli, Methanosarcina barkeri, and *Methanospirillum hungatii.* These organisms reduce carbon dioxide to methane with hydrogen generally serving as the electron donor by the following reaction:

$$4H_2 + CO_2 \rightarrow CH_4 + 2H_2O$$

Therefore, hydrogen is serving as a sole source of reducing power for both methanogenesis and cell carbon synthesis. Other substrates that can be converted to methane include methanol, formate, acetate, and methylamines. With the latter substrates, the carbon substrate serves as both the electron donor and the final electron acceptor. For example, acetate is converted to methane and carbon dioxide as follows:

$$CH_3COO^- + H^+ \rightarrow CH_4 + CO_2$$

The acetate, carbon dioxide, and hydrogen used by these methanogenic bacteria is generated by the fermentation of complex polymers, including cellulose and other polysaccharides, by obligate anaerobes, and by fermentative bacteria (Fig. 3.8).

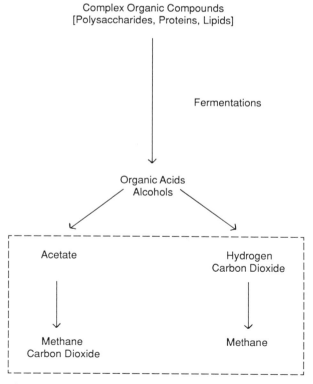

Fig. 3.8. General organic matter transformations associated with methanogenesis. Processes enclosed in the box are catalyzed by methanogens.

Aerobic Heterotrophs and the Soil Ecosystem: The soil microbial community is primarily driven by the energy contained within photosynthetically fixed carbon. In other words, the soil ecosystem is indirectly solar powered. When we consider that most other carbon pools entering soil are products of decomposition of plant carbon (e.g., animal biomass, composts and sludges, as well as petroleum), the intense dependence of belowground life processes on plant productivity is highlighted.

Biomass carbon entering soil can be divided into two primary pools, biodegradation resistant and readily or easily metabolized carbon. Clearly, the biodegradation resistant categorization is a relative term in that these substances (e.g., lignified materials) are decomposed in soil, albeit slowly. The readily metabolized fractions are defined as those fixed carbon materials that provide a carbon and energy source for the soil microbial community with a minimal energy expenditure by the microbe. Included in these energy expenditures are the activities necessary to synthesize catabolic enzymes as well as those required to activate the substrate molecule to enhance its degradation susceptibility. Consider, for example, the phosphorylation of hexoses with adenosine triphosphate to form hexose-monophosphates (Fig. 3.5).

The biodegradation resistant substances may contain sufficient chemical energy reserves to support microbial growth, but a diversity of enzymes are necessary to convert the compounds to forms where that energy can be readily recovered by the microbial community. Thus, the energy expenditures for synthesis of the enzymes that are necessary to convert these substances to forms from which the microbes may recover energy commonly exceed the potential energy that can be recovered. The biodegradation resistant group of compounds includes plant lignin as well as soil humic acids.

It is not unusual to know a priori that a certain metabolic activity must be present in a soil sample, but due to difficulties associated with the analysis of a complex mixture of abiotic and biotic components, detection or quantification of the process is precluded. Fortunately, the properties of comparative biochemistry dictate that metabolic reactions occurring in bacteria or fungi, which can be studied in laboratory culture, are common with minor variations to those occurring in the soil microbial community. Thus, if a fixed carbon substance is amended to a soil sample and its decomposition is documented, it is reasonable to conclude that decomposition is occurring, with minor variations, by commonly described biochemical pathways (e.g., Embden Meyerhoff Parnas and citric acid cycle).

This observation allows construction of a model predictive of the fate of the primary classes of organic carbon compounds entering soil (Fig. 3.9). As in other ecosystems, in soil, most fixed carbon will be catabolized to simple organic acids, which can be activated to acetyl-coA and fed into the tricarboxylic acid cycle. For example, proteins are hydrolyzed by proteases and deaminated to organic acids. Aromatic-ring-containing substances are hydroxylated to facilitate ring cleavage (Fig. 3.10). In this process, molecular oxygen is incorporated directly into the aromatic ring.

Hence, it is logical to predict that, within any soil sample containing viable microbial populations, all the common enzymes to intermediary metabolism associ-

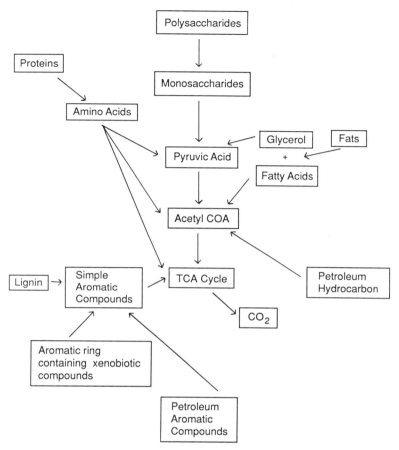

Fig. 3.9. Conceptual model of carbon catabolism by the soil heterotrophic microbial community.

ated with these cycles plus those activities, such as proteases and hydrolases necessary to produce the precursors of these cycles, are present in that soil sample.

Humification: The complexity of the soil system, including the potential for a variety of spontaneous chemical reactions, provides a number of fates for reactive biochemical intermediates, which may not be detected in laboratory culture. Essentially any reactive organic compound entering soil could be incorporated into soil humic acids (i.e., it can be humified). Because many of the reactions associated with incorporation of organic compounds into humic acids are chemically catalyzed, the reaction rate is controlled in large part by the reactivity of the substrates and the probability of interaction of the reactants. In typical soils containing less than 2 percent organic matter, the probability of such reactions is limited, although significant. Because in most systems the controlling factor in humification is the probability of collision of the reactants, the survival time of the molecule in soil becomes

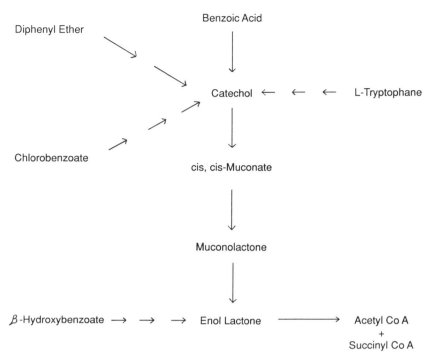

Fig. 3.10. Examples of pathways leading to entry of the carbons of aromatic compounds into the TCA cycle. The specific pathway depicted is the catechol pathway. Other aromatic-ring-degrading processes include the gentisate and protocatechuate pathways.

a major predictor of whether the substance will be decomposed totally or at least partially, or whether it will be humified. The more long-lasting organic components, such as complex polyaromatics in petroleum or lignin, are more likely to be humified than the more ephemeral proteins and simple sugars. See Tate (1986) for further discussion of humification processes.

3.3. CONCLUDING COMMENTS

The soil microbial community is clearly distinguished by its diversity in metabolic capacity. Along with the common variety of fixed carbon substances catabolized by living organisms, soil microbes also oxidize a number of soil minerals (e.g., sulfides), metals (e.g., ferrous ion), and several cations and anions containing atoms capable of existing in numerous oxidation states (e.g., ammonium, thiosulfate). Combined with the capacity to use several different terminal electron acceptors (e.g., molecular oxygen, nitrate, nitrite, sulfate, and numerous organic compounds), microbial life can exist in soil under essentially any chemical and physical conditions that allow for sustenance of cellular integrity.

The societal legacy of this myriad of metabolic capabilities is the potential for the soil microbial community to survive severe chemical insults and return to previously existing conditions. Influxes of acidic solutions of sulfate (acid mine drainage) may be ameliorated by anaerobic reduction of the sulfate to sulfide (i.e., the sulfate is used as a terminal electron acceptor). Toxic organic compounds may be mineralized in the presence or absence of molecular oxygen. Heavy metal contaminants can be removed from interstitial waters by precipitation with the sulfide produced by sulfate reduction. Survivability and sustenance of the soil microbial community is built to a large extent on its vast catalog of catabolic capabilities.

REFERENCES

Aftring, R. P., B. E. Chalker, and B. F. Taylor, 1981. Degradation of phthalic acids by denitrifying, mixed cultures of bacteria. Appl. Environ. Microbiol. 41:1177–1183.

Alexander, M. 1977. Introduction to Soil Microbiology. John Wiley and Sons. New York, 467 pp.

Altenschmidt U., and G. Fuchs. 1991. Anaerobic degradation of toluene in denitrifying *Pseudomonas* sp.: Indication for toluene methylhydroxylation and benzoyl-CoA as central aromatic intermediate. Arch. Microbiol. 156:152–158.

Angers, D. A., A. Pesant, and J. Vigneux. 1992. Early cropping-induced changes in soil aggregation, organic matter, and microbial biomass. Soil Sci. Soc. Am. J. 56:115–119.

Barnhart, C. L. H., and J. R. Vestal. 1983. Effects of environmental toxicants on metabolic activity of natural microbial communities. Appl. Environ. Microbiol. 46:970–977.

Beam, H. W., and J. J. Perry. 1973. Co-metabolism as a factor in microbial degradation of cycloparaffinic hydrocarbons. Arch. Mikrobiol. 91:87–90.

Beam, H. W., and J. J. Perry. 1974. Microbial degradation of cycloparaffinic hydrocarbons via co-metabolism and commensalism. J. Gen. Microbiol. 82:163–169.

Biederbeck, V. O., and C. A. Campbell. 1971. Influence of simulated fall and spring conditions on the soil system. I. Effect on soil microflora. Soil Sci. Soc. Am. Proc. 35:474–479.

Billon-Grand, G., J. B. Fiol, A. Breton, A. Bruyère, and Z. Oulhaj. 1991. DNA of some anaerobic rumen fungi: G + C content determination. FEMS Microbiol. Lett. 82:267–270.

Boyle, M. 1989. The environmental microbiology of chlorinated aromatic decomposition. J. Environ. Qual. 18:395–402.

Christensen, H., D. Funck-Jensen, and A. Kjoller. 1989. Growth rate of rhizosphere bacteria measured directly by the tritiated thymidine incorporation technique. Soil Biol. Biochem. 21:113–118.

Dore, J., and D. A. Stahl. 1991. Phylogeny of anaerobic rumen *Chytridiomycetes* inferred from small subunit ribosomal RNA sequence comparisons. Can. J. Bot. 69:1964–1971.

Espejo, R. T., and P. Romero. 1987. Growth of *Thiobacillus ferrooxidans* on elemental Sulfur. Appl. Environ. Microbiol. 53:1907–1912.

Evans, P. J., D. T. Mang, K. S. Kim, and L. Y. Young. 1991a. Anaerobic degradation of toluene by a denitrifying bacterium. Appl. Environ. Microbiol. 57:1139–1145.

Evans, P. J., D. T. Mang, and L. Y. Young. 1991b. Degradation of toluene and *m*-xylene and transformation of *o*-xylene by denitrifying enrichment cultures. Appl. Environ. Microbiol. 57:450–454.

Federle, T. W., M. A. Hullar, R. J. Livingston, D. A. Meeter, and D. C. White. 1983. Spatial distribution of biochemical parameters indicating biomass and community composition of microbial assemblies in estuarine mud flat sediments. Appl. Environ. Microbiol. 45:58–63.

Focht, D. D., and W. Brunner. 1985. Kinetics of biphenyl and polychlorinated biphenyl metabolism in soil. Appl. Environ. Microbiol. 50:1058–1063.

Gordon, G. L. R., and M. W. Phillips. 1989. Degradation and utilization of cellulose and straw by three different anaerobic fungi from the ovine rumen. Appl. Environ. Microbiol. 55:1703–1710.

Häggblom, M. M., and L. Y. Young. 1990. Chlorophenol degradation coupled to sulfate reduction. Appl. Environ. Microbiol. 56:3255–3260.

Henry, S. M., and D. D. Grbic-Galic. 1991. Influence of endogenous and exogenous electron donors and trichloroethylene oxidation toxicity on trichloroethylene oxidation by methanotrophic cultures from a groundwater aquifer. Appl. Environ. Microbiol. 57:236–244.

Hutchins, S. R. 1991. Biodegradation of monoaromatic hydrocarbons by aquifer microorganisms using oxygen, nitrate, or nitrous oxide as the terminal electron acceptor. Appl. Environ. Microbiol. 57:2403–2407.

Janke, D., and W. Ihn. 1989. Cometabolic turnover of aniline, phenol, and some of their monochlorinated derivatives by the *Rhodococcus* mutant strain AM 144. Arch. Microbiol. 152:347–352.

Kim, J.-H., and J. M. Akagi, 1985. Characterization of a trithionate reductase system from *Desulfovibrio vulgaris.* J. Bacteriol. 163:472–475.

Kludge, C., A. Tschech, and G. Fuchs. 1990. Anaerobic metabolism of resorcyclic acids (*m*-dihydroxybenzoic acids) and resorcinol (1,3-benzenediol) in a fermenting and in a denitrifying bacterium. Arch. Microbiol. 155:68–74.

Koike, I., and A. Hattori. 1975. Energy yield of denitrification: An estimate from growth yield in continuous cultures of *Pseudomonas denitrificans* under nitrate-, nitrite-, and nitrous oxide-limited conditions. J. Gen. Microbiol. 88:11–19.

Kristjansson, J. K., B. Walter, and T. C. Hollocher. 1978. Respiration-dependent proton translocation and the transport of nitrate and nitrite in *Paracoccus denitrificans* and other denitrifying bacteria. Biochemistry 17:5014–5019.

Lowe, S. E., M. K. Theodorou, and A. P. J. Trinci. 1987. Growth and fermentation of an anaerobic rumen fungus on various carbon sources and effect of temperature on development. Appl. Environ. Microbiol. 53:1210–1215.

Mihelcic, J. R., and R. G. Luthy. 1988a. Degradation of polycyclic aromatic hydrocarbon compounds under various redox conditions in soil–water system. Appl. Environ. Microbiol. 54:1182–1187.

Mihelcic, J. R., and R. G. Luthy. 1988b. Microbial degradation of acenaphthene and naphthalene under denitrification conditions in soil–water systems. Appl. Environ. Microbiol. 54:1188–1198.

Nozawa, T., and Y. Maruyama. 1988. Anaerobic metabolism of phthalate and other aromatic compounds by a denitrifying bacterium. J. Bacteriol. 170:5778–5784.

Patrick, W. H., Jr., and A. Jugsujinda. 1992. Sequential reduction and oxidation of inorganic nitrogen, manganese, and iron in flooded soil. Soil Sci. Soc. Am. J. 56:1071-1073.

Postgate, J. R. 1984. The Sulphate-Reducing Bacteria. Cambridge University Press, New York. 108 pp.

Ramanand, K., and J. M. Suflta. 1991. Anaerobic degradation of *m*-cresol in anoxic aquifer slurries: Carboxylation reactions in a sulfate-reducing bacterial enrichment. Appl. Environ. Microbiol. 57:1689–1695.

Schmidt, S. K., and M. Alexander, 1985. Effects of dissolved organic carbon and second substrates on the biodegradation of organic compounds at low concentrations. Appl. Environ. Microbiol. 49:822–827.

Schmidt, S. K., K. M. Scow, and M. Alexander. 1987. Kinetics of *p*-nitrophenol mineralization by a *Pseudomonas* sp.: Effects of second substrates. Appl. Environ. Microbiol. 53:2617–2623.

Schnell, S., F. Bak, and N. Pfennig. 1989. Anaerobic degradation of aniline and dihydroxybenzenes by newly isolated sulfate-reducing bacteria and description of *Desulfobacterium anilini*. Arch. Microbiol. 152:556–563.

Scow, K. M., S. Simkins, and M. Alexander. 1986. Kinetics of mineralization of organic compounds at low concentrations in soils. Appl. Environ. Microbiol. 51:1032–1035.

Simkins, S., and M. Alexander. 1984. Models for mineralization kinetics with the variables of substrate concentrations and population density. Appl. Environ. Microbiol. 47:1299–1306.

Simkins, G., and M. Alexander. 1985. Nonlinear estimation of the parameters of Monod kinetics that best describe mineralization of several substrate concentrations by dissimilar bacterial densities. Appl. Environ. Microbiol. 50:816–824.

Stucki, G., and M. Alexander. 1987. Role of dissolution rate and solubility in biodegradation of aromatic compounds. Appl. Environ. Microbiol. 53:292–297.

Szewzk, R., and N. Pfennig. 1987. Complete oxidation of catechol by the strictly anaerobic sulfate-reduction *Desulfobacterium catecholicum* new species. Arch. Microbiol. 147:163–168.

Tate, R. L., III. 1985. Carbon mineralization acidic, xeric forest soils: Induction of new activities. Appl. Environ. Microbiol. 50:454–459.

Tate, R. L., III. 1986. Soil Organic Matter: Biological and Ecological Effects. John Wiley and Sons, New York. 291 pp.

Teunissen, M. J. H., J. M. Op den Camp, C. G. Orpin, J. H. J. Huis in 't veld, and G. D. Vogels. 1991. Comparison of growth characteristics of anaerobic fungi isolated from ruminant and non-ruminant herbivores during cultivation in a defined medium. J. Gen. Microbiol. 137:1401–1408.

Tiedje, J. M., A. J. Sexstone, T. B. Parkin, and N. P. Revsbech. 1984. Anaerobic processes in soil. Plant Soil 76:197–212.

Traore, A. S., C. E. Hutchikian, J. Le Gall, and J.-P. Belaich. 1982. Microcalorimetric studies of the growth of sulfate-reducing bacteria: Comparison of the growth parameters of some *Desulfovibrio* species. J. Bacteriol. 149:606–611.

Tasaki, M., Y. Kamagata, K. Nakamura, and E. Mikami. 1991. Isolation and characterization of a thermophilic benzoate-degrading, sulfate-reducing bacterium, *Desulfotomaculum thermobenzoicum* sp. nov. Arch. Microbiol. 155:348–352.

Volger, K. G. and W. W. Umbreit. 1941. The necessity for direct contact in sulfur oxidation by *Thiobacillus thiooxidans*. Soil Sci. 51:331–337.

Wang, Y.-S., R. V. Subba-Rao, and M. Alexander. 1984. Effects of substrate concentration and organic and inorganic compounds on the occurrence and rate of mineralization and cometabolism. Appl. Environ. Microbiol. 47:1195–1200.

Webb, J., and M. K. Theodorou. 1991. *Neocallimastic hurleyensis* sp. nov., an anaerobic fungus from the ovine rumen. Can. J. Bot. 69:1220–1224.

Wiggins, B. A., and M. Alexander. 1988. Role of chemical concentration and second carbon sources in acclimation of microbial communities for biodegradation. Appl. Environ. Microbiol. 54:2803–2807.

4 Process Control in Soil

We cannot accurately predict the response of a soil biological activity to environmental perturbation unless we understand the ecosystem properties that control soil processes as well as the nature of the reactions of interest. This foundational prerequisite presents a paradox for those associated with the study of soil biological activity within the confines of environmental and soil sciences. Microbiological principles and metabolic reactions are understood at the individual cellular and molecular levels (i.e., the micro level), whereas soil properties delimiting the expression of microbial potential are assessed at the macro level (e.g., ranging from several grams of soil to a total ecosystem characterization). Furthermore, the environmental scientist requires integration of the microsite and macrosite realms on an ecosystem-wide or terrestrial level. These three spheres of concern (microsite, ecosystem-wide, worldwide) are not always easily merged, but such reconciliation of data is essential.

Macro- and Microsite Reality and Ecosystem Heterogeneity: The primary difficulty associated with merging the results of microbial and soil chemical and physical analyses is the high degree of soil property variability both in a vertical and in a horizontal plane within the soil profile (see chapter 1). A false description of the microsite traits may be gained by studying relative large soil samples. Values for soil properties, such as pH or cation exchange capacity, measured with a 10-g soil sample may diverge widely from the actual situation within which the microbes exist.

Comparable problems are associated with extrapolating from a manageable soil sample to provide an ecosystem-wide model. Unfortunately, as with evaluation of the microbial-world dynamics, soil properties of a large soil site are sufficiently variable that it is difficult, if not impossible, to collect a soil sample that is a reliable representative sample of total site variability. In conclusion, we must remember that there is a primary conflict in evaluating process control of microbially catalyzed processes in complex soil sites. In other words, although microbiology is a science that studies the minute, it must fit into the soil science framework of total ecosystem or even worldwide relationships.

This conflict in scientific purview is exemplified by studies of carbon dioxide evolution from soil. Respiratory carbon dioxide generation measures biological activity at the micro- or macrosite level (i.e., μ^3 or cm^3), but such activity also affects the quantities of greenhouse gases generated and their impact on terrestrial climates, which in turn controls conditions at the soil microsite where the greenhouse gas originated. Soil microbial respiration is directly controlled by such factors as temperature, moisture, and organic matter inputs, which are determined to a large degree by the greenhouse phenomenon. To quantify the contribution of soil microbial respiration to atmospheric levels of greenhouse gases localized assessments must be conducted in a manner that allows their integration into the more comprehensive, world situation. Hence, any real-world-based soil microbiological study must be designed within the limitations and hazards of data extrapolation relating the realm of the microbe to that of higher plants and animals.

The telescoping nature of the application of soil microbiological research reveals a characteristic, and, to some degree, an intellectual trap of our discipline. Data are necessarily a summation of a variety of reactions occurring at highly variable soil microsites. Thus, a composite representation of soil processes must emphasize dominant reactions, even though other processes of major importance to ecosystem integrity and function may be overlooked or the importance of their contribution to system complexity minimized. For example, the bulk of terrestrial soils are aerobic, because soil pores are contiguous with the atmosphere. Molecular oxygen diffuses freely into the soil pores or is transferred deep into the soil matrix through mass action (e.g., the influx of atmospheric gases due to the removal of shallow groundwater or the percolation of oxygen-bearing irrigation or rain water through soil pores). Thus, it is easy to emphasize aerobic processes in conceptual models to the detriment of obligatorily anaerobic reactions. Yet, we know that anaerobic soil bacteria occur and are active in essentially all soils and that the products of anaerobic microbial activity are essential for ecosystem survival. Thus, the potential for a biologically catalyzed reaction to occur must not be excluded simply from the results of an appraisal of overall system traits. Characterization must include a macro- and microsite evaluation.

The Black Box Viewpoint of Soil Systems: An additional conflict in developing soil microbiological studies concerns the different images of ecosystem function, which are derived from evaluating the activity of individuals compared to total ecosystem activities. The biological and structural heterogeneity of soils necessitates developing a dual understanding of the central object of soil biological studies. Evaluation of community activity can be likened to a box that conceals all microprocesses occurring therein, revealing only external effects of manipulation (i.e., a condition analogous to the engineering black box). In this instance, the identities and quantities of only the substances entering or exiting the box can be determined. We can begin to understand the internal process rates developed for modeling the ecosystem by studying this "black box." In contrast, the properties of the individual organisms (reactions and their associated rates within the black box) may be clarified. These sorts of data reveal an intriguing picture of the adaptation of microbes to their environment. Unfortunately, we can evaluate individual enzymes and processes ad infinitum and still not produce a clear means of assembling the data or the concepts into an image of the working whole. Thus, for soil microbiological research, total ecosystem studies must be developed with a clear understanding of the limitations of ignorance of the "minutia," whereas when evaluating the microsite, we must consider its intercalation into the overall system dynamics.

All soil processes are a summation of the activities of a variety of individuals, albeit a cell, multicellular complex, or even collection of isoenzymes. The individual in the soil must be appreciated, and knowledge of the microsite conditions impinging on its existence are essential; yet, appreciating the individual is only valuable for ecosystem evaluation in light of the revelation it provides in discerning overall site community dynamics and their impact on ecosystem function.

Hence, the topics broached in this chapter acknowledge the impact of the individual microbes and the properties of their habitat on total ecosystem dynamics. The initial studies will involve an evaluation of the concept of a limiting factor and experimental quantification of the impact of various environmental properties on biological activities. These studies are followed by examination of the variation of specific soil chemical and physical traits. Soil properties interacting with biological entities to be evaluated herein include nutrient requirements, soil moisture, oxygen tension, redox potential, pH, and temperature.

4.1. MICROBIAL RESPONSE TO ABIOTIC LIMITATIONS: GENERAL CONSIDERATIONS

4.1.1. Definition of Limitations to Biological Activity

The classical definition of a soil-property-controlling biological activity can be summarized by Liebig's law of the minimum: **Under steady-state conditions, the essential material available in amounts most closely approaching the critical minimum needed by a given organisms will tend to be the limiting one"** (Dommergues et al., 1978). This principle was originally developed to describe the impact of the variation of chemical nutrients on microbial growth and activity, but it has been expanded recently to include a variety of growth limitors including temperature and inorganic and organic metabolic inhibitors.

This relationship pertains to communities in steady state. This condition is necessary because the organisms and their interactions can be adequately assessed in relationship to rate-controlling soil conditions only at steady state. Otherwise, populations and/or their activities may be increasing to reach an augmented density or activity level allowed by the new prevailing environmental conditions, or they may be declining to accommodate newly imposed conditions when their activities are assessed.

Although it is conceptually attractive, Liebig's law is limited in its ability to describe the dynamics of soil ecosystems. This inability results from three conditions:

- Soil systems are so complex that microbes rarely have the luxury of dealing with a single stress factor.
- This combined stress produces responses in the microbial community that cannot be anticipated based on any soil property acting singly on the microbial cells.
- The structure of the soil ecosystem makes it difficult to determine the exact conditions at the microsite where the organisms are living.

Implications of the Interaction of Limiting Factors on Microbial Activity: It is rare for one soil property to preclude life processes. To some degree, exceptions can be associated with portions of the dry deserts of Antarctica (Benoit and Hall, 1970; Cameron, 1972) or incidences of extreme anthropogenic intervention (e.g., acidification due to acid mine drainage, nuclear power plant cooling water outflows, or soils receiving massive influxes of heavy-metal-containing mine wastes). Similarly, it is difficult to predict the existence of a terrestrial habitat where no limitations to microbial growth occur; microbial growth and replication occur at will. Essentially, all soil microbes are subject to compromises in the rates at which they function. Most commonly, a variety of soil physical and chemical conditions have an impact on functional microbes in such a way as to cause them to operate at sub optimal levels. The soil microbes must adapt to an array of stresses rather than react to one imposing problem. For example, Wildung et al., (1975) documented the interactive nature of temperature and water content on soil respiration. Soil temperature was shown to alter respiration rates at moisture contents as low as -106 to -88 bar suction. Similarly, Wardle and Parkinson (1990) found that the lowest ecosystem stabilities were most frequently associated with low-temperature and low-moisture conditions.

Population Responses to Stress Factor Interaction: The tolerance that a life process has to alterations of environmental conditions may also vary due to the interaction of two factors. Based on laboratory studies, we can predict that a given microbial population would be unaffected by a particular soil property, yet in native soils, the growth of the organisms is precluded under seemingly acceptable conditions or its range of occurrence is truncated. For example, fungi grow in culture within the pH range of 3.0 to 8.0, but in soil the practical range is closer to 3.0 to 6.5. In this case, the impact of a chemical controller is altered by the microbe's capacity to compete with cohabiting biological populations.

Similarly, some environmental properties may compensate for cellular inadequacies, thus allowing the microbial populations to develop outside the range predicted for this maximum, minimum, or optimal activity. For example, it is not unusual to isolate soil microbes capable of growth at 30°C in culture but not at 37°C in a particular medium. If growth factors are added to the culture medium, the organisms grow quite well at the higher temperature. The inability to grow at the higher temperature without growth factors results from the fact that an enzyme essential for the production of the growth factor is temperature sensitive (i.e., it is denatured at the higher temperature). Thus, availability of this obligatory growth factor in the environment allows the previously inactive organism to proliferate in the seemingly detrimental habitat.

Spatial Heterogeneity and Determination of Limitations to Microbial Growth: Another difficulty with applying Liebig's law of the minimum to conditions in native soil systems is more of a practical concern than a conceptual difficulty. Due to soil spatial and temporal variability, data may imply nonexistent restrictions to microbial growth. Basically, these observations are based on the fact that a microorganism must encounter an inhibitory or stimulating soil component before its growth or activity is affected. Microbes grow as microcolonies in soil microsites where their needs are met. Thus, microbial growth is limited by the distribution of carbon, nitrogen, and other nutrients; aeration; anoxic conditions; and so on. Any troublesome soil property occurring beyond the perimeter of the colony is nonexistent from the viewpoint of the cells. Furthermore, environmental conditions vary on a daily, seasonal, and annual basis. For example, consider the rise in temperature in the surface centimeter of soil on an early Spring day. In the early morning hours, the temperature may be suboptimal for microbial activity, perhaps the soil is frozen. As the hours pass, soil temperature increases, stimulating biological processes until some optimal or maximal rate is reached. On the rare hot days, temperature combined with the resultant increased desiccation rate could present a situation that inhibits the life processes occurring in the very top portion of the soil profile. Similar temporal effects can be described for other variables, such as nutrient concentration, pH, and moisture. Nutrients may be leached through soil during rain fall events, which clearly cause dramatic changes in soil moisture levels, whereas pH variation may occur at the microsite level as a result of acid production by the microbes themselves or the influx of human products, such as acid rain or acid mine drainage waters. Thus, analysis of the soil site at a particular time of day may indicate that the growth-promoting or -inhibiting condition does not exist, whereas if the measurements were taken a few hours earlier or later contrasting results would be obtained.

These observations indicate the need for modifying our concept of a limiting factor to fit the complexities of soil. A more appropriate basis for viewing control of soil microbial processes follows: **The growth of a soil microbiological population at steady state depends on a combination of several limiting factors acting collectively and interdependently at the soil microsite wherein the microbe is functioning.** For example, denitrifiers in soil are limited by variation in easily metabolizable carbon levels, nitrate concentrations, and oxygen concentrations, whereas autotrophic nitrifiers are frequently controlled by soil pH, ammonium concentrations, and oxygen tension.

Clearly, as previously noted, extremes in any of these soil properties may preclude biological activity and hence negate any measurable effect of the other factors. For example, consider the occurrence of denitrification in the presence of free oxygen. Even with optimal nitrate and carbon concentrations, denitrifiers are inactive when molecular oxygen is available as an electron acceptor. But, in the bulk of the situations, the combination of limitations determine the extent of denitrifier population development in a soil site.

4.1.2. Elucidation of Limiting Factors in Soil

Under most circumstances, the complexities of the soil environment, as well as the microscopic and submicroscopic size of the soil microbial community, preclude clear determination of limiting factors in the field. In some situations, field observations are quite useful (e.g., Antarctic dry deserts or anthropogenically impacted sites where extremes in physical or chemical conditions are easily discerned), but for the vast majority of the terrestrial ecosystem, controlled laboratory-based study of the system is required to quantify the contribution of each individual soil property on biological reaction rates. We can determine functions of biological activities in soil by observing properties, singly or combination, in laboratory-incubated soil samples or by assessment of field activities in several diverse systems.

Limitations of Laboratory-Generated Data: Scientific conditioning creates a desire in experimenters to study the effect of varying each individual soil property on biological processes under the controlled conditions in the laboratory. These controls offer the security of simplicity and reproducibility. In laboratory experiments, soil conditions may be varied singly or in concert so that the resultant impact on microbial activity can be assessed accurately. Easily manipulated factors include pH, temperature, moisture, and nutrient levels. These conceptually alluring studies may yield results that describe the capacity of existing microbial populations in the soil system to cope with alteration of their habitat in the short term. The use of such data for predicting the resiliency of the native soil population to long-term perturbation is frequently limited. To achieve the latter objective, soil samples should be incubated for a sufficient time period to allow the microbial community to adjust to the newly imposed conditions, particularly if the chemical or physical properties of the soil are being altered.

We can see the limitations that result from a failure to allow community development in laboratory study by evaluating the impact of varying soil pH beyond ranges normally detected in the soil ecosystem studied. It is frequently desirable to determine the pH range over which an enzymatically or biologically catalyzed process occurs. We can readily observe the impact of varying this parameter by taking one or a few soil samples and adjusting the pH over the range known to be tolerated by general microbial populations. An ecological maxim is that the active microbial population in any soil sample is that which is most capable of coping with the prevailing soil pH. Thus, major changes in the pH of an individual soil sample may result in inhibition of the currently existing microbial strains. Such a study (i.e., adjustment of soil pH in the laboratory) must lead to the conclusion that the biological community in that particular soil sample has a defined pH range in which it can function (Fig. 4.1). This type of experiment fails to reveal the potential for selecting microbial communities that are better adapted to function at the newly imposed soil pH. Since acidiophilic or alkylophilic microbes may exist in low numbers in the original soil sample and generation times of soil microbes are measured in portions of or multiples of days, extended incubation of the soil sample at the newly established pH are usually necessary for acclimation of the microbial community to occur.

Complications of data interpretation associated with adjusting soil pH result not only from population selection but also from the buffer capacity of the soil. Sufficient acid can be added to a soil sample to reach an acidic pH selected for a particular experimental design, but if the pH of the soil sample is measured a few hours after initial adjustment, it is commonly found that the pH level has returned to a value approaching the preamendment level. Both the microbial community and the soil buffer capacity minimize the effect of soil acidification. Thus, not only is it necessary to incubate the soils for a sufficient period to allow selection of acid-

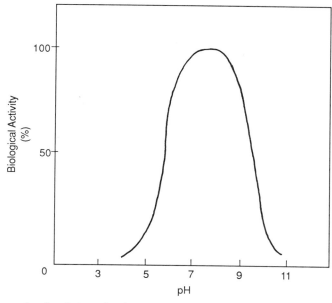

Fig. 4.1. Example of variation of an hypothetical biological activity with changes in soil pH. For this representation, an optimum of approximately pH 7 with rapid declines in activity above and below this value is depicted.

resistant microbial populations, but the natural tendency for the soil to return to native conditions also must be considered. Soil pH changes during the incubation period may allow continued activity of the native population rather than development of acidophiles or acid-tolerant organisms. Thus, to prevent misinterpretation of the laboratory studies, soil pH of the amended soils must be monitored throughout the study. In conclusion, native soil samples may be adjusted to mimic variation in the particular soil property, but selection of microbial populations and the tendency of the soil to modulate external perturbations must be considered during data interpretation.

Evaluation of Soil Properties and Microbial Variability in Native Samples: An alternative available to overcome this potential for laboratory-generated misunderstandings of the complexity of soil community resiliency is to study microbial population dynamics in several native soil samples that exhibit the entire range of variation for the property within the range of interest. Microbial populations in native soils can be reasonable anticipated to have adapted to the soil site. Selecting appropriate soil samples for these studies is essential. Except under fortuitous situations, it is rare to find a series of soil sites where only one or a few soil properties vary in an otherwise constant environment. Thus, when studying native soil samples, it is necessary to assess not only the biological activity of interest, but also all possible chemical and physical soil properties that could impinge upon that activity must be quantified. Data interpretation is most commonly accomplished by analyzing the correlation between the various soil properties and the biological activity or by using regression analysis.

Problems in data interpretation with these field survey experiments are encountered in that statistically significant irrelevant relationships may be obtained. For example, the relationship may be fortuitous in that the variables being measured may correlate with a third factor that has not been considered. A potential solution to this difficulty is that, once significant correlations are determined in the field, the data can be verified under more controlled conditions in the laboratory. Another means of confirmation is to use the field-collected relationships to construct a mathematical model of the biological activity of the soil ecosystems. Field validation of the model may provide evidence that supports the conclusion that appropriate relationships have been elucidated and quantified.

4.2. IMPACT OF INDIVIDUAL SOIL PROPERTIES ON MICROBIAL ACTIVITY

From a somewhat naive point of view, soil appears to present a reasonably accommodating habit for microbial growth. Mineral nutrients, organic matter (including fresh plant material), and growth factors necessary to meet microbial nutrient requirements generally exist within appropriate temperature, moisture, and pH ranges for microbial growth. Also, most soils contain microbial populations with the necessary genotypic capacity to catalyze all requisite reactions for ecosystem development. Unfortunately, for the microbe, soils also contain a variety of growth inhibitors (e.g., heavy metals and organic and inorganic acids). Therefore, to understand biological activity in the soil fully, the needs of the organisms must be considered in relationship to the form and distribution of chemical and physical prerequisites for microbial growth.

4.2.1. Availability of Life-Sustaining Nutrients

Living cells must have an energy source, electron acceptors, trace minerals, a variety of macronutrients, and, frequently, a variety of growth factors (i.e., vitamins and amino acids that are incorporated into cellular substituents intact). Although these substances are generally present in soil, the extent of microbial community development is determined by their distribution and concentrations. Therefore, the following discussion evaluates the biological requirements for growth and reproduction and their availability to soil communities. Details of microbial mechanisms for energy recovery and incorporation of micro- and macronutrients into biomass are presented in chapter 3.

Soil Microbial Nutrient Resources: The primary energy supply for all living systems is solar energy. Photosynthetic (photoautotrophs) higher plants and microbes convert this energy into the chemical energy used by nonphotosynthetic (heterotrophs) organisms to support cellular growth and replication. Because of the diversity of metabolic capacities of microorganisms, a small portion of the soil microbial community can take advantage of nonsolar-derived energy sources available in soil including reduced minerals (e.g., S^0, Fe^{2+}) and products of cellular respiration (e.g., NH_4^+, H_2). Typical energy-yielding transformations of these substances are

$$NH_4^+ \rightarrow NO_2^- \rightarrow NO_3^-$$

$$H_2 \rightarrow H_2O$$

$$S \rightarrow SO_4^{2-}$$

$$Fe^{2+} \rightarrow Fe^{3+}$$

The bulk of the soil community is driven by the oxidation of organic substances. Heterotrophic organisms (i.e., chemoorganotrophs or heterotrophs) primarily use organic carbon that enters the soil ecosystem as root exudates or plant litter, although plant biomass can also be provided through anthropogenic intervention (e.g., through soil amendment with manure or composts) or with a variety of anthropogenically modified and industrially produced waste materials (e.g., petroleum byproducts and industrial wastes). Because the energy contained in newly synthesized plant biomass is easily recovered by soil microbes, sites receiving these substances can be likened to oases within a desert of biological activity.

Recall that a primary chemical maxim is that all oxidative reactions must be balanced by reductions (chapter 3). The biological reactions catalyzed within soil are not an exception to this rule. The most common electron acceptor in soil is molecular oxygen, which is reduced to water. For facultative anaerobic organisms, oxygen or organic carbon can serve as electron acceptors, whereas for strictly anaerobic bacteria, organic carbon acceptors such as organic acids are reduced. Some aerobic organisms growing under anoxic conditions reduce nitrate or nitrite (denitrification) or sulfate (dissimilatory sulfate reduction). (See chapter 3 for further details of these processes.)

Following energy sources, the next most abundant requirement for cellular growth and proliferation in soil is a variety of macronutrients (i.e., carbon, nitrogen, phosphorus, and sulfur supplies). These substances are the major building blocks for cellular biomass. Carbon and nitrogen are generally derived from the mineralization of plant biomass, whereas phosphorus and sulfur not only are components of biomass but are also present in high concentrations in soil minerals. Thus, pathways of carbon and nitrogen mineralization are frequently interlinked, whereas phosphorus and sulfur supplies and transformation in soil is generally independent of organic matter availability and mineralization rate. (See chapter 7.)

Determination of *in situ* Limiting Nutrients: Evaluation of nutrient limitations in soil communities is usually focused on energy or macronutrient supplies. Typically, documentation of such limitations entails amending soil samples with a nutrient suspected of controlling microbial metabolic activity. If the soil microbes use the soil amendment, a variety of responses in the microbial community may be observed. A generalized depiction of the type of data that can be recovered from

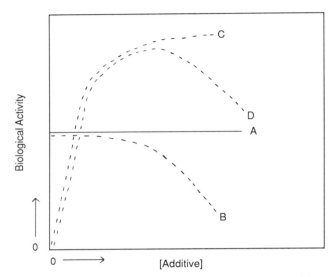

Fig. 4.2. Hypothetical curves describing response of soil biological activity to amendment with proposed growth-limiting nutrients. A. The added substance is not controlling cell replication. B. The additive is nonlimiting and toxic to the biological community. C. The substance was present in the soil sample in growth-controlling concentrations. D. The additive is growth limiting, but it is also toxic in elevated concentrations.

such studies is presented in Fig. 4.2. If the microbial population is not controlled by the soil amendment or an alternative factor limits biological activity, curves of type A are observed. For example, observation that metabolizable fixed-carbon compounds are present in low concentrations in a sample of submerged swamp soil could lead to the hypothesis that the microbial populations are carbon-limited; but when a metabolizable carbon source is added to samples of swamp soil incubated under anoxic conditions, no change in microbial growth of culturable aerobic bacteria occurs. That is, the carbon source is not limiting to microbial growth. In this situation, the absence of the electron acceptor, free oxygen, precluded the rapid catabolism of available carbon and resultant multiplication of the aerobic organisms. Thus, prior to soil amendment, sufficient carbon and energy existed in the low levels of organic substrate existent prior to amendment to support the degree of microbial respiration occurring therein at a maximal level. Oxygenation of the system or provision of an alternative electron acceptor (e.g., nitrate) could result in the development of a carbon-limited system. The primary delimiter of microbial activity (a terminal electron acceptor] is relieved by addition of free oxygen or nitrate. Then the availability of an energy source can become a primary obstacle to maximal microbial activity.

An alternative to this neutral response of soil amendment is the situation described by curve B. These data reflect the condition where soil concentration increases of a hypothetical controlling nutrient results in inhibition of overall microbial respiration. An example of this effect of soil amendment is provided by adding a variety of aromatic compounds to a soil sample. In low concentrations (typically less than 0.1 percent (wt/wt)) microbial respiration may be stimulated by such compounds as catechol or benzoic acid, but as their concentration is increased, microbial respiration is inhibited. Thus, if maximally tolerated levels of these substances already exist in a soil site, further amendment with them causes cellular respiration to decline.

More easily interpreted examples of relief of metabolic limitations are exemplified by curves C and D. In these situations, the test compound controls microbial activity. As its concentration is increased, microbial activity or respiration increases until a plateau is reached. No further increase in activity is detected. At this point, another soil factor limits the microbial community or population studied. At maximum concentration, the amendment may have no further impact on soil activity (curve C), or it may become inhibitory (curve D) as described previously by the model curve B.

The fluctuations in soil microbial populations after clear-cutting a forest provide an excellent example of the impact of nutrients availability of biological activity *in situ* (e.g., Lundgren, 1982). Generally, clear-cutting the forest initially results in an increase in soil microbial biomass. This microbial growth results from the decomposition of excess organic matter, both living and that previously accumulated in soil as plant residues. These substances become available to the microbial community through destruction of the aboveground portions of the plant biomass. Roots die and carbon compounds from the debris produced by cutting the trees leach into the soil profile. Furthermore, machinery and logging traffic on the soil surface disrupts the soil structure, which enhances the availability of any organic matter occluded in soil aggregates. Three microbial communities can be described for this system in transition: the native forest soil community that existed prior to anthropogenic interention, that supported by the release of plant debris during logging (a disturbed soil community), and that which develops in response to the newly established ecosystem properties (bare soil). In a nutrient-controlled situation, the disturbed soil community exhibits maximal population density, whereas the bare soil reflects the denuded, minimally production aboveground ecosystem.

Carbon Limitation of Microbial Function: Examining the occurrence of macronutrients in a soil sample (primarily carbon, nitrogen, sulfur, and phosphorus) and their ratios in soil microbial communities reveals that carbon resources generally control the soil microbial community development. Soil microbial biomass contents commonly correlate with soil organic carbon contents and are generally stimulated by fixed-carbon amendments (e.g., Knapp et al., 1983, Schnürer et al., 1985). The capacity for soil microbes to use efficiently and effectively the pools of metabolizable carbon contained in soil and to be easily stimulated by soil amendment with such materials has been adequately and frequently document. Similarly, as is commonly noted, nitrogen is rarely limiting to soil microbes, but the active incorporation of nitrogen into soil microbial biomass may cause nitrogen to become limiting to the plants growing in the ecosystem (see chapter 10 for a further discussion of soil nitrogen transformations). The low concentrations of sulfur and phosphorus required by soil microbes may be supplied through mineralization of plant biomass or through solubolization of soil minerals so that, although these macronutrients are commonly present in growth-controlling concentrations for higher plants, soil generally contains adequate sulfur and phosphorus pools for microbial development.

Growth Factors: Growth factors are organic compounds required by the cell in small quantities (e.g., vitamins and amino acids, purines, and pyrimidines). They are incorporated into the cell intact. Note that if sufficient concentrations of these materials exist in soil, they may also be used as carbon, nitrogen, and/or energy sources. For example, a large proportion of the bacteria growing in the rhizosphere can mineralize amino acids to meet these nutrient and energy requirements.

Species requiring growth factors are more common in soil than are those capable of synthesizing them *de novo*. This capability implies that common sources of growth factors must be present in soil. Four reservoirs of microbial growth factors can be identified readily.

- *Other organisms growing in soil:* Each microbe growing in soil can synthesize a portion of the growth factors necessary for biomass production (auxotrophs). A microorganism may produce more than is required for internal consumption. Excess growth factors leak from the cell into the suspending milieu. Alternatively, supplies of the growth factor become available to prototrophs (those requiring the growth factor) upon death and lysis of the auxotrophic cell. This cross-feeding of nutrients (use of growth factors leaking from a producer cell to cohabitants of the habitat) is frequently observed when viable cells are grown on petri plates containing vitamin-free or amino acid-free media that support the development of heterotrophic bacterial colonies. Small satellite colonies develop around the auxotropic bacterial colonies.

- *Root exudates and sloughed cells:* Plants commonly leak a mixture of vitamins and amino acids from their roots. Furthermore, as the root extends through soil, packets of nutrients as sloughed cells are released from the root surface and consumed by the soil microbial community.

- *Anthropogenically supplied organic substances:* Organic matter incorporated into the soil (waste disposal, sludges, composts, and accidental spills) may also contain a variety of vitamins as well as macronutrients.

- *Whole plant and animal cells:* Predators and parasites receive their growth factors directly from the host cells.

The dominant organisms present in an ecosystem are usually those requiring a few growth factors, at most. More fastidious metabolic needs are usually associated with the flora of living plants or animals (e.g., rumen or intestinal organisms, root symbionts). In these latter situations, an organism that can scavenge growth factors from its environment clearly has an advantage over those that must expend energy to synthesize the factor. Once the available supply of the vitamin or amino acid becomes limiting, the roles are reversed. In a soil system barren of growth factor resources, auxotrophs are better competitors than are prototrophs. An exception to this generalization is associated with those microbes that can form resting structures, such as endospores. For these creatures, once the easily recoverable supply of growth factors is depleted and their competitive advantage is lost, they form a resting structure wherein they persist until growth conditions improve.

Trace Metal Supplies in Soil: A variety of metallic ions are required for microbial growth. Examples of universally required cations are potassium, iron, and magnesium whereas most living cells need calcium, copper, and zinc. Cobalt is likely required by most organisms for vitamin B_{12}. Molybdenum is required for nitrogen fixation and nitrate metabolism. As would be anticipated by the predominance and variety of mineral components of soils, except for molybdenum, these metals are generally not limiting to soil microbes.

4.2.2. Moisture

Water is an essential contributor to the growth of soil microbes. Water not only provides an essential medium for growth of microbial populations, but it is also a primary participant in a variety of cell processes. The major roles for water in soil can be summarized as follows.

- *Serves as an essential material for flora and fauna:* A number of soil enzymes catalyze the incorporation of portions of the water molecule into biological compounds. Water is a direct participant in hydrolysis reactions (e.g., cleavage of cellulose into glucose and proteins into amino acids) as well as hydroxylations, such as the hydroxylation of picolinic acid and other pyridine-ring-containing compounds (e.g., see Dagley and Johnson, 1963; Hirschberg and Ensign, 1972; Hochstein and Dalton, 1967; or Tate and Ensign, 1974).

- *Affects gas exchange in soil:* Soil pores are filled with air and water. If all pores are water-filled, the soil is water-logged or saturated. Since gaseous diffusion is greater in air than in water, the greater the water content of a soil, the higher the probability of its becoming oxygen-limited or even anoxic.

- *Affects microbial nutrient supply:* Water is the transport medium of microbial nutrients to the cell and of waste materials away from the cell. For non-water-soluble nutrients, water serves as the transport medium for the extracellular enzymes or the solubolizing agent to reach the substrate and convert it to water-soluble products or forms. Extracellular enzymes are generally involved in the cleavage of complex polymers into water-soluble subunits, whereas surfactants increase the availability of hydrophobic materials (e.g., lipoidal substances) to microbial cells.

- *Affects soil temperature:* The higher the soil water content, the more resistant that soil ecosystem is to temperature fluctuations. Hence, soil temperatures rise more slowly during diurnal cycles in moist or water-saturated soils during the Spring than in the desiccated, or air-dried soils of the mid-Summer.

- *Serves as a growth medium for microbial colonies:* Microbes function in the water layer on soil particles or within soil pores. Even in a seemingly dry soil sample, a layer of water must exist surrounding microbial microcolonies and soil particles to support a viable soil community. In reality, although soil is predominantly a particulate system, microbial life forms contained therein live in an aquatic world. This water-saturated system may consist of a microfilm a few microns thick or it may extend throughout a soil micro- or macropore. In either case, for the microorganism to function properly, it must be totally bathed in water.

Measurement of Soil Moisture: Standard techniques for quantification and expression of soil moisture levels are necessary to the development of predictive models (conceptual or mathematical) of the impact of soil moisture variation on soil processes and to the provision of a basis for comparison of moisture effects on biological activity in contrasting soil types. The procedure should be easily conducted, reproducible, and assess an aspect of soil moisture that directly affects soil microbial activity.

Traditionally, soil moisture has been reported as gravimetric water content (ω, g H_2O/g dry soil, usually dried to a constant weight at 105°C) and volumetric water content (θ, cm^3 H_2O/cm^3 soil). These two means of expressing soil moisture are related by the soil bulk density [ρ_b, g solids/cm^3] in the following relationship:

$$\theta = \omega\rho_b \qquad (4.1)$$

A third term encountered on a perhaps all too frequent basis is g H_2O/g wet soil. Use of the latter term is particularly distressing. It provides some means for comparing soil data when a single soil type is considered, but it cannot be used to compare results from different soil types.

Although gravimetric and volumetric water content are probably the most frequently used representations of soil moisture by soil microbiologists, their use is somewhat problematic in that neither value describes the actual availability of water to the soil microbial community. Furthermore, these values do not vary independent of soil type. Thus, comparable gravimetric or volumetric moisture values in two different soil types are not equivalent. Although a relationship between soil moisture and a specific microbial activity can be derived for a single soil type using these values, we cannot determine the generalized principles behind the relationship.

A partial solution to this difficulty is to use water potential as a measure of water availability in soil pores. Water potentials, which are measurements that have been used by soil scientists for many years, are governed by surface tension forces and the radius of the menisci of the water. Thus, values can be assessed by principles of capillary rise. The mathematical expression for this relationship follows:

$$\tau = \frac{2\gamma}{r} \qquad (4.2)$$

where r is the equivalent or effective capillary radius, τ is the suction necessary to absorb the water, and γ is the surface tension of water. Note that as the equivalent capillary radius becomes smaller, less water is present, and the suction necessary to absorb the water increases. Equivalent expressions of suction are as follows:

$$1 \text{ bar} = 10^6 \text{ dyn cm}^2 = 100 \text{ kPa} = 1022 \text{ cm } H_2O = 75 \text{ cm Hg}$$
$$= 0.987 \text{ atm}$$

Common water potential values found in the soil microbiology literature are the moisture content of soil at -0.3 bar (field capacity)

Another useful expression of soil moisture is water-filled pore space (Doran et al. 1988; Linn and Doran, 1984). Soil microbes exist in a mixed soil air–water system. In reality, the degree of displacement of air with water within the soil pore combines with the rate of microbial respiration to produce anoxic or anaerobic microsites within the soil profile. (See Skopp et al., 1991, for a detailed study of this process.) Complete displacement of air with water restricts diffusion and thereby hastens imposition of oxygen-limited conditions. When moisture approaches or exceeds field capacity, the percent soil pore space filled with water or air becomes a better indicator of aerobic vs. anaerobic microbial activity than water content or water potential. Use of percent water-filled pore space to express soil moisture levels is practical because it requires knowledge only of the soil bulk density and the gravimetric soil water content. (The particle density of mineral soils is assumed to be 2.65 Mg/m^3.) The percent water-filled porosity is defined by the following relationship:

$$\%\text{WFP} = \frac{\theta_v}{\text{TP}} 100 \tag{4.3}$$

where θ_v is equivalent to the product of the gravimetric water content ω and the soil bulk density ρ_b and TP is the percent total porosity $(1 - \rho_b/\rho_p)(100)$ where ρ_b is the soil bulk density and ρ_p is the soil particle density. For aerobic processes, maximal activity is detected at a percent water-filled porosity of about 60 percent. Below 60 percent water limits activity (insufficient moisture exists to provide the contiguous surface film required by the soil microbes), whereas above 60 percent aerobic microbial activity decreases (apparently due to reduced oxygen availability). Caution is necessary when using these generalized figures. Neilson and Pepper (1990) found discrepancies when they used percent saturation or water-filled pore space to characterize soil aeration, especially when soils of differing bulk densities or those receiving particulate organic matter amendments were compared.

The effect of water in a heterogeneous system can also be expressed as water activity. This measurement of moisture is most commonly used in food storage to describe water availability. Water activity is related to the equilibrium relative humidity (ERH) as follows:

$$A_w = \frac{\text{ERH}}{100} \tag{4.4}$$

The difficulty with using water activity in soil microbiological research is the complications associated with its measurement. Long incubation times in a chamber where the moisture level of the test sample is allowed to equilibrate with the chamber atmosphere are usually required. Data can be extremely variable. Water availability affects species diversity, survival, movement, and activity of microbes. Of all the expressions that may be used to define the moisture conditions under which the soil microbe is abiding, water activity is probably the most useful. Unfortunately, it is also the most difficult and time-consuming (i.e., impractical) measure of soil moisture.

Impact of Moisture Variation on Soil Microbial Activity: Total microbial activity varies from nearly nonexistent levels at low water availability to a maximal level under optimal soil air/moisture mixtures (Fig. 4.3). Typically, soil bacteria can function under seemingly severe soil moisture tensions. For example, Wilson and Griffin (1975) found that respiration of a mixed bacterial population in soil was maintained at a reasonably high level between −8 and −30 bar but was negligible at −50 bar. Similar results have been reported by Wildung et al. (1975) and Knight and Skujin (1981). As soil moisture levels are saturated and free-oxygen supplies are depleted, overall activity again declines. This generalized curve applies to the totality of microbial respiratory activities and, to a certain degree, to individual processes. In the latter situation, the points of minimal or optimal activity are characteristics of the trait being studied. Obligatorily anaerobic processes clearly are optimized under saturation conditions.

Tolerance to restrictive soil moisture levels varies with major microbial groups as

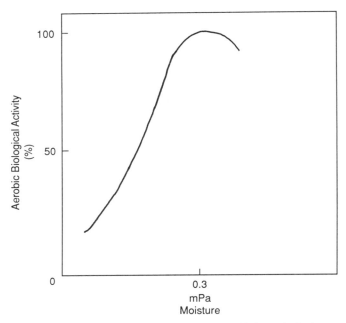

Fig. 4.3. Variation in aerobic biological activity in soil. In this hypothetical example, optimal moisture is approximately 0.3 mPa.

well as with individual microbial species. Generally, bacteria are more exacting than are actinomycetes and fungi. Hence, the latter organisms tend to predominate in the drier soils, because, at least in part, they form resting structures. Robertson and Firestone (1992) postulated that bacteria may use extracellular polysaccharide to enhance survival under moisture-limiting conditions. A purified extracellular polysaccharide from a *Pseudomonas* species contains several times its weight in water at low-moisture tension. Within the realm of the bacteria, a gradient of tolerance occurs. For example, ammonium tends to accumulate in dry soils because the ammonifying microorganisms are less sensitive to the low-water activities then are nitrifiers.

Selection of microbial populations with increased resistance to moisture limitations in chronically arid soils was demonstrated by Knight and Skujins (1981). They used adenosine triphosphate and soil respiration measurements as indicators of microbial biomass and activity and evaluated the impact of variation in soil moisture between -2 and -100 bar tension on microbial activity in two arid soils and a subalpine forest soils. The microbial activity increased in arid soils between -2 and -20 bar soil moisture, whereas the optimal moisture in the chronically wetter forest soils was -2 bar, and activity continually decreased below that level. Thus, a characteristic microbial flora capable of managing the prevailing environmental conditions developed in each of the extreme environmentally different soils studied.

It is important for students of microbiology to understand what these observations reveal about the function of the soil microbial community and, at least as importantly, what is not stated. The positive side of this observation is that as long as the soils are not so dry as to preclude biological function, an appropriate microbial population develops. The negative aspect is that the microbial population that grows under restricted soil moisture is not as active as a comparable population growing in less stringent conditions. In other words, were microbial respiration measured under limited soil moisture and at optimal soil moisture, even if the organisms in each situation were absolutely optimized, less respiration would be found in the low-moisture soil. From a practical viewpoint, this means that we should not anticipate as much organic matter to be decomposed in a desert field without irrigation as could occur in the same field with irrigation.

The impact of excessive soil moisture (flooded or chronically water-logged conditions) on soil microbial activities is more predictable than is the situation where there is a shortage of free water. Interactions of soil moisture levels with availability of molecular oxygen become evident in the flooded or chronically water-logged soil sites. A traditional division of the microbial realm has been separation by oxygen tolerance (anaerobes, facultative anaerobes, and aerobes). Flooded soils are not necessarily oxygen-free, but should available free-oxygen supplies be depleted, anaerobic and facultative microbial populations are stimulated while obligate aerobes die or enter resting stages. The most dramatic effect of instituting anaerobic conditions in soil is the induction of facultative (e.g., fermentations), anoxic aerobic metabolism (e.g., denitrification), or strictly anaerobic processes (e.g., methanogenesis). Organic acids may accumulate, nitrogen oxides could be evolved, and ultimately, methane generation or sulfate reduction may be instituted. From a practical viewpoint, anoxic soils may be recognized by precipitation of metal sulfides or evolution of an amine or sulfide odor. Characteristics of the microflora are dependent upon the duration of the flood period and the nature of electron acceptors present. Methanogenic bacteria require the development of highly reducing conditions, whereas fermentative organisms are less stringent.

As stated previously, a flooded soil is not necessarily either anoxic or highly reducing. Water movement through the environment as well as diffusion of atmospheric gases into the system may prevent exhaustion of free oxygen. This phenomenon was shown in a study of the muck soils of south Florida (Tate, 1979). When the soils were flooded, an increase in aerobic bacterial populations was initially observed (Fig. 4.4). The soils were flooded to maximum depths of about 31 cm. As the free oxygen and easily metabolized carbon source levels declined a parallel reduction of aerobic bacterial populations occurred. Thus, had the experimenter assumed that flooded and anoxic conditions were equivalent, the soil microbia behavior contrary to that which occurred *in situ* during the initial portion of the study would have been expected.

Soil Systems with Fluctuating Soil Moisture: Additional assumptions or perhaps attitudes that must be developed before the effect of soil moisture on the microbial community can be understood are a) the moisture level of a particular site tends to be continually changing and b) moisture effects on biological activity are interactive with other system properties.

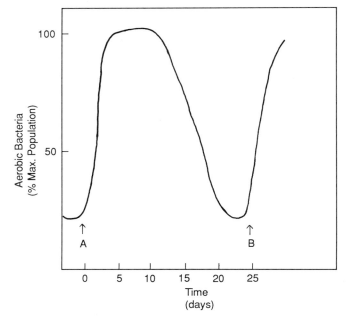

Fig. 4.4. Effect of flooding a dry histosol on aerobic bacterial populations. Soils were flooded to a maximum depth of approximately 31 cm commencing on day 0 (arrow A). Arrow B indicates the time of drainage of the flood waters. (Adapted from Tate, 1979.)

Moisture fluctuation may be on a short-term, localized basis as would be found a) with the diurnal impact of temperature variation and water evaporation on the top centimeter of an exposed soil surface, b) after a heavy rainfall, or c) as a result of seasonal flooding, such as occurs in a wetland rice soil. For example, microbial biomass fluctuations can reveal the short-term effect of soil moisture levels induced by precipitation events (Clarholm and Rosswall, 1980). Precipitation caused increases in microbial biomass lasting 1–2 days, even in non-moisture-limited soil. This observation supports the conclusion that although data for macrosite soil moisture tensions indicate that the microbial populations are not limited by this soil property, microsite discontinuity in moisture films covering soil particles determine the rate of change in and the ultimate level of microbial population densities. In flooded rice field soils, anoxic conditions induce anaerobic microbial populations during the somewhat long term of flooding, but draining of the site after harvest or to establish seedlings induces aerobic and facultative microbial populations. Strict anaerobes may be further limited by the flow of oxygenated water through the system. In that situation, a mosaic develops in the soil with pockets of anoxic soil interspersed with oxygenated soil.

Note that soil systems are described as containing a mixture of aerobic and anaerobic micro- or macrosites, *not* "a partially aerobic system." By definition, each individual soil site is either anaerobic (anoxic) or aerobic. "Partially aerobic" is an inappropriate description in that *aerobic* indicates the presence of oxygen in the system, which is an all or none state. Even if only a few molecules of oxygen are present per cubic centimeter of soil, the site is aerobic.

The interactive nature of soil moisture with other soil properties is documented by examining organic matter availability and osmotic pressure variation in soil. Varying soil moisture with time not only affects the nature of the microbial population present in the soil but also may alter the quantity of nutrients or the physical structure of the microbes' environment. Cycling of soil moisture liberates occlude organic matter through such processes as disruption of soil granular structure or death of the more stringent microbial populations. Therefore, microbial oxidation of soil organic matter is greater in soils with fluctuating moisture than in constantly wet or dry conditions (e.g., see Lund and Goksoyr, 1980; Orchard and Cook, 1983; Reddy and Patrick, 1975; and Sørensen, 1974). Jager and Bruins (1975) noted that the loss of carbon due to microbial respiration in a soil experiencing approximately 60 wet/dry cycles was also proportional to drying temperature, with greatest loss (31.2 percent) in the soil dried at 85°C and least loss (18 percent) in soil dried at 35°C. Drying resulted in an ultimate reduction in microbial population densities, but the highest population densities were found in soil dried at 85°C, suggesting that more organic matter was made available to support microbial growth by the more drastic treatment. A practical implication of this observation relates to maintenance of soil organic matter levels in mineral soils and the minimization of subsidence of histosols (see Tate, 1980). In both situations, maintenance of the native organic matter levels is desirable for preservation of the ecosystem and/or encouragement of optimal soil structure (see Tate, 1987).

An indirect effect of declining soil moisture is the impact of osmotic pressure on biological activity. Generally, the osmotic potential of water is less important than water availability, but in some localized or transient situations variations in osmotic potential emerge as significant controllers of microbial growth. Potentially lethal or inhibitory salt concentrations may be the product of microbial metabolism (e.g., mineral dissolution through acidification of microsites), chemical equilibria (e.g., ionization of water-soluble salts), as well as physical phenomena (evaporation). Localized problems are associated with areas of salt encrustation due to water evaporation, saline waters such as are found in some irrigation systems and around salt water bodies, or saline soils. Transient difficulties may be associated with severe desiccation of soil.

The major impacts of osmotic pressure are likely to be averted in soil because of the fact that soil moisture reduction results in the induction of cellular survival adaptations (e.g., sporulation). Also, in regions with chronically saline soils, selection of organisms with increased resistance to pressure changes may occur. For a discussion of microbial adaptations to osmotic changes, see Csonka and Hanson (1991).

4.2.3. Aeration

The primary role of molecular oxygen in soil is as a terminal electron acceptor. As indicated in chapter 3, energy is produced during electron transport to oxygen. In fact, oxygen is the highest energy-producing electron acceptor. Thus, the greatest cell yields are produced by the total oxidation of an organic-carbon-containing substance to carbon dioxide and water. This important function of molecular oxygen in aerobic metabolism is augmented by its direct incorporation into molecular structures of such compounds as aromatic-ring-containing substances and alkanes by oxygenases. The significance of aerobic metabolism to the soil microbial community is underscored by the observation that most microbes in surface soils are aerobes. Exceptions are soils that are consistently flooded, such as bog soils.

Soil oxygen tension is controlled by gaseous diffusion rates as well as by the respiratory activity of the soil biota. Because diffusion is a relatively slow process, it is not unusual for the rate of consumption of this essential gas to be greater than its rate of replacement. Thus, free-oxygen tensions are rarely too high for soil aerobes, but clearly soil oxygen levels may be too low for obligate aerobes.

Indirect limitors of soil oxygen are soil moisture and the concentrations of metabolizable organic matter. As previously indicated, moisture impacts soil oxygenation in that the greater the soil moisture, the less air-filled pores exist and the slower diffusion rate. Decomposable organic carbon availability controls the rate of oxygen consumption indirectly in that free oxygen may be depleted as this fixed carbon is oxidized. With an influx of easily metabolized organic matter, it is not unusual for anoxic conditions to develop. Examples of practical consequences of oxygen depletion due to soil microbial respiration follow.

- Oxygen depletion in cropped soils after a heavy rainfall event can result in crop damage.

- After a heavy rainfall, with the presence of nitrate and suitable carbon sources, significant losses of fixed nitrogen from soil can result from the induction of denitrifiers.
- Spills of easily decomposable organic matter in soil or incorporation of readily metabolizable organic matter can result in the depletion of free-oxygen supplies, thereby causing ecosystem disruption through oxygen starvation of plant roots, death of small animals, earthworms, nematodes, mites, and so on.

As the free oxygen in soil is depleted, a number of predictable changes in microbial activity occur. When the soil oxygen tension has been reduced to less than 1 percent (v/v), the microbial population appears to shift from being predominantly aerobic to anaerobic. With the development of a reducing atmosphere, growth yields decline because the energy yielded per mole of fixed carbon oxidized anaerobically is far less than that produced from aerobic respiration (see chapter 3). (Note that with dissimilatory respiration of aerobic bacteria [denitrification and sulfate reduction], cell yields approximate biomass production with oxygen as the terminal electron acceptor, because a cytochrome-based electron transport process occurs and the carbon substrates are oxidized completely to carbon dioxide, water, and ammonia.) Fermentation results in accumulation of incompletely oxidized carbonaceous substrates (e.g., organic acids, amines). When highly reducing conditions develop so that methanogenesis and sulfate reduction occur, complete mineralization of fixed carbon substrates again occurs.

4.2.4. Redox Potential

It is difficult to separate the impact of redox potential from oxygen tension effects experimentally. To raise the redox potential artificially in the absence of free oxygen, it is necessary to add oxidizing agents, which are generally toxic to microorganisms. Thus, changes in soil redox potential are related to changes in oxygen levels. If organic matter is added to soil, oxygen is depleted and the potential drops —at times quite precipitously. This is a microbial reaction, because inhibitors of microbial activity prevent both oxygen depletion and the development of reducing conditions.

As alluded to previously, the occurrence of a variety of microbial processes is related to specific redox potentials. Some of these are as follows:

Aerobic carbon oxidation	>0.2 V
Denitrification	0.15 to 0.2 V
Methanogenesis	-0.2 to -0.1 V
Sulfur reduction	-0.2 to -0.1 V

Although these data suggest that the nature of microbial activity in soil is mutually exclusive, it is not unusual to observe these processes occurring concurrently across a landscape of limited size. Again, this reaction results from the heterogeneity of the soil system. Two examples of this situation (variation of redox

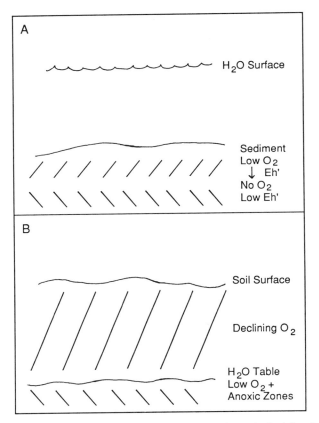

Fig. 4.5 Redox potential and oxygen tension patterns in an hypothetical flooded soil (A) and a soil with an elevated water table (B).

potential in a flooded soil and within the profile of a drained histosol) are provided in Fig. 4.5. In both situations, an oxygen-rich, high-redox-potential region overlays an anaerobic portion of the soil profile. A region of variable conditions generally occurs between these two extreme environments.

4.2.5. pH

The microbiologist's bias tends to favor conducting experiments with soils that have approximately neutral pH and to study microbes capable of functioning therein. Our laboratory media developed for the culture of microorganisms are so constituted, and the majority of the soil systems studied tend to fall in a pH range of approximately 5.5–7.5. Yet, acidic soils constitute a major portion of the world's soil resources, especially as exemplified by soils of tropical regions and temperate forests, and acidification is a major reclamation problem in the normally near-neutral soils of the temperate region. For example, with many mining processes, metal-sulfide-bearing slags are produced. Biological activity in these waste materials results in oxidation of the sulfide to sulfate. Subsequent drainage of these acid-bearing waters presents major problems for downstream soils and receiving waters. Hence, understanding the effect of soil pH on soil microbial community development and function is imperative.

Acidic conditions present a particularly stressful situation for the microbial cell. Organisms existing in acidic habitats can be described as being in a somewhat schizophrenic state. The internal cell pH must be highly buffered at or near neutrality, the pH optimum for internal enzymes. Thus, the effect of varying the external pH necessitates developing cell-surface- and membrane-associated mechanisms to control the internal pH. Furthermore, metabolic processes occurring at the cell surface must be adapted to cope with the acidic conditions.

A variety of pH-associated metabolic problems must be surmounted by the cell. For example, cellular surface charge varies with environmental pH. This variation controls the interaction with soil particles as well as with charged nutrients. Furthermore, surface-bound enzymes as well as extracellular enzymes (see chapter 5) must also be capable of functioning at the elevated hydrogen ion concentration. Maximal activity must either be expressed at the predominant pH, or the enzyme synthesis rate must be modified to compensate for the reduced enzymatic activity occurring under the acidic conditions. Also, adjustment of enzyme synthesis rates may be necessary for those enzymes that are denatured under acidic conditions. If the enzyme product is essential for cell function, then the organism must adjust the enzyme synthesis rate so that it is always present at effective levels.

Indirect Mechanisms for pH Limitation of Biological Activity in Soil: The nutrient concentrations and the toxicity of environmental substituents are both affected indirectly by soil acidity. Trace mineral availability provides an excellent example of pH interactions with nutrient availability. Both iron and manganese are more water soluble at low pH, whereas molybdenum is precipitated at low pH.

Similarly, the capacity to transport charged carbonaceous substances such as fatty acids and amino acids into the cell is dependent upon the charge of the molecules. Reduction of the soil pH below the pK_a of acids results in conversion of the negatively charged entity to a neutral compound. Zwitter ions may have a negative or positive charge or be uncharged depending upon the pH of the suspending medium. Each of these states would alter the capacity of the cell or its catabolic enzymes to catalyze transformation of these materials.

Under acidic conditions, organic acids are toxic to microbial growth. This condition is not a pH effect but rather an impact of the presence of the organic acid itself in an uncharged or neutral state. Thus, accumulating organic acids under anaerobic conditions in acidic environments could be a mechanism for limiting microbial activity. This interaction of pH and anaerobic conditions may contribute to organic matter accumulation in acid swamps.

A final indirect effect of acidic soil pH conditions on biological activity relates to the potential for solubolization of toxic compounds or elements. Aluminum is more available under acidic conditions. It is not uncommon for toxic levels of this cation to be detected in acidic tropical soils.

Direct Modifications of Biological Activity by pH Variation: The direct effect of pH relates to the fact that each microbial strain has an intrinsic range of pH within which it can function. The diversity of soil life is exemplified by the fact that microorganisms grow from about pH 1 to 11. Similarly, because, to a large degree, an organism's enzymes determine the diversity of processes catalyzed, a comparable reaction to environmental pH occurs for these cell components. Although there is microbial life at essentially all naturally occurring soil pH levels, each individual species (and to some extent strain) has a characteristic pH range and optimum pH for growth. For example, *Thiobacillus thiooxidans,* a key organism in oxidation of reduced sulfur-containing compounds, grows only at extremes of acidity (pH 2 to 4), whereas *Nitrosomonas* sp., a key player in biological nitrification, has an optimum pH in the alkaline range and is not active below about pH 5.5. Similarly, another sulfur-oxidizing bacterium, *Thiobacillus thioparus,* grows only at neutral pH.

Some selection of overall contributors to biological activity results from variation in soil pH. In acid soils the dominant activity is generally the result of the fungal community, because most heterotrophic bacteria grow poorly at acid pH. An exception is with water-logged acid soils. Here the aerobic fungi are inhibited by oxygen limitations; therefore, the acid-tolerant bacteria are better able to compete.

Some microbially catalyzed processes occur in soils with a specific pH range. For example, nitrification (oxidation of ammonium through nitrite to nitrate) is suggested to occur primarily in soils with a pH greater than 5.5. This observation results from the fact that nitrification is catalyzed primarily by a small group of autotrophic bacterial species which are pH sensitive (Dommergues et al., 1978). Exceptions to this pH limitation occur in forest soils, but an unequivocal microbiological explanation of the data has not been presented.

In general, if a process occurs over the entire pH range conducive to the development of living cells, the transformations are a result of the activity of a variety of microbial species or strains with varying pH optima. For example, carbon and nitrogen mineralization occur in essentially any soil ecosystem where life is possible. Thus, a measure of the diversity of the microbial community is revealed by the breadth of conditions under which the organisms can function. This maxim relates not only to soil pH but also to other properties, such as moisture and aeration.

Conflicts Between Laboratory Observations and Actual Field Activities: Some anomalous situations have been encountered. In these cases, laboratory experiments with axenic microbial cultures yield data that suggest that the only known microbes capable of catalyzing a particular process cannot function at the prevailing soil pH of the site under study, yet the products of the reaction are detected. For example, as indicated previously, acidophilic autotrophic nitrifiers have not been detected in acidic forest soils, yet nitrate is commonly found in these soils. A variety of explanations may be proposed and should be investigated when such situations arise (Dommergues et al., 1978). The microbes may be growing in soil microsites with a favorable pH. In other words, the acid-limited microbes may be constrained to islands of more moderate conditions contained within a sea of restrictive soil. This situation would be more likely to occur in soils where the overall pH is near the extremity of function of the microorganism. That is, an overall soil pH of approximately 5.0 could be measured, yet microsites 0.5 pH units above or below that value may exist.

An alternative explanation to the anomalous occurrence of a pH-limited process may involve our capability of culturing the responsible microbes. A previously unknown population of microbes with a pH optimum in the range existent in the ecosystem could be present and functioning in the site under study. Only a small portion of the microbial community in any soil can be cultured and studied in the laboratory. Techniques for the study of the vast majority of this diverse community remain to be developed. Thus, the occurrence of processes in the field may be a better indicator of microbial capabilities than is the library of information collected from axenic cultures.

Lastly, a spontaneous chemical conversion of the reactants may occur outside of the pH range optimal for biological activity, as is exemplified by denitrification under acidic conditions. In this situation, bacteria that catalyze reduction of nitrate to nitrous oxide and dinitrogen under acidic conditions can be cultured, but their function in the native acidic soil is questioned. The kinetics of the spontaneous chemical reaction favor that process over the biologically catalyzed nitrogen oxide reduction.

A variety of interactions of soil hydrogen ion concentration and soil biological process have been examined. Life does occur in extremely acid or alkaline soils. The participants may vary from those under less stringent conditions, and the range of processes occurring therein may be somewhat truncated. Stewardship of acid soils and reclamation of acid-impacted soils require consideration of the reactions occurring therein and the limitations of the system as well as ameliorative procedures necessary to adjust the acid-impacted site to conditions more favorable to above- and belowground community development or restoration. Due to the intrinsic buffer capacity of soil, management of an acidophilic community may be necessitated by the inability to adjust the pH of many soils to a more neutral status (e.g., See Magdoff and Bartlett, 1985).

4.2.6. Temperature

The greatest resistance to temperature extremes among living entities is found in the microbial community. The typical growth range for soil inhabitants is more restrictive than the extremes associated with other ecosystems. The typical breadth of temperatures that support microbial growth in soil is from approximately 0 to 70°C, which compares on the upper side of the range for growth of bacteria in hot springs. These organisms are capable of growth at 100°C (Heinen and Lauwers, 1981). The record for maximum growth temperature is attained by sulfate-reducing, barophilic thermophiles located in deep Pacific hydrothermal vents. These organisms are capable of growth at temperatures over 250°C under pressure (Baross and Deming, 1983). Minimal growth temperatures have been shown to be slightly below 0°C in Antarctica soils. With either maximum or minimum growth temperatures, the limiting factor appears to be free water. As long as the conditions allow the existence of free water (either through boiling temperature elevation by augmented pressure or freezing point reduction by salt concentration), microorganisms appear to be capable of growth, albeit slow at the reduced temperatures.

Reduction of soil temperatures below freezing results in significant declines in the microbial populations (Biederbeck and Campbell, 1971). The cause of the cell death is not the freezing of the cell, in that freezing itself caused only a small reduction in microbial numbers. The lethal factor appears to be associated with slow thawing of the soil. As soils thaw, ice crystals form in the microbial cells and disrupt cellular integrity. Sensitivity also relates to the growth status of the microbial cells in that recently cropped soils exhibited greater population declines due to freezing than was detected in inactive soils (Biederbeck and Campbell, 1971). Furthermore, a differential sensitivity of the microbial populations was noted in that bacterial population reduction was greater (92 percent) than that of fungi (55 percent) which was greater than that of actinomycetes (33 percent). Similar differential sensitivity of soil bacteria to freezing and slow thawing was noted by Nelson and Parkinson (1978) in a study of a *Pseudomonas* sp., *Bacillus* sp., and *Arthrobacter* sp. Survival was dependent on soil moisture, storage time, and the thaw rate.

More typically for soil microbes, growth can be depicted as an activity range (Fig. 4.6). The growth rate increases until a maximal temperature, characteristic of the organism, is reached. The maximum growth-supporting temperature is determined by the sensitivity of the most fastidious enzyme, which is essential for cell replication. This temperature sensitivity curve is a good example of the **law of**

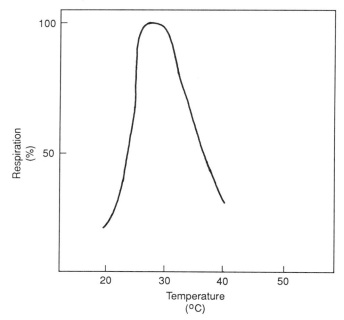

Fig. 4.6 Typical response of a mesophilic bacterial population to temperature variation. An optimum temperature of 25–30°C is depicted.

tolerance, which states that a microorganism will have an ecological minimum and maximum, with a range in between, that represents the limits of tolerance.

Tolerance to temperature can be used to divide microorganisms into three classes. Psychrophiles are defined to grow optimally at temperatures less than 20°C. Mesophiles grow best between 15 and 45°C, whereas thermophiles are active at temperatures greater than 45°C. Few if any psychrophiles are found in most soils. Exceptions are associated with Arctic and Antarctic sites. Mesophiles comprise the bulk of the soil microbial community, whereas most soils contain some thermophiles, even if the soils never achieve the high temperatures.

The presence of these temperature-associated classes of microbes in soil ensures the occurrence of metabolic activity over a wide temperature range. Typically, soil biological activity increases from a minimum at or near 0°C to a maximum and terminus point near 70°C. Differences between soils relate primarily to the proportional increase in reaction kinetics with temperature augmentation.

This rate increase has been represented by several mathematical relationships. (See Tate, 1987, for a discussion of modeling of microbial temperature responses.) Among these models are the Arhenius and Q_{10} relationships. Historically, the latter is probably the most commonly used parameter for soil microbiological studies. The Q_{10} is equal to the reaction rate at temperature one divided by the rate at temperature one minus 10°C; that is,

$$Q_{10} = \frac{K_T}{K_{T-10}} \tag{4.5}$$

Q_{10} values for biological processes range from 1.5 to 3.0 with a mode of approximately 2.0. In general, if the Q_{10} is much greater than 3.0, an environmental parameter besides temperature is controlling the reaction rate. For example, Q_{10} values for petroleum decomposition at low temperatures (10–20°C) are commonly greater than 5. At the low temperatures, catabolism of the substrate is controlled by its water solubility and viscosity. These parameters dramatically change as the temperature is raised, thereby a disproportional impact of temperature elevation of the microbial oxidation is observed.

4.3. MICROBIAL ADAPTATION TO ABIOTIC STRESS

From an analysis of the diversity of soil properties affecting microbial activity, it almost appears that life in soil would be limited at best. Yet, nearly every site on earth—ranging from hot desert soils to the sediments receiving hot spring effluent—supports active microbial populations. Even in situations where a priori considerations would predict that two factors, such as moisture and temperature, exist at limits that preclude microbial growth, microorganisms have been shown to occur in extremely specialized niches. For example, in cold dry deserts of Antarctica, cyanobacteria-dominated communities have adapted to growth between the crystals of porus rocks (cryptoendolithic) where some modulation of the extreme environmental stress occurs (Friedmann, 1982; Johnston and Vestal, 1991). This occurrence portends a variety of mechanisms of community adaptation to these stresses. Three reactions to stressful conditions should be highlighted.

First, if the stress is extreme, there is no adaptation, as exemplified by soils with temperatures greater than 70°C around hot springs or adjacent to nuclear reactor water outflows, some extremely acidic mineland reclamation sites, and soils containing excessively high metal concentrations. The existence of such ecosystems underscores the fragile nature of the biotic system and highlights the fact that at least some growth is necessary for implementing the next two survival mechanisms.

If growth is possible, microbial strains that are resistant or at least tolerant to the selective factor develop. This growth has been observed with osmotic pressure, temperature, oxygen tension, and moisture stress. A reduction of species diversity may occur in these situations, especially if the stress is severe, but a functional community develops. Reducing the microbial diversity, and thereby the genetic diversity, of the community could portend enhanced vulnerability to ecological insult. Soil presents a somewhat unique exception to this principle in that the potential always exists for invasion by propagules from adjacent soils. Thus, the genetic potential required to overcome chemical or physical stress in a particular site could exist in less-stressed neighboring soils. In this case, passive or active transfer of propagules from the neighboring soils may result in the establishment of a new community in a site that previously lacked the resources necessary for survival.

In some situations, the microbial community may possess the capacity to alter the stress factor, thereby relieving the impediment to growth or survival. For example, microorganisms may modulate inhibitory acidic or alkaline conditions or decompose toxic substances. Organic acids or even antibiotics may be decomposed to carbon dioxide by microbial strains resistant to their toxicity. Even nonbiodecomposible toxins such as heavy metals may be transformed to nontoxic states. They may be complexed, precipitated, oxidized or reduced, or even chelated. The products are rendered non-water-soluble, converted to water-soluble but nontoxic forms, or incorporated into a chelate complex that can be leached from the microbial habitat. Inhibitory inorganic acids may also be rendered ineffective. For example, sulfate is reduced to hydrogen sulfide, which is precipitated by metal ions, such as iron, which are common in soil. In each of these examples, a homeostasis is returned by activity of resistant populations conferring benefits to the survivors of the susceptible organisms.

4.4. CONCLUDING COMMENTS

The nature, activity, and future of any soil microbial community is determined by the capacity of its individual members to adapt to or modify negative soil properties. Those members of the community best equipped to cope with the combination of ecological stresses are active while more stringent organisms succumb or enter resting stages. Modifying the physical, biological, or chemical properties of the site results in the induction of microbial processes to restore the homeostasis of the community. This adaptation may result in the reemergence of an ecosystem resembling that existing prior to perturbation, development of a new community better suited to the newly developed conditions, or total destruction of the biological activity. The ultimate fate depends upon the genetic diversity of the populations, the severity of the stress imposed, as well as the time allowed for recovery. Some situations may allow recovery in hours to days, whereas in other situations full recovery may not be evident for decades or centuries.

REFERENCES

Baross, J. A., and J. W. Deming. 1983. Growth of 'black smoker' bacteria at temperatures of at least 250°C. Nature (London) 303:423–426.

Benoit, R. E., and C. L. Hall. 1970. The microbiology of some dry valley soils of Victoria Land Antarctica. *In* M. W. Holdgate (ed.), Antarctic Ecology, Vol. 2, pp. 697–701. Academic Press. New York.

Biederbeck, V. A., and C. A. Campbell. 1971. Influence of simulated Fall and Spring conditions on the soil system. I. Effect on soil microflora. Soil Sci. Soc. Am. Proc. 35:474–479.

Cameron, R. E. 1972. Microbial and ecological investigations in Victoria Valley, Southern Victoria Land Antarctica. *In* G. A. Llano (ed.), Antarctic Terrestrial Biology. Antarctic Research Series, Vol. 20, pp. 194–260. Am. Geophys. Union, Washington, DC.

Clarholm, M., and T. Rosswall. 1980. Biomass and turnover of bacteria in a forest soil and a peat. Soil Biol. Biochem. 12:49–57.

Csonka, L. N., and A. D. Hanson. 1991. Prokaryotic osmoregulation: Genetics and physiology. Annu. Rev. Microbiol. 45:569–606.

Dagley, S., and P. A. Johnson. 1963. Microbial oxidation of kynurenic, xanthurenic, and picolinic acid. Biochem. Biophys. Acta 78:577–587.

Dommergues, Y. R., L. W. Belser, and E. L. Schmidt. 1978. Limiting factors for microbial growth and activity in soil. *In* M. Alexander. (ed.) Advances in Microbial Ecology. 2:49–104. Plenum Press. New York.

Doran, J. W., L. N. Mielke, and S. Stamatiadis. 1988. Microbial activity and N cycling as regulated by soil water-filled pore space. Paper No. 132. Proc Int. Soil Tillage Res. Org. ISTRO, 11th. 11–15 July 1988.

Friedmann, E. I. 1982. Endolithic microorganisms in the Antarctic cold desert. Science 215:1045–1053.

Heinen, W., and A. M. Lauwers. 1981. Growth of bacteria at 100°C and beyond. Arch. Microbiol. 129:127–128.

Hirschberg, R. L., and J. C. Ensign. 1972. Oxidation of nicotinic acid by a *Bascillus* species: Source of oxygen atoms for the hydroxylation of nicotinic acid to 6-hydroxynicotinic acid. J. Bacteriol. 108:757–759.

Hochstein, L. I., and B. P. Dalton. 1967. The purification and properties of nicotine oxidase. Biochem. Biophys. Acta 139:56–64.

Jager, G., and E. H. Bruins. 1975. Effect of repeated drying at different temperatures on soil organic matter decomposition and characteristics, and on the soil microflora. Soil Biol. Biochem. 7:153–159.

Johnston, C. G., and J. R. Vestal. 1991. Photosynthetic carbon incorporation and turnover in Antarctic cryptoendolithic microbial communities: Are they the slowest-growing communities on Earth? Appl. Environ. Microbiol. 57:2308–2311.

Knapp, E. B., L. F. Elliott, and G. S. Campbell. 1983. Microbial respiration and growth during the decomposition of wheat straw. Soil Biol. Biochem. 15:319–3232.

Knight, W. G., and J. Skujins. 1981. ATP concentration and soil respiration at reduced water potentials in arid soils. Soil Sci. Soc. Am. J. 45:657–660.

Linn, D. M., and J. W. Doran. 1984. Effect of water-filled pore space on carbon dioxide and nitrous oxide production in tilled and nontilled soils. Soil Sci. Soc. Am. J. 48:1267–1272.

Lund, V., and J. Goksoyr. 1980. Effects of water fluctuations on microbial mass and activity in soil. Microbial Ecol. 6:115–123.

Lundgren, B., 1982. Bacteria in a pine forest soil as affected by clear-cutting. Soil Biol. Biochem. 14:537–542.

Magdoff, F. R., and R. J. Bartlett. 1985. Soil pH buffering revisited. Soil Sci. Soc. Am. J. 49:145–148.

Neilson, J. W., and I. L. Pepper. 1990. Soil respiration as an index of soil aeration. Soil Sci. Soc. Am. J. 54:428–432.

Nelson, L. M. and D. Parkinson. 1978. Effect of freezing and thawing on survival of 3 bacterial isolates from an arctic soil. Can. J. Microbiol. 24:1468–1474.

Orchard, V. A., and F. J. Cook. 1983. Relationship between soil respiration and soil moisture. Soil Biol. Biochem. 15:447–453.

Reddy, K. R., and W. H. Patrick, Jr. 1975. Effect of alternate aerobic and anaerobic conditions on redox potential, organic matter decomposition, and nitrogen loss in a flooded soil. Soil Biol. Biochem. 7:87–94.

Robertson, E. V., and M. K. Firestone. 1992. Relationship between desiccation and exopolysaccharide production in a soil *Pseudomonas* sp. Appl. Environ. Microbiol. 58:1284–1291.

Schnürer, J., M. Clarholm, and T. Rosswall. 1985. Microbial biomass and activity in an agricultural soil with different organic matter contents. Soil Biol. Biochem. 17:611–618.

Skopp, J., M. D. Jawson, and J. W. Doran. 1991. Steady-state aerobic microbial activity as a function of soil water content. Soil Sci. Soc. Am. J. 54:1619–1625.

Sørensen, L. H. 1974. Rate of decomposition of organic matter in soil as influenced by repeated airdrying—rewetting and repeated additions of organic matter. Soil Biol. Biochem. 6:287–292.

Tate, R. L., III. 1979. Effect of flooding on microbial activities in organic soils: Carbon metabolism. Soil Sci. 128:267–273.

Tate, R. L., III. 1987. Soil Organic Matter. Biological and Ecological Effects. John Wiley & Sons. New York. 291 pp.

Tate, R. L., III. 1980. Microbial oxidation of organic matter of histosols. *In* M. Alexander (ed.), Advances in Microbial Ecology, Vol. 4, pp. 169–201. Plenum Press. New York.

Tate, R. L., and J. C. Ensign. 1974. Picolinic acid hydroxylase of *Arthrobacter picolinophilus*. Can. J. Microbiol. 20:695–702.

Wardle, D. A., and D. Parkinson. 1990. Interactions between microclimatic variables and the soil microbial biomass. Biol. Fert. Soils 9:273–280.

Wildung, R. E., T. R. Garland, and R. L. Buschbom. 1975. The interdependent effects of soil temperature and water content on soil respiration rate and plant root decomposition in arid grassland soils. Soil Biol. Biochem. 7:373–378.

Wilson, J. M., and D. M. Griffin. 1975. Water potential and the respiration of microorganisms in the soil. Soil Biol. Biochem. 7:199–204.

5 Soil Enzymes as Indicators of Ecosystem Status

Logic suggests that soil enzymatic activity should be a useful measure of soil biological potential. Enzymes are the mediators for biological transformations of native and foreign soil organic and mineral components. Thus, we can predict that these catalysts provide a meaningful assessment of reaction rates of soil processes. In several instances, a variety of enzyme activities have been shown to relate to soil biological potential (e.g., Dick, 1984; Nannipieri et al., 1979; Tate et al., 1991), but frequently little or no correlation between apparently essential enzymatic activities and soil respiration, microbial biomass or population density, and microbial activity is observed. In soil systems, expression of enzymatic activity, total enzyme molecules present, and substrate levels are frequently uncoupled.

For example, a reaction may be catalyzed in a soil microsite yet provide no metabolic benefit for the soil biological community. This situation results from the fact that for an enzymatically catalyzed process to benefit growing organisms, its product(s) must be available to the cells at the appropriate time and place. With cell-linked enzymes (those contained in the cytoplasm or associated with cell membranes and walls, including periplasmic enzymes), this concurrence is not a significant difficulty. However, with the large number of extracellular enzymes in soil, substrates may be transformed by the latter enzymes into products of potential utility for cellular function, yet be of no real value for living systems, because the products and needy cells do not encounter each other.

Interpretation of soil enzymological data is further complicated by the fact that a particular soil enzymatic activity does not necessarily represent a homogenous entity. The rate of the enzyme-catalyzed process is the result of cell-associated as well as cell free enzyme molecules. However, in contrast to pure culture analyses where the cell may produce a single enzyme type to catalyze a particular reaction, a large number of different species, may synthesize enzymes with contrasting capacities to catalyze the same reaction in soils. Enzyme sources in soils include plant, animal, fungal, and bacterial cells. Thus, the quantity of any soil enzymatic activity detected in a particular soil sample is the sum of active and potentially active enzymes. Furthermore, because of the variety of different sources of enzymes, the basic properties of the specific enzymes catalyzing an individual reaction may differ (i.e., several isoenzymes for each enzymatic activity occurring in soil may exist). Thus, not all the enzymes catalyzing the same reaction are assessed under their optimal conditions when a generalized assay is used for their quantification, thereby resulting in an underestimation of the enzymatic activity of the soil sample.

These introductory observations reveal the complexity of the ecosystem in which enzymes must function in soil. Soil enzymes are essential in assessing ecosystem function; yet, the complexity of the soil system complicates interpretation of data involving these catalysts. In this chapter a basic understanding of the nature of soil enzymes and their role in ecosystem function is developed.

5.1. PHILOSOPHICAL BASIS FOR THE STUDY OF SOIL ENZYMES

Soil enzymology has been an accepted scientific study for nearly 100 years. A large body of published reports documents extensive scientific effort directed at a) the evaluation of the variation in the quantities of numerous enzymatic activities with soil and ecosystem type as well as b) the optimization of the procedures for measuring these activities. These data provide a basis for developing a conceptual model of the soil interactions affecting enzymatic activity. As a starting point in developing an understanding of the diversity of soil enzymatic activity, it is reasonable to assume that all enzymes necessary to support growth and replication of indigenous soil populations are present in any soil sample. The impact of soil particulates (e.g., adsorption), chemical properties (e.g., pH, salinity), or the presence of inhibitors (e.g., heavy metals) may preclude direct quantification or result in under estimation of the total quantity of a metabolic activity present in a soil sample, but the principles of comparative biochemistry predict that the enzymes must occur in the soil sample. Levels of enzymes detected in a soil sample and their utility are affected by the diversity of natural and xenobiotic organic compounds of the soil system, the chemical and physical complexity of the system, and the extreme heterogeneity of the microsite wherein the enzymes must be active.

Research involving quantification of soil enzymes is exemplified by the following examples of fundamental research.

- *Evaluation of basic properties of enzymes in soils.* This evaluation includes determining the source of the enzymatic activity, the kinetics of the process catalyzed, the physical limitations to enzymatic activity, the stability of the enzymatic activity in soil, and the variation of activity with soil and ecosystem type (e.g., Duxbury and Tate, 1981; Sarker et al., 1989; Tateno, 1988; Zantua and Bremner, 1975).
- *Enzymes as a measure of soil productivity:* Soil fertility relates to a composite of soil traits including biological and abiotic soil properties. Considerable research has been done to explain the relationship between basic enzymatic activities and overall soil fertility (e.g., Moore and Russell, 1972; Verstraete and Voets, 1977). Contrasting conclusions have been reached regarding the use of soil enzymes in estimating soil fertility. Moore and Russell (1972) found little linkage between dehydrogenase activity and soil fertility. Verstraete and Voets (1977) noted that yields of winter wheat related positively to phosphatase

activity, whereas they observed a negative relationship of this enzymatic activity to sugarbeet yields.

- *Enzymatic activity as an indirect measure of microbial biomass:* Considerable effort has been expended in determining the relationship between the levels of particular enzymatic activities and total microbial biomass or soil respiration. Theoretically, any metabolic activity that is expressed in active cells, is present in concentrations proportional to the number of cells, and can be accurately quantified in soil samples, can be used to measure microbial biomass. This activity must be directly associated with the microbial cells and rapidly inactivated or destroyed when cellular integrity is disrupted or with cell death. A common enzymatic activity used to estimate microbial activity is dehydrogenase (e.g., Casida, 1977). Dehydrogenase activity may be proportional to soil respiration, but direct correlation with microbial population density is unusual (Frankenberger and Dick, 1983; Tabatabai, 1982). Frankenberger and Dick (1983) found that alkaline phosphatase, amidase, α-glucosidase, and dehydrogenase activities significantly correlated with microbial respiration in glucose-amended soils. No correlation was observed in unamended soils. Alkaline phosphatase, amidase, and catalase activities did correlate with microbial biomass.

- *Indicator of capability to conduct biogeochemical cycling:* Phosphatases, sulfatases, and nitrogen oxide reductases have been used to estimate potential to catalyze essential portions of the carbon, phosphorus, sulfur, and nitrogen cycles, respectively (e.g., Häussling and Marschner, 1989; Tarafdar and Jungk, 1987; Tate, 1984).

- *Indicator of the negative effects of pollutants:* General overall ecosystem stability is related to the general health of the soil microbial community. Disruption of soil microbial activity as shown by changes in levels of metabolic enzymes can serve as an estimate of ecosystem disruption. This relationship is clearly shown when soils are polluted with heavy metals. The metal may combine with exposed enzymes through association with sulfhydryl groups, thereby altering the tertiary structure of the protein sufficiently to reduce or destroy the enzymatic function (e.g., Cole, 1977; Doelman and Haanstra, 1979, 1989; Mathur and Sanderson, 1980; Tyler, 1981). For example, Doelman and Haanstra (1979) found increasing inhibition of soil respiration and dehydrogenase activities with increasing lead concentrations from 375 to 1500 μg Pb/g dry soil. Similarly, Cole (1977) noted inhibition of amylase, α-glucosidase, and invertase in lead-amended soils.

- *Enzymes as a predictor of bioremediation potential and success:* As the biological community is reestablished in damaged soils, associated enzyme activity increases. Similarly, in organic chemical-contaminated soils, increases in enzymatic activities involved in mineralization of the pollutant serves as a predictor of capacity of native populations to return the soil to the previously existing conditions (e.g., Burnes and Edwards, 1980; Dick et al., 1988; Klein et al., 1985; Stroo and Jencks, 1982).

With each of these examples, research results that demonstrate successful application of enzymological data to the achievement of the study objectives can be presented, but numerous failures have also been recorded. These conflicting results are frequently the product of a failure to consider fully the complexities of the soil environment. For a selected enzymatic activity to reflect the level of soil biological processes, the following criteria should be met:

- The enzymatic activity must be associated with living cells, or the products must be accessible to such cells.
- The enzyme must be a participant in a biological reaction essential for the process of interest.
- The activity measured must be proportional to the quantity of enzyme present. Active enzyme must not be obscured by large pools of soil-stabilized abiontic enzyme or spontaneous chemical transformations.
- The enzyme activity must be essential for the process of interest (i.e., alternate synthetic pathways must not be major contributors to the conversion or the proportion of the reaction rate contributed by the alternate route must not change as the soil condition is altered).

These observations regarding the use of enzymes as indicators of soil biological activity as well as the considerations regarding valid interpretations of the data reveal an underlying conceptual viewpoint of the soil enzymologist. The philosophical understanding of the soil enzymologist resembles that of a biologist evaluating the activity of a tissue more closely than it does that of a bacteriologist studying axenic cultures. For the soil enzymologist, activity of the individual microbial species is less important than the sum of all metabolic activities occurring in the soil system. Each cell, enzyme, animal, or plant tissue piece is considered merely to be a part of the larger entity, the soil sample. From the view of a soil microbiologist (or perhaps a better term—ecosystem microbiologist), the function of the whole is more important than the intricacies of the parts. This is not to say that the identity of the individual has been trivialized. Indeed, to gain a complete understanding of the contributors to system homeostasis, as well as ecosystem potential, the potential and expressed communities of each individual life-form must be understood, but it is their interactions as a team rather than their performances as individuals that are generally important to the sustenance of the total ecosystem.

5.2. BASIC SOIL ENZYME PROPERTIES

Note that the basic biochemical concepts relating to enzyme function and kinetics are unchanged in the soil ecosystem. Enzymes are proteins whose activity is characterized by parameters that reflect the interaction among the enzyme, its substrates, and the products of the reaction. Synthesis rate of *de novo* enzymatic activity are controlled by cellular metabolic processes. The apparent variation in enzyme properties measured in soil that appears to contradict the data collected using axenic cultures or partially purified or characterized enzyme preparations is a result of the interaction of enzyme molecules and biological cells with the variety of chemical and physical components in the soil microsite. Thus, this section delineates the common points among *in situ* function of enzymes in soil microsites and properties of these catalysts in purified or semipurified preparations studied in the laboratory as well as the divergence associated with understanding these entities in the laboratory and field.

Soil Enzymes as Proteins: Enzymes are proteins that combine with their substrates in a stereospecific manner to catalyze biochemical reactions by modifying the electron configuration around the chemical bonds at the site of catalysis, thereby lowering the activation energy of the reaction. This rather simple definition provides a basis for predicting modifications in enzymatic activity and limitations to the longevity of the activities imposed by soil properties.

Initial considerations relate to the proteinaceous nature of enzymes. These proteins are folded into a three-dimensional structure (tertiary structure) that optimizes interactions with the substrates for the reaction. Thus, any soil properties that modify the protein tertiary structure alter the enzymatic reaction rate. The tertiary structure of the protein can be modified by interaction with soil metals, by the ionic changes induced by varied soil pH, by changes in hydration due to desiccation, as well as by a covalent linkage to soil humic substances. The enzymatic reaction rate in soil is therefore dependent upon the substrate concentrations, pH, ionic strength of the interstitial water, temperature, as well as presence or absence of a variety of enzyme inhibitors and activators.

Extreme disruption of the enzyme three-dimensional structure is termed denaturation. It can be likened to the changes that we observe when egg white proteins are heated to high temperatures (i.e., the egg white proteins are denatured by the high temperature into a substance that a major percentage of the population considers more palatable than the uncooked egg). Denaturation of enzyme proteins in soil may be induced by pH extremes, high temperatures, as well as interaction with a variety of polyvalent metalic anions (e.g., mercury, copper, and cadmium).

Enzyme vs. Enzymatic Activity: A basic aspect of describing soil enzymes relates to the terminology used to designate that activity assayed in the soil sample. It is not unusual to hear soil enzymologists state that a particular enzyme (e.g., phosphatase) was measured in a soil sample. Yet, it must be understood that the name of an enzyme can be used only when it is clear that the activity of a single enzyme type has been measured. For example, the phosphatase produced by a specific soil bacterial species, say that of a *Pseudomonas* species, can be quantified. In that situation, the experimenter can state that *Pseudomonas* phosphatase was examined. But, with a soil sample, a wide variety of bacteria, fungi, as well as higher animals and plants that produce distinct phosphatase enzymes can exist. Should the conversion of organic phosphate to mineral phosphate be assayed in that soil sample, the technician would find it difficult to state which of the specific phosphatase enzymes were quantified. Thus, the most accurate description of the procedure would be to state that phosphatase *activity* was measured.

A further qualification of the description of the soil process measured involves indicating whether potential or actual field activity is being estimated. This point relates to the enzyme kinetics parameters discussed later in this chapter and the specific assay conditions used to characterize the soil enzyme. Enzyme activities are generally measured in the laboratory under optimal chemical conditions with non-limiting substrate levels. However, rate-limiting substrate concentrations in soil are generally available, and the enzyme must cope with such difficulties as moisture, pH, or salinity limitations. Thus, the activity measured in a test tube is the maximum rate at which the enzyme activity can be expressed. This value is termed the potential enzyme activity.

Role of Enzymes in Soil: The most clearly delineated enzymatic functions in soils are those cytoplasmic activities associated with intermediary metabolism. Less distinct is the contribution of extracellular enzymes to the metabolism of living cells. Extracellular enzymes are secreted by soil biomass to solubolize or metabolize substrates external to the cell in order to produce nutrients needed by the microbes, to detoxify environmental substituents, or to modify the microenvironment of the living cell in a manner to improve the probability of cellular survival. These enzymes are highly valuable for total ecosystem function as long as the products of their activity are available to living cells. Note that a positive benefit for ecosystem function may be credited to extracellular enzymes as long as *any* living entity benefits. Thus, the cell producing the enzyme may be at a disadvantage, but the community gains. Should these products or the cells not concurrently exist, the extracellular enzymes could be classed as having no ascribable role in the soil ecosystem. Similarly, a variety of endo- or extracellular enzymes may be stabilized in native soil organic matter. Although the enzymes may continue to catalyze transformations of potential benefit to the living soil community, any gain by the cells would be fortuitous.

Stability of Soil Enzyme Activity: Because enzymes, be they within the living cell or existing free in soil interstitial water or bound to collodial soil organic matter and clay particles, are proteins and because proteases (enzymes that hydrolyze proteins to amino acids) are ubiquitous in soil, enzymes are reasonable labile soil substituents. Some enzymes may be stabilized in soil organic matter and therefore have a half-life of decades or even centuries (Skujins and McLaren, 1969), but most enzymatic activities are reasonably short-lived. Those enzymes contained within cellular protoplasm (i.e., associated with intermediary metabolism) turnover at rates dictated by normal cellular metabolism. The change in enzyme concentration within the cell with time (dE/dt) can be described by the following general equation:

$$\frac{dE}{dt} = \frac{DE_{synthesis}}{dt} - \frac{DE_{decomposition}}{dt} \quad (5.1)$$

Within the cell, these rates are controlled by the metabolic status of the cell. In a starving cell or even a resting cell, the decomposition rate may be much larger than the synthesis rate—especially of nonessential activities—so that the quantity of a particular enzyme activity rapidly declines.

External to the cell, some enzymes may be stabilized by covalent linkage to soil humic acids. The physical attachment of the enzyme to these substances limits access to the enzyme by proteolytic enzymes, thereby reducing the potential for proteolysis. This covalent linkage of the enzyme proteins to soil organic matter also stabilizes the enzyme tertiary structure in a manner that reduces the potential for denaturation of the protein. For enzyme activities which are a combination of endo- and extracellular activities, the apparent change in total activity in soil is modulated by the quantity of stabilized enzyme. Total enzyme activity E_T is described as

$$E_T = E_{\text{stabilized}} + E_{\text{cellular}} \qquad (5.2)$$

where the quantity of cellular enzyme and its turnover rate is described by equation 5.1.

Stabilization of Enzyme Activities in Soil: Because a significant portion of the enzymatic activity detected in soil samples may be the result of enzymes covalently bound to colloidal soil organic matter or associated with clay particles, consideration of the nature of these stabilized proteins and the processes leading to their long-lived state is worthwhile. There are two major fates of enzymes lost from cells due to cell lysis or active secretion of the proteins into the extracellular milieu; the proteins may be decomposed or they may be stabilized through linkage to soil humic materials or adsorption to clay minerals.

Decomposition of proteins in soil interstitial waters may be preceded by denaturation. Proteins whose native environment is the cell cytoplasm, the cell membrane, or the periplasm generally find the environment external to the cell harsh. These proteins are vulnerable to suboptimal pH, unfavorable ionic strength of soil interstitial water, as well as the presence of denaturing agents such as heavy metals. Sensitivity to the external milieu varies with the nature of the enzyme protein. For example, Frankenberger and Bingham (1982) found a different sensitivity of a variety of intra- and extracellular enzyme activities to soil salinity. Dehydrogenase activity was severely limited by salinity whereas a variety of hydrolases were more resistant. The authors concluded that osmotic desiccation of the microbial cells may have resulted in the release of the intracellular enzymes into the soil interstitial water and their subsequent proteolysis.

Adverse conditions tend to disrupt the protein tertiary structure in such a manner that increases susceptibility to proteolytic activity. Within the cell, proteins may be protected by physical separation of the proteins and proteases as well as existence of the protein in a tertiary structure that limits or perhaps even precludes proteolytic attack. For example, should the terminal residues be protected by being embedded within the protein molecule, C- or N-terminal specific proteases would not be capable of initiating hydrolysis of the protein. This protection could be lost once the protein normally contained in the cytoplasm enters the soil environment.

Those enzymes that are synthesized by the cell and secreted for function external to the cell wall or membrane (extracellular enzymes) generally have a tertiary structure that is more strongly held together (because of the covalent, hydrogen bonding, and ionic linkages) than that of intracellular enzymes. Thus, extracellular proteins are more resistant to denaturation and proteolytic activity than are enzymes whose prime site of function is the protected cove of the cellular protoplasm.

As indicated previously, enzymatic activity can be stabilized in soil through linkage to the soil humic materials. These enzymes are stabilized through ionic, hydrogen, or covalent bonding to soil humic materials. In situations where the linkage is through ionic or hydrogen bonds, the enzymes can be removed easily from the soil by washing them in buffer suspensions. For example, peroxidases frequently can be easily extracted from soil suspensions (Bartha and Bordeleau, 1969). Few enzymes can be extracted from soil with buffer solutions, so it is logical to conclude that this type of association between enzyme proteins and soil organic matter is somewhat uncommon. Similarly, although there are numerous reports of separation of soil enzyme activities from mineral matter (e.g., Ladd, 1972; Nannipieri et al., 1980; Perez-Mateos et al., 1988), these purified enzyme preparations still contain humic substances associated with the proteins. Gosewinkel and Broadbent (1986) separated about 3 percent of the phosphatase activity from a field soil with electron-donating substances, suggesting a disruption of the enzyme–humic acid covalent linkage. The proteins may be bound to humic acids through quinones by nucleophilic substitution, to sulfhydryls, or to terminal and ε-amino acids. In these situations, enzymatic activity of the proteins is maintained if the protein configuration is maintained and the bonding does not obscure the active site. Should either of the conditions not be met, then the enzyme would enter the class of humic-acid-bound proteins rather than be a soil-stabilized enzyme.

The quantity and type of enzymatic activities stabilized in particular soil samples may be distinctive for the individual soil site, such that their analysis could be of value for forensic investigations. Thornton et al. (1975) found that phosphatase, arylsulfatase, urease, invertase, and trypsin activities in three soil series were sufficiently characteristic of the soils that soil samples collected within close proximity could be distinguished by their enzymatic patterns. The Michaelis constants of the enzymes can be used to characterize the soil samples to greater detail.

5.3. PRINCIPLES OF ENZYME ASSAYS

To quantify a specific enzymatic activity in soil samples, knowledge of a variety of properties of the enzyme is essential, even if the activity of interest has been studied previously. In the latter situation, the enzyme assay must be optimized for the particular soil of interest and perhaps modified to increase the probability of achieving the experimental objectives. Contrasting goals for a study may be a) to quantify total activity of a particular enzyme present in a soil sample or b) to determine the amount of that activity expressed *in situ*. These values may differ. For example, using a buffer adjusted to a pH that is optimal for a particular enzymatic activity may allow determination of the highest amount of substrate transformed by an enzyme in a soil sample, but soil enzymes rarely function at their optimal pH. Thus, activity values detected in buffered assays may show little relationship to real field catalysis rates. The assay procedure may be modified so that the *in situ* soil pH and therefore the native enzyme activity are more accurately reflected. Some of the properties of the enzyme that should be evaluated when developing an assay procedure follows. (See Tabatabai, 1982, for a more detailed discussion of this point.)

- The reaction catalyzed should be understood in stoichiometric detail. We need to know about the reactants and the products as well as understand the chemical and physical properties of each of these chemicals. The latter traits are useful in developing enzyme assays because they may provide a useful measure when they are quantified. For example, a strong adsorption maximum could be exploited for a quick assay of product formation or substrate disappearance.
- The requirement for chemicals aside from the substrates for optimal enzyme reaction should be understood. This understanding includes any necessary metal requirements for the proteins and electron acceptors. This is less of a problem for assaying enzymes in a crude milieu such as soil, but it becomes more difficult as the enzyme activity is "purified."
- The kinetics of the reaction in relationship to substrate concentration must be elucidated. Enzyme assays are commonly conducted with saturation concentrations of the substrate. This procedure is necessary because the assay must be proportional to the quantity of enzyme present and not limited by the substrate levels. An exception to this generalization is encountered when saturation levels of the substrate inhibit the enzyme. In this situation V_{max} of the reaction can be used to estimate the quantity of the enzyme present. (See section 5.4.)
- Even though the experimental objective may be to quantify the enzyme activity under field conditions, the optimum pH, temperature, and ionic strength of the enzyme activity should be understood. These optima are particularly important when a variety of soil types is to be assayed. Similar levels of enzyme protein may be present in each soil, yet the expressed activity could vary significantly due to suboptimal chemical or physical conditions.
- A suitable means of monitoring changes in the substrate or product must be developed. This is particularly important in the soil system because the altered conditions of the soil in the assay vessel could result in synthesis of new enzymatic activity or even, in the rare situation, in a decline in enzyme levels. Generally, it is more expeditious to measure changes in product levels, because it is generally more accurate to quantify increases in the level of a compound from zero, than it is to measure slight changes in a substrate that is present in high concentrations.

Methodological Difficulties of Assessing Soil Enzymatic Activity: Even after developing an apparently ideal assay for the enzyme activity of interest, applying the procedure to soil samples may be precluded or special considerations may be necessary in interpreting the results of the procedure. As stressed previously, these difficulties arise primarily because of the complexity of the soil system and the reactions occurring therein. Four specific problems illustrate these difficulties: a) biologically vs. chemically catalyzed processes, b) complications due to microbial growth or enzyme synthesis during the assay time, c) stability of the enzyme products in the reaction mixture, and d) impact of sample treatment prior to enzyme assay.

Biologically vs. Chemically Catalyzed Processes: Within the reasonably defined confines of a microbial culture vessel, it is somewhat easy to assign cause to a particular process (i.e., biological and chemical reactions are usually easily distinguished). In soil, because of the potential for catalysis of organic transformations by clay particles, abiontic enzymes, and spontaneous chemical reactions, such distinctions are much more difficult. It is not unusual for a chemical conversion to be catalyzed both biologically and chemically within the same soil sample and, perhaps even, the individual microsite. Although the process is catalyzed by intracellular enzymes, denitrification provides an excellent example of the overlap between biologically catalyzed and spontaneous chemical reactions at acidic pH values. Denitrification is predominantly the result of enzymological catalysis of the reduction of nitrogen oxides in agricultural soils with mildly acidic (5.5–7.0) pH levels. As the soil pH declines, nitrogen oxides may be denitrified through spontaneous chemical reactions. At the acid pH extremes (<3.5) although it has been shown that biological denitrification can occur (Muller et al., 1980), it is generally assumed that chemical reactions predominate. At intermediate pH ranges, both chemical and biological reactions can occur.

The question thus becomes, How can biological and chemical processes be distinguished in soil? Three reaction properties, which may separate abiotic or biotic contributions to a soil process, are the impact of soil sterilization on the reaction, the heat stability of the enzymes, and the reaction kinetics.

The most commonly used procedure for distinguishing biological processes from spontaneous chemical reactions is to assess activity in sterile soil controls. Biological activity may be destroyed through heat sterilization (generally steam sterilization at elevated temperatures and pressures [121°C, 15 psi], but a variety of chemical sterilants have been used) as well as radiation sterilization. The assumption underlying these methods is that only the biological aspects of the soil are altered by the sterilization procedure and that the chemical processes continue at unaltered rates in the treated soil.

Ideally, an all or none effect on the reaction rate by sterilization is observed. Unfortunately, such easily interpreted data are not often produced; that is, the reaction rate is reduced by sterilization but not precluded. Contributions to the intermediate effect include the possibility that the soil samples may not have been rendered sterile by the sterilization procedure, the process may actually be occurring as a result of chemical processes, at least in part, or sterilization of the soil created conditions where the chemical reaction could occur when it would not necessarily be anticipated to occur in the native soil sample. Radiation-sterilized soil samples are particularly vulnerable to the latter problem. Radiation sterilization creates free radicals in the native soil organic matter. For transformations, such as polymerization of aromatic-ring-containing compounds, formation of free radicals increases the probability of incorporating aromatic-ring-containing pesticides, for example, into native soil organic matter as well as polymerization of the test compounds. In any case, a sterility control must be conducted. Samples of the treated soil should be examined for surviving microbial populations, or any other populations that developed subsequent to sterilization but prior to use in the experiments, by culturing soil samples in a variety of media. Testing for microbial growth in a single medium does not exclude the existence of viable microbes that are unable to grow in the test medium but capable of growth in an alternate medium. Thus, a variety of microbial growth media should be used.

Heat stability of the catalysts can also be used as a measure of enzyme-catalyzed reactions. Proteins are denatured quickly (within minutes) at temperatures above 55°C. Thus, maintaining a soil sample at 55°C for approximately 30 minutes should destroy an enzyme-catalyzed reaction. Again, interpretation of soils data must be tempered with the realization that proteins stabilized in native soil organic matter may become heat-resistant. Thus, a portion of the enzyme activity may survive the heat shock.

A further consideration in differentiating chemical and biological soil processes relates to reaction kinetics. As discussed later, saturation kinetics are observed with enzyme-catalyzed processes. As the substrate concentration is increased in the reaction mixture, a level is reached where no further increase in reaction rate is detected (see Fig. 5.1). The reaction becomes proportional to the quantity of catalyst present rather than the substrate concentration. In contrast, increased concentrations of reactants in a chemically catalyzed mixture generally lead to increased product production until another rate limitation, such as substrate solubility, is imposed. Therefore,

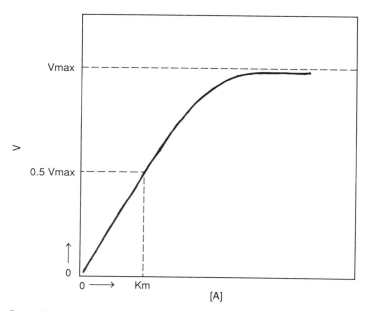

Fig. 5.1. Generalized rate curve for an enzyme-catalyzed reaction where V_{max} is the maximum reaction rate, v is the reaction rate, and [A] is a substrate for the enzyme.

an evaluation of the reaction kinetics can frequently be used as an indicator of biologically vs. chemically catalyzed reactions.

Complications Due to Microbial Growth or Enzyme Synthesis: Another methodological difficulty is associated with the assay procedure itself. Again, with a defined mixture of enzyme substrates, cofactors, and buffer, it is reasonably easy to keep the principles (the enzyme concentration and substrates, including any requisite cofactors) at constant or saturating levels throughout the assay period. This balancing is frequently difficult, if not impossible, when evaluating soil enzyme activities. Soil samples are a complex mixture of substances that may allow replication of enzyme-producing cells as well as *de novo* enzyme synthesis by preexisting microbial populations. Conditions are generally ideal for growth of the enzyme-producing microbial cells or synthesis of nascent enzyme during enzyme assays.

Because the objective with any assay procedure is to determine the quantity of enzyme activity present in the soil sample prior to assay, not that produced during the quantification procedure, methods must be implemented to prevent enzyme synthesis. For assays of a few minutes duration, enzyme synthesis is not a significant difficulty, but some procedures require incubation times of several hours for adequate product to be synthesized for accurate quantification. For example, phosphatase and some protease activity measurements require 4–6 hours or longer. A variety of microbial growth or protein synthesis inhibitors may be incorporated into the enzyme assay mixture. Inhibitors of microbial growth commonly used in soil enzyme assays are toluene, γ-irradiation, and antibiotics. Toluene (10–25 percent (vol/vol) concentrations) is the oldest and is still a commonly used procedure. The inhibitory mechanism of toluene is primarily the dissolution of cell membranes. Therefore, toluene inhibition of cell growth is most useful for assessing extracellular enzyme activity and for those activities where disruption of the cell membrane does not affect the activity. It has been used for phosphatase and invertase activities. Dehydrogenase activity—a nondescript enzyme activity associated with intact cell membranes—is inhibited by toluene inhibition.

In contrast, irradiation of cells with 5–10 Mev electron beam, hard X-rays, or γ-rays is useful for assaying intracellular as well as extracellular enzymatic activity. The differential sensitivity of soil biomass and the potential to effect spontaneous chemical reactions must be considered when using irradiation. (See Skujins, 1967, for a more complete discussion of radiation sterilization applicability to enzymatic assays.)

Antibiotics have also been used to prevent *de novo* protein synthesis during soil enzyme and respiration measurements. When interpreting data derived from using antibiotic inhibitors, we need to look at the differential sensitivity of soil biomass (bacteriocides vs. fungicides), inactivation of the antibiotic due to adsorption to soil components, as well as biodegradation susceptibility of the inhibitors. The degree of inhibition and specificity of the antibiotic should be examined for each soil type to be used in the individual study.

Stability of Enzyme Products in the Assay Mixture: Further methodological complication to be considered when developing enzyme assays involves the fate of the reaction product. As indicated previously, the most sensitive enzyme assays involve quantification of the reaction product. With defined systems, it is relatively easy to choose a substrate or modified substrate that yields products that are stable in the reaction mixture (e.g., a product that is not further metabolized within the reaction mixture). This goal is frequently not possible with an ill-defined soil sample. The product of the enzyme activity may be an ideal substrate for other organisms or enzymes existent in the soil mixture. For example, it is common to measure carbohydrases by quantifying the monosaccharide production rate, but monosaccharides are highly degradable and frequently short-lived in soil. Thus, a control for such assays must involve a determination of the longevity of the monosaccharide yielded by the enzyme.

Impact of Sample Treatment Prior to Enzyme Activity Assessment: The final methodological problem to be analyzed herein relates to soil sample storage techniques. Few enzymatic activities can be measured immediately upon selection of representative soil samples. Soil samples must be collected, transported to the laboratory, generally stored for some period of time, and then assayed. Procedural details that may impact the accuracy of the enzyme assay include conditions during sample storage (temperature, $-21°C$ or $4°C$ or room temperature, or moisture— field moisture or air dried), sterilization procedure for the assay, reaction conditions (e.g., pH, temperature), and static vs. agitated incubation of the reaction vessel. Controls must be recorded to estimate changes in field enzyme levels by each of these procedural variations.

Along with the controls indicated previously, any enzyme assay procedure must include a no soil control (to demonstrate the stability of the reaction mixture and, with colorimetric assays, to indicate background absorbance of the reaction mixture) plus a no substrate control. With the chemical complexity of soils, the possibility exists for compounds that mimic the product to be assayed to occur in the soil sample or for sufficient substrate to preexist in the soil sample for the synthesis of significant quantities of the reaction product in the absence of amended substrate.

5.4. ENZYME KINETICS

Evaluation of the impact of substrate concentration on rate of product formation of an enzyme-catalyzed reaction (i.e., measurement of the kinetics of the enzymatic catalyzed reaction) is a valuable means for elucidating the properties of the soil microsite in which the enzyme is functioning, the number of enzymes catalyzing the reaction of interest that are present in the soil sample, the total amount of enzyme in a soil sample, and the impact of soil properties on expression of the enzyme activity. This type of assessment generally involves determining process reaction rates with increasing substrate concentration, under otherwise constant assay conditions. Rates of soil biological processes are generally either zero-order (the reaction rate is constant and independent of the substrate concentration) or first-order (the rate is proportional to the substrate concentration). Modification of incubation temperature, pH, ionic strength of the incubation solution, plus alteration in concentration of minor elements or cofactors required for the reaction may have a significant impact on the reaction rates.

Zero-Order Biological Processes in Soil: The disappearance of substrate (dS/dt) for zero-order reactions is described by the following simple equation:

$$\frac{dS}{dt} = k \tag{5.3}$$

The constant k is independent of substrate concentration and, thus, is proportional to total enzyme. Although first-order kinetics are more commonly recorded with soil enzyme-catalyzed reactions, zero-order kinetics are frequently encountered, especially in situations where the substrate concentrations used in the assay procedure are considerably higher than would normally exist in the soil site.

Observations of zero-order kinetics for soil processes commonly reflect availability of the rate-limiting substrate for the transformation rather than a basic property of the enzyme protein. Note that in the first-order reaction curve depicted in Fig. 5.1 the reaction rate is independent of increased substrate amendment in the presence of high substrate concentrations. Thus, any enzyme reaction may appear to be zero-order if rate-saturating quantities of substrate a) are used in the reaction mixture to quantify the enzymatic activity or b) already exist in the soil sample due to the presence of high indigenous levels of reactants. (This saturation velocity will not be apparent with enzymes that are inhibited by high substrate concentrations. With inhibition of the enzyme, the reaction rate actually declines when the substrate concentration is too high.) Thus, it is highly probable that some reactions considered to be zero-order in native soil samples are in reality first-order processes functioning at saturating levels of substrate.

Alternatively, zero-order kinetics are also observed when access to the enzyme by the substrate is limited. In this situation, the reaction rate is controlled by the rate of encounter of the enzyme and protein. This situation occurs when the substrate must diffuse to the active enzyme or where the substrate availability is controlled by its limited water solubility. In the latter situation, the effective concentration of the substrate is controlled by its dissolution rate.

First-Order Processes in Soil: In contrast, rates of first-order processes are substrate-dependent. The most commonly used mathematical representation for first-order reactions is the Michaelis–Menten equation. This relationship and some common transformations are presented in Fig. 5.2. Other mathematical models have been developed for describing first-order reaction kinetics, but most soil enzymology is still based on the Michaelis–Menten equation. Note that Michaelis constant K_m is proportional to three rate constants in the general enzyme reaction depicted and that the maximum reaction velocity V_{max} is proportional to the total quantity of enzyme present in the reaction mixture. (Those unfamiliar with the derivation of the basic Michaelis–Menten relationship should consult an elementary biochemistry text.)

Plotting the reaction rate vs. substrate concentration yields a hyperbolic equation (Fig. 5.1). Note that the Michaelis constant is that concentration of substrate that allows the reaction to proceed at one-half the maximum velocity ($0.5\ V_{max} = K_m$), whereas $0.1\ K_m$ and $10\ k_m$ are approximately equivalent to 0.1 and $0.9\ V_{max}$.

It is difficult to derive rate constant values from a hyperbola. The maximum velocity is an asymptote, and it is reasonably difficult to distinguish with precision the substrate concentration yielding half the maximum velocity for determining the Michaelis constant. Therefore, a variety of mathematical transformations have been derived for the reaction that results in straight lines when plotted (Fig. 5.2). With the Lineweaver Burk Transformation, the slope of the line m is equal to the Michaelis constant divided by the maximum velocity K_m/V_{max}, whereas the x-intercept is the reciprocal of the maximum velocity. These values can thus be easily determined by linear regression of the reciprocals of the reaction velocity and substrate concentration data.

Compromises Associated with Determining Michaelis Constants in Whole Soils: Michaelis constants that are derived from study of whole soils rather than purified extracts in defined mixtures are best termed apparent Michaelis constants K_{app}. Michaelis constants are a parameter characteristic of purified enzymes, whose

$$A + E \underset{K_2}{\overset{K_1}{\rightleftharpoons}} X \underset{K_4}{\overset{K_3}{\rightleftharpoons}} P + E$$

Michaelis - Menten Relationship:

$$v = V_{max}\frac{[A]}{K_m + [A]}$$

where $V_{max} = K_3 E_t$

$$K_m = \frac{K_2 + K_3}{K_1}$$

Lineweaver Burk Transformation:

$$1/v = 1/V_{max} + (K_m/V_{max})(1/[A])$$

plot $1/v$ vs $1/[A]$

slope $= K_m/V_{max}$; intercept $= 1/V_{max}$

Hanes - Wolf Transformation:

$$[A]/v = K_m/V_{max} + (1/V_{max})[A]$$

Eadie - Hofstee Transformation:

$$v = V_{max} - K_m\ v/[A]$$

Fig. 5.2. Michaelis–Menton equation and some conversions that are useful in calculating kinetic constants.

reaction kinetics have been assessed under carefully defined conditions. Therefore, values derived for enzyme activities assessed in suspensions of soil particulates may not reflect the same properties that would be derived from studying the same enzyme molecules without the undefined soil matrix. Biological complications with measuring Michaelis constants of soil enzyme activities can result from a) the existence of multiple enzymes capable of catalyzing the reaction in the soil system, b) competing reactions for the substrate (resulting in a lower concentration of substrate available to the enzyme than assumed by the experimenter), or c) the existence of a rate-limiting step before the substrate reaches the enzyme of interest (e.g., permease activity or diffusion limitations of the substrate into the microsite of enzyme activity). Abiotic difficulties are associated with substrate adsorption (thereby becoming unavailable to the enzyme); in addition, suboptimal conditions for the occurrence of the reaction (e.g., presence of inhibitors) may also cause inaccuracies in assessing Michaelis constants of soil enzymes.

The importance of site variation in substrate concentration on enzyme reaction rates can be seen in data published by Tabatabai and Bremner (1971). The arylsulfatase and phosphatase activities in a variety of Iowa surface soils was assessed. These workers found that the Michaelis constant varied with soil type (as would be expected) as well as with incubation procedure. Continuous mixing of the soil samples resulted in lower Michaelis constants than were detected in static samples. The elevated Michaelis constant in the static samples reflected reduced availability of the substrates to the enzyme. Soil particles containing the enzyme activities of interest settled to the base of the test tube during the incubation period in the static soil samples. The reaction rate was dependent upon the quantity of substrate present in the base of the tube and was vulnerable to depletion of the substrate. Once the localized substrate was exhausted, its restoration became diffusion-limited. Thus, the Michaelis constants derived from the static systems were a product of the assay procedure and not necessarily related to the actual enzyme capability of the soil samples.

Application of K_m Values to Detection of Occurrences of Isoenzymes in Soil: Apparent Michaelis constants can be used to demonstrate multiple enzyme forms in soil. Nannipieri et al. (1982) measured a variety of enzyme activities in soil and plotted the results using Eadie-Scatchard plots. Generic examples of the types of data form this analytical procedure are shown in Fig. 5.3. With a single enzyme, a linear relationship is observed (Fig. 5.3A), whereas with two distinct enzymes a break in the kinetics curve may be detected. The data appears to be nonlinear, but in reality it fits two straight lines (Fig. 5.3B). In the situation for activity resulting from two enzymes, the reaction rate is described by the following equation:

$$v = \frac{V_{max1}\, S}{K_{m1} + S} + \frac{V_{max2}\, S}{K_{m2} + S} \qquad (5.4)$$

As the number of different types of enzyme proteins increases, clearly the complexity of the reaction increases. The procedure cannot be used for all combinations of enzymes. Problems are encountered with isoenzymes having highly divergent kinet-

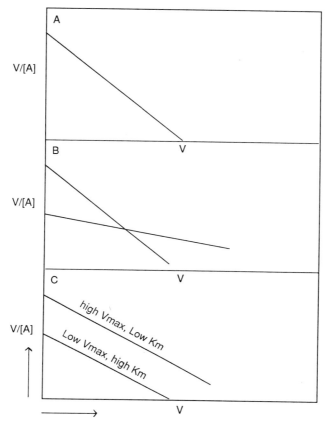

Fig. 5.3. Use of Eadie-Scatchard plots to distinguish multiple forms of enzymes in soil. (A) Plot of data resulting from activity of a single enzyme; (B) plots demonstrating occurrence of two distinct enzymatic enzyme activities; (C) situation where multiple forms of an enzymatic activity are obscured by divergent kinetics of the two isoenzymes.

ics constants (Fig. 5.3C). The rate curve for an enzyme with a high maximum velocity and a low apparent Michaelis constant would totally overshadow the curve descriptive of an enzyme with a low maximum velocity and a high apparent Michaelis constant. In this situation, the reaction would appear to result from the activity of a single enzyme. This is an excellent means of analyzing multiple forms of enzymes in soil under the simple conditions described, but frequently the data are too complicated for such an analysis.

5.5. DISTRIBUTION OF ENZYMES IN SOIL ORGANIC COMPONENTS

Experimental design for studying soil enzymological activities and interpreting data derived from such studies must be based on a thorough understanding of the com-

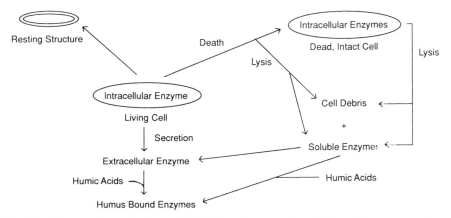

Fig. 5.4. Conceptual model of the distribution of enzymatic activity within major soil enzyme reservoirs.

plexities of distribution of the enzyme proteins within the soil matrix. As indicated previously, enzyme reaction kinetics are controlled by their degree of exposure to a variety of inhibitors and limiting chemical conditions in the soil. Differential protection is derived from containment in cellular structures, suspension in soil interstitial water, or linkage to soil colloids. A conceptual model of enzyme pools in soil and the movement of these materials between the various organic matter pools is presented in Fig. 5.4. Central to the concept of enzyme function in soil is the living cell. From a teleological viewpoint, it is for the function of the living cells that enzymes are initially synthesized. Hence, the cell must occupy the central position of the model. Two classes of enzyme molecules are linked to the living cell. For this discussion the classic definition of a living cell will be used (i.e., a living cell is one that is capable of reproduction.) The cell contains intracellular enzymes and is the primary source of extracellular enzymes.

Intracellular enzyme activities are those found in living microbial, plant and animal cells. In most studies of soil enzymes, those activities associated with animal and plant cells are of less significance than those of fungal, actinomycete, and bacterial cells simply because of the mass of each cell type and their contribution to total soil respiratory and metabolic activity. Most intracellular enzymes are involved in the various aspects of cellular metabolism (e.g., glycolysis, Krebs cycle). These enzymes cannot function outside of the cell due to cofactor needs and sensitivity to variation in pH, redox potential, heavy metals, and other inhibitory physical conditions. For example, a variety of polyphenyl oxidases contain ferrous iron in their active center. Exposure to molecular oxygen results in oxidation of the ferrous ion to ferric ion and inactivation of the enzyme. Some intracellular enzymes are capable of maintaining their activity upon cell lysis (e.g., some proteases), but generally intracellular enzymes are short-lived outside of the protective environment of the cell.

Other enzymes directly associated with respiring microbial cells and therefore not classed as extracellular enzymes per se are periplasmic (the enzymatic activities of gram negative bacteria and enzymes attached to the outer surface of viable cells). The activity of the latter enzymes extends into the ambient medium. In culture, periplasmic activities are distinguished from cytoplasmic enzymes by converting the cell to spheroplasts. This determination is clearly difficult to accomplish with soil samples, but it can be concluded that enzymes that exist in the periplasmic space in culture would also exist in that form in soil. Examples of this class of enzymes include alkaline phosphatase and penicillinase. Leakage of periplasmic enzymes into the external milieu results in their being classified as extracellular enzymes.

Enzymes attached to the outer surface of viable cells are usually defined as extracellular since their activity is directed toward external functions. These enzyme activities are exemplified by a variety of bacterial polysaccharidases, but enzymes embedded in the extracellular gum of plant roots (mucigel) may also be included in this category. These enzymes catalyze the conversion of large nonpermeable compounds into small permeable molecules.

True extracellular enzymes are those enzymes that are secreted by living cells during normal cell growth and division. These activities are found in soil interstitial water. Most extracellular enzymes have low molecular weights (20,000 to 40,000) and are produced abundantly in soil by gram positive bacteria, fungi, and plant roots. An easy means of differentiating extracellular enzymes is by their function. If their substrates cannot enter the cell, then it may be concluded a priori that the enzymes catalyzing their hydrolysis must function external to the cell membrane. Examples of such enzymes include those responsible for hydrolysis of high molecular weight or water-insoluble polymers (cellulose, hemicellulose, etc.), those involved in the destruction of toxins, and those associated with dissolution of host tissue prior to invasion by a pathogen.

Other pools for activities normally classed as being intracellular are those enzymes contained in nonproliferating cells (fungal spores, protozoan cysts, plant seeds, bacterial endospores), attached to or contained within entire dead cells, and those linked to cell debris. Enzymes of nonproliferating cells are generally surrounded by thick, biodegradation-resistant walls. Thus, these activities are not detected in enzyme assays as commonly conducted unless a procedure that disrupts the cell structure is included. This task can be accomplished through sonic oscillation. Cell-debris-associated enzymes are those linked to cell membranes, cell walls, plus internal structures of eucaryotic cells. These structures are rapidly decomposed, but the continuing turnover of microbial biomass assures a constant, low level of enzymes existent in this class. These enzymatic activities mostly include hydrolytic activities. Longevity of enzymes in this grouping is dependent upon the degree of protection afforded by and the lifetime of the cellular structure in which the enzyme is contained (cell membrane or wall) as well as the sensitivity of the enzyme to the soil environment.

Enzymes that have become stabilized in humic matter and associated with clays (abiontic enzymes) are more problematic in developing enzymes study protocols and in interpreting experimental results than any of the enzyme pools previously outlined. Synthesis of the former enzymes was induced in the cells to meet an existing need, but the enzyme activities were uncoupled from the soil biomass through association with abiotic soil components. Thus, although these enzymatic activities may be significant contributors to the total activity detected in a soil sample, their presence in the soil sample may not relate to actual biological processes occurring therein. Soil-stabilized enzyme activities are a major contributor to weak or negative correlation of enzymatic activity with soil biogeochemical processes. They provide a background from which it may be difficult, if not impossible, to distinguish the modulations in cell-associated enzymatic activity. The stability of extracellular enzymes relates primarily to the longevity of humic polymers, and any association with cellular function may be fortuitous. Benefit to soil biomass from these activities results entirely from the potential for enzymatic-catalyzed products to diffuse to active cells.

An example of a practical implication of this dynamic distribution of soil enzymes among biotic and abiontic soil fractions is the interpretation of data derived from the analysis of air-dried soils. Drying the soil may result in the redistribution of enzyme proteins among these intracellular, extracellular, and abiontic pools. Desiccation may result in death of some viable cells, lysis of viable cells and whole dead cells, as well as disruption of soil aggregates where humic- or clay-associated enzymatic activity may have been occluded and therefore not able to be an active participant in soil biochemical activities. Hence, although an enzymatic activity level can be measured in air-dried soils, extrapolation of the results to actual field conditions may be difficult, if not impossible. In reality, drying of the soils creates a new ecosystem with some relationship to that native site from which the soil was collected but with new and perhaps unique properties resulting from the altered physical state of the soil.

5.6. ECOLOGY OF EXTRACELLULAR ENZYMES

If we may take the liberty of personifying the soil microbe, we may note that the microorganism residing in soil is faced with a major dilemma. Considerable energy resources in soil are not readily consumed as carbon or energy sources by microbes because they are not water soluble. Thus they can not be moved to the microbial cell and readily transported across the cell membrane. A survival advantage is derived by any cell capable of using these substances as a carbon and energy source. The difficulty encountered is that extracellular enzymes must be used in transforming the external energy sources into units that can be consumed. These enzymes are an inefficient but essential means of procuring this energy because the probability of their functioning in soil and the subsequent recovery of the product of their action by the enzyme-producing cell is limited.

As indicated previously, soil is an unfavorable environment for any enzyme. Even if the protein is not denatured, adsorbed, or inactivated, it may not encounter its substrate because of the heterogeneity of soil and the tendency of soil organic components to be occluded by soil mineral particulates. Furthermore, if the substrate and enzyme are found in the same microsite, the substrate concentrations may not be sufficient for optimal enzyme activity. Even if the enzyme encounters substrate in sufficient quantities to produce beneficial levels of the product, the transformation may be hindered by soil physical or chemical conditions. For example, enzyme reactions occur in a water matrix. Drying the soil may result in too little moisture for continued enzyme activity. Even if all impediments to expression of an extracellular enzyme activity are avoided, one final difficulty must be overcome before the cell that originally synthesized the enzyme benefits from expenditure of energy resources to produce the extracellular enzyme protein. The product of the reaction must reach the cell. Since soil is a three-dimensional entity, the water-soluble product of extracellular enzyme catalysis may diffuse or be leached in paths that lead it away from the "hungry" cell. The product of the extracellular enzyme may also be consumed by cells other than the extracellular enzyme-originating cell. Clearly, the capacity to produce extracellular enzymes is of survival importance to the growing cells, and the proteins are synthesized in native soil sites and microbial growth due to their activity is readily demonstrable.

A variety of adaptations increase the probability of the success of extracellular enzymes. Synthesis of the extracellular enzyme may be dependent upon the presence of the substrate in the vicinity of the cell. This phenomenon reduces the probability of synthesis and excretion of enzymes with no possibility of energy gain. One means for this adaptation to be effective is for the cell to synthesize continuously a low level of the enzyme. Full enzyme production occurs only when the enzyme's substrate is present in the region of the cell. When the extracellular enzyme synthesized during the noninduced period encounters its substrate, a product that induces full enzyme synthesis is yielded. The reduced enzyme production in the absence of the enzyme substrate reduces energy expenditure and is hence of survival advantage to the microbe. This mechanism would be advantageous only to the growing cell in sites where there is a reasonable probability of encountering the substrate. Otherwise, the synthesis of even low levels of the enzyme would be a waste of metabolic energy.

A variation on the limited induction mechanism involves organisms for which the synthesis of the extracellular enzymatic activity is totally inducible (i.e., no extracellular enzyme is synthesized in the absence of its substrate). This condition is advantageous in that no energy is wasted in the synthesis of unnecessary enzyme molecules but a price is extracted in that the response to an influx of metabolizable substrate is slow because de novo enzyme synthesis must be induced. Enzyme induction would involve one of the following mechanisms: a) Production of soluble, induced molecules could be the result of activities of coresident microbes in the microsite. This would be an example of commensulism. b) Inducers could be yielded by nonbiological catalysis of the substrate. Some substrates may be decomposed chemically into intermediates that could stimulate enzyme synthesis. c) The inducer could be produced by death and lysis of a proportion of the population. In this case either the inducer or enzymes necessary to yield it are released from the cell. Lastly, d) the inducer could be produced through action of the humic acid fraction (i.e., catabolism of the enzyme substrate by the stabilized enzymes present in the soil microsite).

Other potential adaptive mechanisms involve physical limitation of the distance the enzyme could migrate from the cell. This adaptation includes enzymes that, by definition, are not classified as extracellular enzymes, but they certainly are synthesized for external activity. Activities grouped in this category include periplasmic enzymes as well as those linked to cell walls or retained in cell slime layers. The cell linkage of externally active enzymes require cellular adaptations to optimize the potential for the cell to encounter the substrate. For this mechanism to be successful, the microbe must be in direct contact with the substrate. Generally, this mechanism is advantageous to the motile organism capable of a chemotactic response to available substrates. This adaptation is commonly observed with plant cell wall decomposers (e.g., colonies of cellulytic bacteria growing on the surface of plant debris).

5.7. CONCLUDING COMMENTS

Soil enzymes have been shown to be useful, if not essential, tools in evaluating soil life processes. When properly applied, enzyme assays reveal the nature of the biological processes occurring in a particular soil and the impact of external forces on these transformations. Interpretation of data is limited by our capacity to relate activity measured in the laboratory manipulated soil sample to conditions at the actual field site; yet, with a proper appreciation for soil heterogeneity and the limitations to biological processes occurring therein, such analyses provide a real depiction of the soil system.

REFERENCES

Bartha, R., and L. Bordeleau. 1969. Cell-free peroxidases in soil. Soil Biol. Biochem. 1:139–143.

Burnes, R. G., and J. A. Edwards. 1980. Pesticide breakdown by soil enzymes. Pestic. Sci. 11:506–512.

Casida, L. E., Jr. 1977. Microbial metabolic activity in soil as measured by dehydrogenase determinations. Appl. Environ. Microbiol. 34:630–636.

Cole, M. A. 1977. Lead inhibition of enzyme synthesis in soil. Appl. Environ. Microbiol. 33:262–268.

Dick, W. A. 1984. Influence of long-term tillage and crop rotation combinations on soil enzyme activities. Soil Sci. Soc. Am. J. 48:569–574.

Dick, R. P., D. D. Myrol, and E. A. Kerle. 1988. Microbial biomass and soil enzyme activities in compacted and rehabilitated skid trail soils. Soil Sci. Soc. Am. J. 52:512–516.

Doelman, P., and L. Haanstra. 1979. Effect of lead on soil respiration and dehydrogenase activity. Soil Biol. Biochem. 11:475–479.

Doelman, P., and L. Haanstra. 1989. Short- and long-term effects of heavy metals on phosphatase activity in soils: An ecological dose-response model approach. Biol. Fert. Soils 8:235–241.

Duxbury, J. M., and R. L. Tate III. 1981. Enzyme activities in cultivated Histosols. Soil Sci. Soc. Am. J. 45:322–328.

Frankenberger, W. T., Jr., and F. T. Bingham. 1982. Influence of salinity on soil enzyme activities. Soil Sci. Soc. Am. J. 46:1173–1177.

Frankenberger, W. T., Jr., and W. A. Dick. 1983. Relationship between enzyme activities and microbial growth and activity indices in soil. Soil Sci. Soc. Am. J. 47:945–941.

Gosewinkel, U., and F. E. Broadbent. 1986. Decomplexation of phosphatase from extracted soil humic substances with electron donating reagents. Soil Sci. 141:261–267.

Hüaussling, M., and H. Marschner. 1989. Organic and inorganic soil phosphates and acid phosphatase activity in rhizosphere of 80-year-old Norway spruce [*Picea abies* (L.) Karst.] trees. Biol. Fert. Soils 8:128–133.

Klein, D. A., D. L. Sorensen, and E. F. Redente. 1985. Soil enzymes: A predictor of reclamation potential and progress. *In* R. L. Tate III and D. A. Klein (eds.), Soil Reclamation Processes. Microbiological Analyses and Application. pp. 141–171. Marcel Dekker. New York.

Ladd, J. N. 1972. Properties of proteolytic enzymes extracted from soil. Soil Biol. Biochem. 4:227–237.

Mathur, S. P., and R. B. Sanderson. 1980. The partial inactivation of degradative soil enzymes by residual fertilizer copper in Histosols. Soil Sci. Soc. Am. J. 44:750–755.

Moore, A. W., and J. S. Russell. 1972. Factors affecting dehydrogenase activity as an index of soil fertility. Plant Soil 37:675–682.

Muller, M. M., V. Sundman, and J. Skujins. 1980. Denitrification in low pH spodosols and peats determined by the acetylene inhibition method. Appl. Environ. Microbiol. 40:235–239.

Nannipieri, P., B. Ceccanti, S. Cervelli, and C. Conti, 1982. Hydrolases extracted from soil: Kinetic parameters of several enzymes catalyzing the same reaction. Soil Biol. Biochem. 14:429–432.

Nannipieri, P., B. Ceccanti, S. Cervelli, and E. Matarese. 1980. Extraction of phosphatase, urease, proteases, organic carbon, and nitrogen from soil. Soil Sci. Soc. Am. J. 44:1011–1016.

Nannipieri, P., F. Pedrazzini, P. G. Arcara, and C. Piovanelli. 1979. Changes in amino acids, enzyme activities, and biomass during soil microbial growth. Soil Sci. 127:26–34.

Perez-Mateos, M., S. Gonzalez-Carcedo, and M. D. Busto Nuñez. 1988. Extraction of catalase from soil. Soil Sci. Soc. Am. J. 52:408–411.

Sarkar, J. M., A. Leonowicz, and J.-M. Bollag. 1989. Immobilization of enzymes on clays and soils. Soil Biol. Biochem. 21:223–230.

Skujins, J. J. 1967. Enzymes in Soil. *In* A. D. McLaren and G. A. Peterson (eds.), Soil Biochemistry, Vol. 1, pp. 371–414. Marcel Dekker. New York.

Skujins, J. J., and A. D. McLaren. 1969. Assay of urease activity using ^{14}C-urea in stored, geologically preserved, and in irradiated soils. Soil Biol. Biochem. 1:89–99.

Stroo, H. F., and E. M. Jencks. 1982. Enzyme activity and respiration in minesoils. Soil Sci. Soc. Am. J. 46:548–553.

Tabatabai, M. A. 1982. Soil enzymes *In* A. L. Page (ed.), Methods of Soil Analysis. Part 2. Chemical and Microbiological Properties. American Society of Agronomy, Inc. Madison, WI.

Tabatabai, M. A., and J. M. Bremner. 1971. Michaelis constants of soil enzymes. Soil Biol. Biochem. 3:317–323.

Tarafdar, J. C., and A. Jungk. 1987. Phosphatase activity in the rhizosphere and its relation to the depletion of soil organic phosphorus. Biol. Fert. Soil. 3:199–204.

Tateno, M. 1988. Limitations of available substrates for the expression of cellulase and protease activities in soil. Soil Biol. Biochem. 20:117–118.

Tate, R. L., III. 1984. Function of protease and phosphatase activities in subsidence of Histosols. Soil Sci. 138:271–278.

Tate, R. L., III, R. W. Parmelee, J. G. Enrenfeld, and L. O'Reilly. 1991. Enzymatic and microbial interactions in response to pitch pine root growth. Soil Sci. Soc. Am. J. 55:998–1004.

Thornton, J. I., D. Crim, and A. D. McLaren. 1975. Enzymic characterization of soil evidence. J. Forensic Sci. 20:674–692.

Tyler, G. 1981. Heavy metals in soil biology and biochemistry. *In* E. A. Paul and J. N. Ladd (eds.), Soil Biochemistry, Vol. 5, pp. 371–414. Dekker. New York.

Verstraete, W., and J. P. Voets. 1977. Soil microbial and biochemical characteristics in relation to management and fertility. Soil Biol. Biochem. 9:253–258.

Zantua, M. I., and J. M. Bremner. 1975. Comparison of methods of assaying urease activity in soils. Soil Biol. Biochem. 2:291–296.

6 Microbial Interactions in Soil and Community Development and Resilience

Soil is an harsh habitat for microbial community development. There are a variety of common physical and chemical soil properties that singly or in combination with other soil properties could preclude or minimize the extent of microbial growth and activity. Although these physical and chemical soil traits appear to provide a sufficient barrier to the invading microbial pioneer to prevent colonial establishment, their impact is commonly augmented by the inherent resistance of the native biological community to invasion by alien organisms.

Interestingly, as long as invading microbes do not encounter extremes in physical and chemical properties, the biological community frequently determines the potential for successful colonization. This conclusion is based on the long understood fact that microbes do replicate when introduced into sterile soil samples (e.g., Conn and Bright, 1919; Katznelson, 1940). Although the microbial population may be functioning at the margins of their acceptable physical and chemical limitations in a particular sterilized soil sample, growth occurs if all the metabolic and physical requirements of the microorganisms are met.

Typical data are provided by evaluating the behavior of *Escherichia coli* in a sterile organic soil, Pahokee muck (Fig. 6.1) (Tate and Terry, 1980). With initial populations of approximately 1.0×10^5 propagules g^{-1} dry soil, more than a hundred-fold increase in population density resulted. Note that when large populations of bacteria were added to the soil, growth was limited or prevented, as was shown when bacterial populations of greater than 3.0×10^7 propagules g^{-1} dry soil were added to muck samples. In this case, the population density of the added *E. coli* in the sterile soil remained stable over the 10-day incubation period. Limited cellular replication when excessive population densities are added to soil samples may result from space limitations (i.e., the physical space for cellular replication is not available) or from biological interactions between the bacterial propagules inhibiting or preventing growth. Furthermore, over the duration of the study (10 days) a significant decline in the microbial populations did not occur in sterilized soil.

These laboratory observations should not be surprising, because colonization of sterile soil habitats or sites with low population densities must be assumed a priori to occur naturally. The most obviously related situation involves the invasion and colonization of nascent soils, such as volcanic ashes. Conditions may be extremely stressful especially in newly formed volcanic soils, but as the physical and chemical

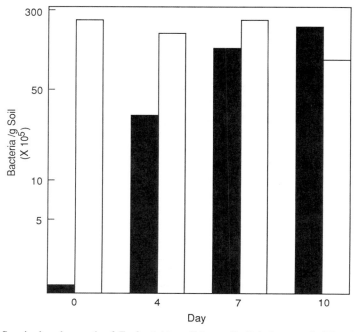

Fig. 6.1. Survival and growth of *Escherichia coli* in sterile Pahokee muck. The shaded bars represent soil samples amended with approximately 1.1×10^5 bacterial propagules g^{-1} soil, whereas the open bars depict soil samples receiving approximately 3.0×10^7 cells g^{-1} soil.

limitations to microbial growth are ameliorated (e.g., surface area increased through fragmentation of rocky structure, incorporation of fixed organic matter as the result of growth, and death of pioneer microbial communities), extensive microbial population densities and the intricate microbial communities characteristic of mature soil ecosystems develop.

These somewhat simplistic laboratory observations and rare-case field situations do not represent the most common experience involving the behavior of invader populations in native field soils. When mature field soils are amended with foreign microbes (such as through soil application of sewage effluent), the introduced organisms quickly die. Data exemplifying this natural phenomenon are easily recovered from an examination of the public health literature. In a recent survey, Reddy et al. (1981) noted half-lives for fecal coliforms amended to a variety of soil samples of 2–150 hours, for *Salmonella* sp. 2–185 hours, and for poliovirus half-lives of 7–416 hours were found. This short-term survival of microbes foreign to the soil community is fortuitous for purification of pathogens from societal waste materials and has provided the scientific justification for approval of land-based sewage and similar waste material disposal methods.

In contrast, the failure of soil-amended microbial populations to survive for extended periods is more troublesome in situations where the development of new microbe-based catabolic activity in a soil site is desirable. For example, with the current desire to develop genetically engineered or laboratory-selected bacterial strains for use in soil renovation procedures, limited survival and/or activity of the derived, axenic cultures when inoculated into the contaminated soil site raises a significant concern regarding the practicality of such procedures.

Since alien organisms readily survive when added to sterile soil samples, we conclude that the primary barriers to establishing exogenous microbial populations in soil relates to a limited capacity for the introduced organisms to cope with the preexisting soil populations. This minimal survival of alien propagules in soil may involve a variety of biological exclusionary mechanisms including the inability of the invader organisms to compete successfully with indigenous populations, sensitivity to toxicants existing in soil, and susceptibility to common soil predators or parasites.

These contrasting observations between native and sterile soil samples underscore the importance of biological interactions in soil microbial community composition and function. To understand situations where insufficient die-off of undesirable microbes occurs in soil or the contrasting inability to establish desirable microbial species in a soil site, a clear delineation of the interactions within the total soil community (i.e., viruses, bacteria, fungi, protozoa, nematodes, and higher animals) is required. Thus, the objective of this chapter is to analyze the general traits or types of microbial interactions in soil and to determine the impact of these interactions on community adaptation and resilience following ecological insult.

6.1. ELUCIDATION OF COMMON CONCEPTS OF MICROBIAL COMMUNITY FUNCTION

In order to explain the biological associations controlling soil microbial communities, we must develop a concept of the total ecosystem. Thus, we must consider the terms delimiting the system under study and the life contained therein. First, we must come to a common understanding regarding the meaning of the statement "ecology of the organism." The word *ecology* is commonly misused by both the general public and the scientific community. By definition, **ecology is the study of the relationship of organisms to their environment.** This environment includes all abiotic and biotic factors impacting upon the cell. Biotic properties of an ecosystem include not only those traits commonly classified by a society as being "natural" but also any consequences of anthropogenic intervention or interaction with the site. Although the effects of human activities on a particular soil ecosystem are by definition natural (because people are commonly a component of the aboveground ecosystem), anthropogenic factors are usually distinguished in evaluation of ecosystem function because human interactions commonly alter the nature and properties of the biological community.

Anthropogenic interactions modify the properties of the soil biological community when they sufficiently modify a controlling property of the soil ecosystem to shift the equilibrium of the site. This alteration of system descriptors can be exemplified by examining the conditions that must be met for amendment of a chemical to a soil site to shift the biological equilibrium. There are three preconditions for effect.

- *The chemical must be added in concentrations that significantly change the levels preexisting in the site.* For example, nitrogen contained in sewage effluent could be a major nitrogen source to plant communities in land disposal systems. An exception is the drained histosols of Southern Florida. The soils of the Everglades agricultural area are composed of approximately 85 percent organic matter, which is in part biodegradable. Decomposition of this organic soil results in the mineralization of about 1400 kg N ha^{-1} annually (see Tate, 1980). Thus, adding a few kilograms of mineral nitrogen to these soils through sewage effluent disposal would be not make a meaningful change in the soil nitrogen concentrations.
- *The chemical must be added in sufficient quantity to alter the steady-state equilibrium.* For example, there is considerable concern regarding the impact of acid rain on environmental systems. In well-buffered, calcareous soils, the impact of the acid content of the rain water is modulated by the buffering action of the calcium carbonate and soil organic matter.
- *The chemical may be toxic to the soil organisms.* Biodegradable toxicants temporarily inhibit microbial processes. Examples of biodegradation-susceptible compounds commonly encountered by soil communities are pesticides and petroleum components contained in accidental spills or used as carriers during pesticide application. In contrast, a nonbiodegradable compound would have a more lasting impact on ecosystem function if toxicant-resistant microbial communities are not developed.

The impact of an environmental insult on the basic properties of an undisturbed site are commonly of interest at the ecosystem level, but full explanation of the processes occurring therein require specific targeting of a particular portion of the soil biota. Interactions may be evaluated at the individual organism, population, community, or total ecosystem levels. For most soil inhabitants, definition of the individual—the single bacterial, algal, or protozoan cell—is reasonably simple to accomplish. This entity is more difficult to define for filamentous organisms, fungi, and some actinomycetes. For most of these organisms, a fragment of the mycelial structure may at time be considered to be the basic entity of this ecological hierarchy, the individual.

Above the individual level, which is studied infrequently with soil microbiological research, is the more easily evaluated population. **A population is defined as being constituted of all the individuals of the same species or function.** More rigorous definitions require a species definition, but in soil microbiology it is not unusual to discuss populations of nitrifiers, denitrifiers, or diazotrophs as well as populations of *Bacillus* or *Pseudomonas* sp. The summation of the populations in a particular ecosystem constitute the community, whereas, as indicated previously, the ecosystem is the combination of both the living and nonliving components of the environment.

6.2. CLASSES OF BIOLOGICAL INTERACTIONS COMMON IN SOIL COMMUNITIES

In simple situations (e.g., an axenic culture growing in a laboratory), the microbial population can be said to be controlled by its nutrient supply, moisture (for colonies growing on solid media), energy source, and a variety of other chemical and physical traits defining the ecosystem. But, even in this simple laboratory culture, cell–cell interactions impact the development of the individuals. The microbes are not alone in their ecosystem. Each individual must deal with the physical presence and activity of their progeny. Even if a microbe enters a new habitat as a single spore and no other organisms are present—an highly unlikely situation—the primary product of the pioneer is its progeny. That is, for successful colonization of the site to occur, the growing, respiring microbe must divide. At the least, the presence of this new cell results in competition for nutrients and space as well as an increase in waste materials accumulated in the vicinity of the developing microcolony. These parent–progeny interactions are simple in that the phenotypic and genotypic differences between the cells are minimal. Except for any mutations that may have resulted during the division cycle, each cell is essentially identical. Thus, both may be said to have equal opportunity in the competition for available resources.

As the complexity of the community increases, this simplicity is replaced by a variety of more complex, positive and negative interactions. There are few pure cultures in terrestrial ecosystems. Microbes must cope with the actions of a variety of individuals with greater or lesser capacity to cope with cohabitants of their microsite. A summary of the classical designation of the types of interaction affecting soil microbial community development is presented in Table 6.1. Each of these

TABLE 6.1. Biological Interactions Occurring in Soil Ecosystems.

Type of Interaction	Species A	Species B
Positive Interactions:		
Neutralism	0	0
Commensulism	+	0
Protocooperation	+	+
Symbiosis	+	+
Negative Interactions:		
Competition	−	−
Amensalism	0	−
Parasitism and Predation	+	−

processes is discussed individually from the view of their importance to soil microbial community as well as total ecosystem development.

6.2.1. Neutralism

Neutralism is included on the list of interactions of soil microbes for completeness, but it is unlikely that such minimal impacts of an organism on its neighbors would occur in a soil ecosystem. By definition, in a neutralistic association the organisms co-occurring in the same habitat have absolutely no effect on the cohabitants. Due to limitations to growth afforded by soil physical structure (space limitation) as well as restrictions resulting from supplies of growth factors, carbon resources, and the generally harsh conditions of soil, such a degree of noninteraction of adjacent populations is unlikely.

6.2.2. Positive Interactions Between Soil Microorganisms

Commensulism: In this opportunistic relationship, one organism can grow or function as the result of the action of a second organism. The second organism gains no benefit from the relationship. Commensulism is probably the most common soil microbial interaction. Examples include such critical soil processes as existence and function of anaerobes, organic matter decomposition, cross-feeding of nutrients, provision of growth factors, and inactivation of toxins.

Anaerobic bacteria cannot function in the presence of free oxygen. Many of these organisms are highly sensitive to even low levels of molecular oxygen and die quickly in its presence. Clearly, adaptations are necessary for these organisms to exist in an aerobic world. Facilitation of anaerobic activity by respiration of aerobic populations is demonstrated while evaluating microcolony development in a soil aggregate. Soil aggregates occur in soil with a predominance of aerobes growing on their surface and with anaerobes replicating internally. Facultative organisms may exist and function throughout the soil aggregate. For a benefit to be accrued by the internal, oxygen-sensitive anaerobes, populations of aerobes on the aggregate surface must use molecular oxygen at rates faster than it can diffuse into the interior of the aggregate, thereby indirectly creating the conditions allowing for the development of the anaerobic microsite. Thus, the anaerobes functioning within the soil aggregate do so at the behest of the rapidly respiring surface-residing populations, whereas the latter organisms do not necessarily gain any benefit from the internal populations.

The decomposition of plant tissue in soil triggers similar beneficial, gratuitous associations. Many soil microbes can decompose the easily metabolized substituents of cellular cytoplasm (e.g., amino acids, proteins, amylose, simple sugars), but those materials are separated from the general soil microbial community by the cellulosic cell wall. Cellulose-degrading microbes metabolize the cell wall, thereby rupturing the cell and making the internal contents available to noncellulytic microbial populations. Because, in general, cellulytic microbes are poor competitors for easily metabolized substrates, the cellulose degraders gain no or minimal benefit from the activity of those populations simulated by the release of internal cellular substituents.

Similar metabolically based commensul relationships are associated with production of metabolic by-products and growth factors. A variety of waste products are yielded from the respiring microbes. These substances may include ammonium, produced from the dissimilatory reduction of nitrate, as well as a variety of partially oxidized carbonaceous substances, such as a variety of organic acids. These may be used by cohabitants of the soil community as carbon and energy sources.

Bacteria requiring vitamins (prototrophs) are commonly isolated from soil. In a study of 499 soil bacterial isolates, Lochhead and Burton (1957) found that 27.1 percent of the organisms required one or more vitamins for growth whereas 63 percent of these more fastidious organisms needed more than one vitamin. The most common essential vitamins required by soil bacteria were thiamine and biotin (Table 6.2). By definition, any organism that can grow without an external vitamin source must be synthesizing that vitamin (auxotrophs). Thus, the latter organisms provide the essential substances to the prototrophic population. Many organisms synthesize only the quantities of vitamins required to meet their own metabolic needs. These vitamins become available to prototrophic members of the community upon death and lysis of the auxotrophic cell. Perhaps of more significance to the active soil community is the fact that many auxotrophs can excrete vitamins. More than 50 percent of the isolates studied by Lochhead (1957) excreted one or more vitamins. These latter organisms encourage the development of commensual relationships with other soil inhabitants.

A variety of toxic substances that enter an ecosystem through anthropogenic intervention or are produced internally may have a generalized inhibitory effect on microbial respiration. For example, acidification of a soil site by acid mine drainage or the internal production of sulfate from elemental sulfur or sulfide oxidation (see chapter 14) results in the inhibition of all acid-sensitive organisms. Reduction of sulfate by sulfate-reducing bacteria causes a generalized improvement of ecosystem conditions for the members of the soil community.

TABLE 6.2. Vitamin-Producing and Vitamin-Requiring Populations in Soil.

Vitamin	Require* (Percent Isolates)	Producers** (Percent Isolates)
Thiamine	19.4	35.5–56.7
Biotin	16.4	15.2–32.7
Vitamin B_{12}	7.2	29.9–34.6
Pantothenic Acid	4.6	—
Folic Acid	3.0	—
Nicotinic Acid	2.0	—
Riboflavin	0.6	35.2–67.3

*From Lochhead and Burton (1957).
**From Lochhead (1957).

Protocooperation: Protocooperation involves a temporary association in which both organisms benefit. Any microbe that can catalyze the requisite reaction can function in the combination. Thus, the combination is neither obligatory nor specific. A generalized example of such a cooperation could be found to occur in any environment in which neither generically defined growth factors A or B are present. Microbes requiring either of these growth factors would not be expected to grow in the system. Yet, both groups of organisms can be present in the absence of their prerequisites for replication. For this combination of organisms to exist in the absence of the requisite nutrients for their growth, the microbial population that requires substance A must produce compound B whereas the B auxotroph must produce nutrient A. Neither of these organisms are capable of growth in the absence of the coexisting population, but both can function in combination. The combination of organisms existing in this bi-member community is determined by metabolic capacity not species designation. That is, a third species with the requisite traits could just as easily participate in the development of the community. Specifically, such a system could be envisioned to occur in soils lacking both fixed nitrogen and carbon (newly formed volcanic soils). Successful biological associates in the pioneer community could include a green alga (a photoautotroph) and a nitrogen-fixing bacterial species (a diazotroph).

Symbiosis: To some, symbiosis is a more extreme example of protocooperation, whereas, for others, it constitutes a separate class of biological associations. Symbiosis is defined to be an obligatory, nontransitory, mutually beneficial interaction. This association is either essential for the existence of the organism in the ecosystem or for the occurrence of a particular processes, such as nitrogen fixation. The nature of this association does not mean that either of the partners cannot occur singly, under different circumstances, such as where all of the nutrients necessary for growth are supplied or where competition with other inhabitants of the ecosystem is relieved. For example, *Rhizobium* sp. and legumes combine in a symbiotic relationship to fix nitrogen (see chapter 11). In native systems, the association is obligatory for nitrogen fixation to occur, but legumes and rhizobia are capable of growth separately as long as all their nutrient requirements are met.

Commonly encountered symbiotic associations in soil systems are the bacterial–higher plant association just presented; an actinomycete–higher plant nitrogen-fixing interaction involving actinomycete species, *Frankia* sp. (chapter 11); and a variety of higher plants and mycorrhizal associations involving plant roots and a variety of fungi (chapter 7).

6.2.3. Negative Biological Interactions

Negative interactions or antagonisms are interesting from a variety of viewpoints. These detrimental associations may explain the failure to inoculate soil communities successfully with propagules of beneficial organisms (including genetically engineered microbes). Also, the principles of antagonistic associations underlie the use of soils as purifying agents for pathogen-containing waste materials and may be exploited in development of biological control methods for control of plant pathogens.

Competition: Competition, rivalry over a limiting factor, is generally described in relationship to a less active species being suppressed by a second, more vigorous, species. That is, in theory, species A derives full benefit of the available resources, whereas as development of the second population or species B is reduced. Should the competition be sufficiently intense, the poorer competitor is eliminated (**competitive displacement**). Generally, with competition for limiting nutrients, if the supply is continuous, the poorer competitor is eventually displaced.

This all-or-none effect of the competition clearly does not represent the true field situation. If the weaker population replicates at all, it will exhaust at least a small portion of the object of competition (e.g., carbon, nitrogen, molecular oxygen, phosphorus, space). Thus, the ultimate population density of the more successful as well as its growth rate must be reduced. Furthermore, in natural ecosystems, competition based on the availability of a single resource is unusual. A variety of ecosystem properties tend to mitigate the intensity of competitive interactions, thereby allowing coexistence of both competitors (Fredrickson and Stephanopoulos, 1981).

An obvious question regarding competition is; What makes a microorganism a good competitor? An understanding of these factors is useful in such environmentally important projects as developing genetically engineered organisms for bioremediation. An organism with exceptional ability to mitigate environmental hazards is of limited or no value if it cannot compete with indigenous soil microbes. Thus, selection of desirable genetic traits in bioremediation inoculants should include survival as well as the desired metabolic capacities.

Several explanations of competitive advantage have been proposed, but none fully explains population interactions and community development in all situations. Most likely, no single phenotypic microbial trait is essential for the successful domination of a habitat, but rather a variety of capabilities contribute and are expressed depending upon the properties of the competitor, the habitat in which they are growing, and the population-density-controlling condition. Four properties that are clearly instrumental in successful competition are a) growth rate, b) efficiency of nutrient utilization, c) capacity to catabolize a variety of metabolic substrates, and d) motility.

Growth rate may be effective in competing for nutrients and space, the organism with the shortest generation time being the most effective competitor. This property is actually linked to the second mechanism in that efficient use of nutrients is a necessary contributor to an augmented growth rate. The lesser the quantity of energy that must be expended for growth, the higher the growth rate. The best means of preserving energy for the microorganisms is to avoid the unnecessary synthesis of enzymes. For example, an organism that synthesizes its own growth factors even in the presence of adequate levels of those substrates would necessarily have its growth yields reduced by the quantity of energy used in synthesizing the growth factor and the requisite enzymes.

Substrate versatility and motility both offer contrasting advantages to the competing microbes. Once the original substrate has been exhausted, the capacity to use a second carbon or energy source that is present in the habitat has a competitive advantage for molecular oxygen and space competition. The motile organisms possess the capacity to escape the competitive situation by migrating to a less stressful situation, whereas the less fastidious members of the community escape competitive limitations by catabolizing alternate carbon and energy resources.

Environmental factors can determine the outcome of competitive interactions (Rosenzweig and Stotzky, 1979). In a competition between the fungus *Aspergillus niger* and the bacterium *Serratia marcescens,* amendment of soil with 3 percent kaolinite reduced the antagonism, and higher concentrations of the clay totally eliminated it. The antagonism was also influenced by soil pH and carbon availability (Rosenzweig and Stotzky, 1980). The data suggested a direct correlation between the degree of inhibition and the rate of glucose utilization by the bacteria, indicating that the antagonism resulted from competition for carbon. This competition was influenced by the clay content (kaolinite) of the soil and the soil pH.

Amensalism: Amensalism is the suppression of the growth of one organism by products of growth of a second. This may result from a situation as simple as the alteration of the soil pH or the production of a growth-inhibiting or lethal biological product. *Thiobacillus* sp. commonly reduces soil pH through the oxidation of sulfide to sulfate. Because the pH may reach values as low as 2, the growth of any pH-sensitive microbes is inhibited.

Two major types of biological inhibitors or toxins are produced by soil microbes: those effective at high concentrations (organic acids, chelators) and those that are effective at low concentrations (antibiotics). The growth-controlling impact of the former compounds in soil has been reasonably well accepted, because the substances can be quantified easily in soil samples and their interactions with soil microbial populations can be shown. The role of antibiotics within the soil ecosystem is more problematic.

There is a conceptual barrier that has, for a considerable time, prevented appreciation of the impact of antibiotics in complex, heterogeneous systems such as soil. When a soil system is examined as a whole, it is difficult to demonstrate either the presence or activity of any antibiotic *in situ*. Although the antibiotic may be present in inhibitory concentrations in a soil microsite, averaging this localized concentration with that portion of the soil sample containing little or no antibiotic leads to an underestimation of the potential impact of these microbial products *in situ*.

Further conceptual difficulties are encountered when we note that, in culture, antibiotics are commonly produced in a resting stage following a period of rapid growth. Because it is reasonable to hypothesize that the primary benefit of antibiotics in soil is during that intensively competitive period associated with cellular growth, conceiving of a role for these substances in native soil sites is difficult. As has been shown for streptomycin production by *Streptomyces griseus* (Szabo et al., 1985), this lack of coincidence between time of need and synthesis should not preclude occurrence of a benefit of the antibiotic to microbial growth. Streptomycin was synthesized by vegetative mycelia and bound by spores. Release of the antibiotic from the spores during germination is seen as a competitive advantage for the young hyphae in the microenvironment.

A simple argument in favor of the role of these substances in soil relates to the fact that a major portion of the soil community can produce antibiotics. The complexity of the molecules suggests a priori that the substance should not be synthesized without some selective advantage being conferred on the producer. But, when the total ecosystem is examined, no apparent advantage resulting from antibiotic synthesis or sensitivity is evident.

As suggested previously, benefits of antibiotic synthesis may be accrued at the microsite level. When soil ecosystems are evaluated at the point of interaction of the individual microbes, then a potential benefit of antibiotic synthesis can be gleaned. The role of these inhibitors in native ecosystems is still conjectural, but their potential has been shown clearly in laboratory, greenhouse, and field studies. For example, *Streptomyces olivocinereus,* which produces the antibiotic heliomycin, was shown to be antagonistic to *Arthrobacter crystallopoites* in laboratory medium and when both populations were introduced simultaneously into sterile soil (Polyanskaya et al., 1983). Also several antibiotic-producing bacterial strains tentatively identified as *Bacillus subtilis* and a fungus, *Penicillium nigricans,* protected onions from infection with *Sclerotium cepivorum* in native muck soil in a controlled environment chamber and reduced the occurrence of the resulting disease, onion white rot, when used as a seed treatment in the field (Utkhede and Rahe, 1980). Similarly, amendment of the antibiotic streptomycin or streptomycin-producers to the *Rhizobium* inoculum for soybeans (*Glycine max* L. Merrill) and alfalfa (*Medicago sativa* L.) increased nodulation and plant biomass (Hossain and Alexander, 1984; Li and Alexander, 1986, 1988). Thomashow et al. (1990) found that inoculation of wheat roots with antibiotic-producing (phenazine-1-carboxylic acid) *Pseudomonas fluorescens* resulted in antibiotic synthesis in the rhizosphere of wheat plants and a reduction in the occurrence of take-all disease caused by *Gaeumannomyces graminis.*

Another group of biologically synthesized compounds that appear to be useful in reducing plant disease through antagonism of pathogens are siderophores. (See Leong, 1986, for a review of this topic.) These substances appear to be active at higher concentrations than is characteristic of antibiotics, but when they result in suppression of microbial growth at low concentrations, they can be classified as antibiotics. Siderophores are extracellular, low-molecular weight (500 to 1000 dalton) iron-transporting compounds synthesized by a variety of microorganisms growing under low iron conditions. These substances selectively complex ferric ion with a high affinity, thereby reducing iron availability to competing organisms. Most commonly studied siderophore-synthesizing microbes from the view of controlling plant pathogens are members of the *Pseudomonas fluorescens–Pseudomonas putida* group. Inoculation with a variety of *Pseudomonas* strains appears to be useful in controlling *Gaeumannomyces graminis* var. *Tritici* (Hamdan et al., 1991; Thomashow and Weller, 1990) and *Rhizoctonia solani* (De Freitas and Germida, 1991) on wheat. Siderophores may also be instrumental in reducing pathology in disease-suppressive soils. (See Kloepper et al., 1980, as an example of these studies and Hornby, 1983, as a review of suppressive soils.)

Parasitism and Predation: Predators and parasites, organisms that feed upon living biomass, play a key role in the soil ecosystem. Soil bacterial and fungal populations are capable of producing resting structures or of entering resting stages (metabolic states where respiratory and metabolic activity are reduced to the minimal level necessary to maintain cellular integrity). Thus, without a means of encouraging the recycling of these nutrients contained in the resting cell biomass through the biogeochemical cycles, essential nutrients (e.g., carbon, nitrogen, phosphorus) as well as space could be retained in or by these structures at the detriment of more active soil organisms. Furthermore, facultative predators also encourage the cycling of growth substances and liberation of space through saprobic consumption of microbial biomass. Parasites and predators maintain the soil bacterial and fungal populations in an active state and enhance nutrient movement between soil reservoirs through consumption of microbial biomass. This feeding activity of predators and infectivity of parasites maintains a younger, more active, soil microbial population.

Essentially all types of predators or parasites are present in the soil ecosystems. Bacteria, which prey on other bacteria (e.g., Casida, 1980, 1983), bdellovibrios, bacteriophages, protozoa, as well as nematodes, are all active in soil ecosystems. These organisms may ingest their nutrients by consuming intact cells (**holozoic feeding**), as is commonly described for protozoa, or extracellular enzymes that lyse other bacteria, fungi, or algae may be produced. This latter process is illustrated by the predatory bacterial populations.

A key consideration in evaluating predator or parasite behavior in any ecosystem relates to the observation that both the host and parasites or prey and predators coexist in the same ecosystem. Thus, we can ask, Why do predators and parasites not totally destroy their prey and hosts? A variety of laboratory studies has been conducted to uncover the causes for the existence of prey in the presence of viable predators. These studies generally involve incubation of a suscept plus a predator (e.g., Danso and Alexander, 1975) or plus a parasite (e.g., Wiggins and Alexander, 1985) in a liquid culture. Most of the analyses of the specific survival adaptations outlined next were derived from studying this type of simplified microcosm-type interactions. In these studies, the prey populations commonly achieve an equilibrium density of approximately 10^6 propagules/ml of culture. In the Wiggins and Alexander (1985) study, a minimal host population of approximately 10^4 colony-forming units was required for bacteriophage replication. If population densities significantly greater than this value are amended to the system, their populations decline to the lower density. In studies where low prey densities were incubated in growth-supporting media in the presence of a predator protozoan population, the prey population increased to the limiting population density of about 10^6 propagules/ml culture, which was subsequently maintained.

Clearly some mechanisms must preclude total elimination of susceptible populations, or in the long run both groups of organisms will cease to exist. The continued survival of a "food" reservoir is essential for obligate predators, but facultative organisms are freed of this limitation in that they can use living prey or consume organic matter saprobically. Some relief from this limitation to parasitic or predatory behavior is also gained by the organism that can consume a variety of host or prey species. Once one, perhaps even the prime, food source becomes limiting, consumption of the alternate food source not only spares the original prey population but also reduces the energy expended by the predator in search of food, because we can assume that selection of the host is driven, to at least a significant degree, by the probability of encounter.

A variety of mechanisms have been proposed to explain the balanced attack of parasites or predators (for a detailed review of this subject, see Alexander, 1981). Examples include the following survival adaptations.

1. *Interactions among predators.* Increased predator density due to localized consumption of the prey may result in competition among the protozoa. This interference with grazing due to frequent contact among the animals could actually result in cannibalism of the predator population. This means of population control may occur in native sites, but it is not the primary mechanism in that, in laboratory studies, amendment of stabilized bacterial–protozoan systems with more bacteria causes further increases in the protozoan population densities.

2. *Predators under biological control.* With biological control of predator/prey or parasite/host associations, the predator and parasite populations may be controlled by populations that feed upon them directly or their population densities (and thereby their feeding capacities) may be delimited by the presence of biologically formed toxins. Because of the complexity the biological community in soil, organisms that feed upon bacterial predators or parasites are likely to occur commonly. Again since stabilization of predator–prey populations occurs readily in model systems with one protozoan and bacterial species present, feeding on the bacteriovorus organisms by higher organisms cannot be the sole explanation of population stabilization.

   Toxic interaction is an attractive hypothesis, but insufficient data have been accumulated to allow its evaluation.

3. *Genetic feedback.* The primary assumption behind this mechanism is that a spontaneous mutation and population selection that allows development of host cells with reduced susceptibility (desirability per se) to the predator) can occur. This is certainly a plausible adaptation of the prey population, but at least in short-term laboratory cultures, such selection appears not to occur. In studies of the interactions of *Rhizobium meliloti* and the protozoan *Naegleria,* the surviving population appeared to be as susceptible to protozoan attack as the original population encountering the predator (Danso and Alexander, 1975). Development of bacteriophage-resistant cultures through modification of receptors on the bacterial cells is commonly observed. Thus, it is logical to assume that modification of the bacterial cell wall or even production of chemo-attractants reduces interactions with predator populations, but this has yet to have been demonstrated in laboratory studies of predator–prey interactions.

4. *Refuge.* Native systems are not homogeneous. Both soils and waters are comprised of mixtures of particulates (detritus in aquatic systems) intermixed with a somewhat contiguous aquatic system or water layer. In such systems, the

prey cells may be screened from contact with the predators by the intervention of pores through which the larger predator cell cannot pass. Thus, even though its food source may be within a few microns of its cell, the predator may die. The more heterogeneous the environment, the greater the probability that refuges occur. This is particularly true of soils wherein a vast diversity of pores of varying sizes interconnect the various microsites suitable for microbial growth. Because survival of bacteria occurs in both soils and liquid cultures (where no particulates are present to provide refuge), refuges are not the sole mechanism for prey survival in soil; however, this mechanism could logically be of some importance in soil.

5. *Switching.* Most predators can consume a variety of prey species, although one or a few may appear to be more desirable. For example, Casida (1989) noted that nonsoil bacteria (*Escherichia coli* and *Bacillus mycoides*) amendment to soil samples resulted in protozoan population density increases, whereas no response in the predator population was detected following amendment of soil samples with the native bacterial species *Arthrobacter globiformis*. Should the predator exist in a complex community as found in soil, the random attack of a variety of bacterial species could result in less stress on each individual species or strain. No volition of the predator is implied in this process. The statistical probability of random encounter with the various species present in the system reduces the destruction of any single population.

An exception to this probabilistic evaluation of the occurrence of this switching-prey concept was suggested by Mallory et al. (1983). They observed that if alternative prey population densities exceeded the threshold for active predation and the growth rate of the prey was less than the predation rate, then elimination of the prey species could occur in their microcosm study.

Again the data to support switching as a mechanism for survival of prey cells are scant, but logic suggests that there may at least be a contribution of this probabilistic sparing of microbial cells from attack in a complex system such as native soil where both microbial growth and predation rates are low and limited microbial population densities of all species generally occur.

6. *Density dependence.* As with the sparing mechanism, density dependence is dependent upon the probability of the encounter of the predator and the prey. The higher the density of each population, the greater the probability of feeding. Danso and Alexander (1975) noted that at high population densities of *Rhizobium meliloti* cells in a defied cultures, final densities of the predators (*Hartmanella* sp., *Naegleria* sp., and *Vahlkampfia* sp.) were proportional to the initial prey populations. The rate of elimination of the prey cells was proportional to initial abundance of the predators, but the final populations of both predators and prey were independent of this number. This observation leads to the conclusion that although many parameters of the feeding curve are determined by participant population densities, the population density of surviving bacteria is not. Similarly, in a study of *Bdellovibrio* feeding on a bioluminiscent bacterial prey, Varon and Zeigler (1978) found that a minimal

population of 3.0×10^6 prey/ml culture was required for *Bdellovibrio* cells to have at least a 50 percent chance of survival. A prey density for development of population equilibrium in their system was 7×10^5 cells/ml culture. Thus, once the prey density has been reduced to the point that energy expenditure in food acquisition exceed that returned by prey consumption, sparing of the prey population occurs. That is, the prey population density must be sufficient to provide the requisite energy for the predator or parasite to replicate and search for new food supplies. A population of 10^6 cells/ml of liquid culture apparently approaches that minimal density for adequate energy return to the feeding organisms. Density-dependent sparing of prey cells likely contributes at least a portion of the survival of suscepts in native environments.

7. *Replication to compensate for killing.* With high population densities of host cells, the probability of an encounter between them and predators or parasites is reasonably high. Thus, a reasonably rapid population decline is anticipated to occur. As the host populations recede, the rate of destruction is reduced simply because of the lower probability of encounter of feeders and host or prey cells. Should the latter populations be replicating, a point is reached where the rate of predation or parasitism is equivalent to the rate of growth of the prey. Considering the low bacterial growth rates commonly occurring in soil, replication to compensate for killing would be significant only in situations with low predator or parasite stress, such as would be associated with low feeder populations.

It is clear, from analyzing these hypothesized means of controlling suscept population densities in soil, that no single explanation is adequate to answer our initial question: Why do predators and parasites not totally destroy their prey and hosts? All the proposed mechanisms are logical, and although specific ecosystems can be cited where they are not functional, environmental situations can be proposed wherein they may be important. Data invalidating an all-inclusive role of a particular survival adaptation in sparing of suscept populations have generally been derived from studying two- or three-member liquid cultures. These microcosms, per se, are special cases and in themselves lack the complexity of a soil microsite wherein a particular survival adaptation dismissed in the liquid culture may be crucial for continued existence of individual, predator-stressed microbial populations. Thus, the final conclusion regarding host or prey survival in soil is that a variety of properties of both the feeding populations and their food source contribute to the equilibrium population densities reached and that the exact mechanisms contributing to this process vary with species involved and the nature and properties of the individual microsites wherein the interactions are occurring.

6.3. TROPHIC INTERACTIONS AND NUTRIENT CYCLING IN SOIL

To this point, this analysis of microbial interactions has been concentrated at the somewhat simplistic level of single- or bilevel trophic interactions, that is, bacterial population vs. bacterial population or bacterial populations vs. protozoan community. Clearly, all trophic levels of the soil biological community contribute to ecosystem development. This loosely associated cooperation is best illustrated by examining the trophic interactions involved with oxidation of plant biomass and some of the factors limiting the process rates and population densities. (For a more comprehensive evaluation of trophic level interactions in soil, see Anderson, 1988, and Anderson et al., 1985 invertebrates; Bamforth, 1988 protozoa; Freckman, 1988 nematodes; Visser, 1985, invertebrates; and Usher, 1989, arthropods.)

Although microbiologists tend to concentrate their studies on the biochemical contributions of bacterial and fungal species (and to some degree actinomycete populations) to organic matter decomposition in soil, the overall rate of this process is dependent upon viability of the protozoan, nematode, microarthropod, and animal populations indigenous to soil. Although considerations of specific biochemical transformations and their rates are generally emphasized in evaluation of organic matter decomposition in soil, without the contributions (many of which are only indirectly related to organic matter transformation) of higher organisms, nutrient cycling in most ecosystems would be severely limited. This is best exemplified by listing and evaluating some specific roles for protozoa, nematodes, mites, and other animals in soil ecosystems.

For decomposition of plant debris to occur at rates essential for the return of mineral nutrients contained therein to the aboveground biomass, it must be populated with soil microbes. This introduction of decomposer propagules occurs through mixing the debris within the soil profile as well as inoculating the plant litter retained on the soil surface. Mixing organic matter within the soil profile is accomplished by earthworms, ants, termites, and even burrowing animals. For example, by collecting surface litter and storing it within their nests, termites and ants may create islands of plant debris within a sea of nearly litter-free soil. The biodecomposition of this termite and ant sequestered organic matter alters plant nutrient availability, soil structure, and soil humic and fulvic acid composition (e.g., see Anderson and Wood, 1984; Gupta et al., 1981; Pomeroy, 1983; Wood, 1988; and Wood et al., 1983). This localized improvement of soil properties is clearly shown by studies of tropical savanna soils (Lee, 1974; Ofer et al., 1982). Ofer et al. (1982) found that harvester ants in semiarid pastures enhanced plant growth in the vicinity of the mound, increased soil humic and fulvic acids and altered the distribution of organic nitrogen within the various colloidal soil organic matter fractions. For a more detailed evaluation of the role of ant and termite activities in tropical soils, see Anderson and Flanagan (1989).

A contrasting situation is found in ecosystems where the plant litter remains distributed in a surface organic layer such as is commonly observed with temperate forest soils. Since the primary decomposers of plant debris are soil microbes, stimulation of the decomposition rate results from inoculation of the litter by soil animal movement between the A horizon soil and the litter layer. This transfer of microbial propagules may be accomplished by a variety of organisms including earthworms, mites, and ants. Generally the inoculation of the plant debris results from passive transport of the microbial propagules on the surface of the animal body.

Soil animals may also modify the soil structure in a manner that stimulates organic matter decomposition. Aerobic soil microbes may be limited by the rate of oxygen diffusion into soil pores or by saturation of the pores with moisture due to low water infiltration rates. Each of these limitations may be relieved by earthworm casts and ant hill construction. This soil structural modification by earthworms and the resulting impact on plant residue decomposition is shown in the studies of Zachmann and Linden (1989). Treatment of surface-applied corn residues with earthworms (*Lumbricus rubellus* Hoffmeister) resulted in more rapid residue decomposition and altered water balance of the soils.

As indicated in chapter 4, surface area can be a primary controller of biological decomposition. This is particularly acute with plant biomass. A primary indirect means of stimulating microbial decomposition of leaf tissue and fras is to increase its surface area. Soil animals both directly (through their own feeding) and indirectly (through physically breaking the plant structure by their passage through the litter layer) accelerate the breakup of the biomass physical structure. Clearly, even higher animals, including human activity, accelerate the breakdown of the plant physical structure.

A variety of soil animals directly catabolize soil organic matter. The contribution of these organisms to total organic matter respiration in soil is generally minuscule compared to the activity of soil bacterial and fungal populations, but it may be quite significant. Elliott et al. (1988) suggest that protozoans, nematodes, and microarthropods may in concert contribute up to 40 percent of the nitrogen mineralized in North American grasslands. Populations instrumental in this process include earthworms and isopods. For example, the woodlouse *Porcellio scaber* feeds on decaying pine needles. Their activity is primarily limited by the structural toughness of the needles themselves (Soma and Saito, 1983).

Probably, the major contribution of trophic levels above soil bacteria, fungal, and actinomycete populations to soil organic matter cycling results from their use of the primary feeders (bacteria, fungi, actinomycetes) as food sources. These secondary feeders include protozoa and nematodes. This feeding on the primary feeders has two effects on soil microbial activity:

- As exemplified by nitrogen mineralization (Woods et al., 1982), nutrients yielded from mineralization of organic nitrogen sources in soil is usually incorporated into the microbial biomass first. These sequestered nutrients are not made available to other participants of the biological community until the microbial biomass decays. This is accomplished by the action of soil predators.
- The secondary feeders stimulate microbial growth thereby keeping the primary organisms in a more active state (e.g., see Anderson et al., 1981). The activity of the secondary feeders can increase the rate of metabolism of soil organic matter by keeping the bacterial population young and active. Remember, however, that the total amount of carbon mineralized does not change; all that is altered is the time frame over which the organic matter is decomposed. The quantity mineralized is controlled more by its degradation susceptibility and the quantities present.

Alteration of the activity of secondary feeders can also result in a reduction in the rate of organic matter decomposition. In this situation the primary population is reduced sufficiently that mineralization is reduced. Santos et al. (1981a; 1981b) examined the metabolism of creosote litter in desert by a community composed of microarthropods (tydeid mites), bacteriophagic nematodes, and bacteria. Destruction of the mite population resulted in an increase in their food source, the nematodes. This increased nematode density created greater stress on the bacterial population upon which they were feeding. The resultant decline in bacterial density was expressed as a reduction in litter decomposition. In contrast elimination of both secondary feeders, nematodes and mites, increased bacterial numbers and litter decomposition. Therefore, it could be said that the mites were effective in controlling litter decomposition by indirectly affecting the soil bacterial population. Lower mite populations lead to reductions in the nematode density, which in turn increased bacterial populations densities and organic matter mineralization. Similarly, Grant et al. (1983) found that destruction of the invertebrate ostracod populations in wetland rice fields caused a tenfold increase in nitrogen rates, which results from the parallel threefold increase in blue-green algal populations upon which the ostracods were feeding. In previous work, Grant and Alexander (1981), suggested that the selective feeding traits of the ostracod could be exploited to increase nitrogen fixation in flooded soils. They found that animal size and temperature affected feeding rates of the animals as did cell age of the cyanobacteria. (See Ingham et al., 1985, for a further analysis of bacterial, fungal, and nematode interactions in the soil community.)

6.4. IMPORTANCE OF MICROBIAL INTERACTIONS TO OVERALL BIOLOGICAL COMMUNITY DEVELOPMENT

In the preceding analysis, as in the evaluation of physical and chemical properties impacting soil ecosystem processes, a dissection of system function has occurred. Truly, these essentials of the soil system cannot exist, and most cannot be studied, singly, or even as isolated pieces of a puzzle, per se, but they must be assembled into a conceptualization of the working whole. Indeed, the ultimate importance of any biotic or abiotic trait or interaction is its contribution to community stability and resilience, that is, the capacity of the ecosystem to be maintain and its resistance to disruptive perturbations. Classically, this stability of the soil community is defined as **homeostasis (the tendency of a biological system to resist change and to remain in a state of equilibrium)**.

Maintaining the equilibrium that defines the existing ecosystem type is most commonly accomplished by **negative feedback (a change in one or more populations induced by a modification of some environmental variable that brings about a response in other populations in such a manner that the original fluctuation is opposed or dampened)**. Negative feedback can be seen as a biological equivalent to the physics action–reaction principle. A change in one portion of the system must result in an opposing alteration in a section part in order for the

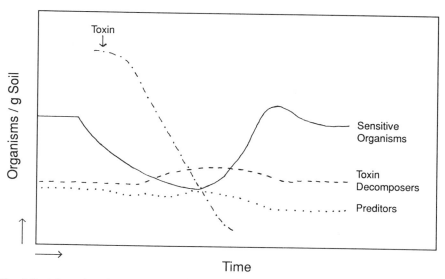

Fig. 6.2. Adaptations in a generalized microbial community to reestablish conditions existent prior to input of a biodegradable toxicant.

existing situation to be maintained. Applicability of the principle of negative feedback to maintain community stability can be exemplified by a hypothetical situation wherein a biodegradable toxin is added to a soil community (Fig. 6.2). The system is defined to be composed of three biological groups—toxin-sensitive bacteria, protozoa capable of feeding upon the bacterial populations, and bacteria that decompose the toxin. For this example, it is not necessary that the toxin be mineralized, but rather a simple detoxification would be sufficient to allow return to the steady-state population. Note that the initial reaction of the population is a decline in population densities of those organisms that are sensitive to the toxin. Because of the time necessary for induction of the enzymes associated with detoxifying the aggravating amendment, a delay in toxin destruction is commonly noted. Increases in the population of toxin degraders can result from using the toxin for carbon or energy, in which case the toxin would be totally mineralized. Similar increases in toxin decomposer population densities may result if they derive their energy from the biomass of the killed toxin-sensitive populations. In either situation, a finite quantity of carbon and energy is available for growth of the toxin-degrader populations. Once the toxin has been removed from the system, surviving toxin-sensitive populations may reproduce. At this point, conditions favor return to preinsult equilibrium population densities, but a further induction or negative feedback process must occur—protozoan populations increase to "assist" in returning toxin-degrader population density to preamendment levels. With the loss of the nutrient source (exhaustion of the toxin or depletion of the biomass of cells killed by the toxin), the toxin decomposers cease to grow. Protozoan activity facilitates the return to the steady-state levels by consumption of the excess biomass. With time the preamendment equilibrium can be reestablished. This is, in reality, an idealized situation in that it relies upon a) the toxin being totally destroyed, b) sensitive populations not being totally eliminated by the toxic material, and c) no permanent change in the controllers of population densities. The overall manifestation of soil microbial populations may appear to represent a complete return to the preexisting equilibrium, but, in reality, it would be more common for the system to move to a new equilibrium condition approximating the pretoxin insult conditions. Obvious ecosystem alteration would not occur unless the newly established community differed sufficiently from that which preceded it to alter overall system properties.

6.5. MANAGEMENT OF SOIL MICROBIAL POPULATIONS

With anthropogenic intrusions into native sites as well as intensive management of developed soil systems, alteration or control of the nature of the soil microbial community is becoming more desirable, if not obligatory, for system survival. Intervention into population density control as well as the nature of the species present can be as simple as selecting *Rhizobium* strains with improved nitrogen fixation efficiency compared to native populations or as complex as enhancing the activity of an array of decomposers of petroleum substituents. In either situation, the underlying question regarding the success of the inoculation or site management procedure rests at least as much if not more on the competitive abilities of the alien organism as on possession of the phenotypic capacity to achieve the desired ends of the project.

Most commonly, generalized soil inoculation has been a failure. Whether it is the use of free-living nitrogen-fixing microbes (e.g. *Azotobacter* sp.) or of bacteria highly efficient in the transformation of organic phosphorus to inorganic phosphorus (*Bacillus megaterium*), negative results predominate in the literature. A common explanation for stimulation of crop yields by soil inoculation with either of these organisms (i.e., the reported success of soil inoculation) is that sufficient population densities of the alien organism were added to the soil that the death and decay of their biomass significantly increased soil fixed nitrogen and phosphorus reserves.

Even with the scattered successes regarding generalized soil inoculations, the preceding evaluation of microbial interactions in soil suggests why colonization by alien organisms should not be expected. Soil is a complex ecosystem in which a vast array of microbial populations have developed. It is reasonable to assume, that if a niche exists for a microbe to function in a mature (climax) system, organisms will probably already have been selected to fill it. The occupant of the niche would be the most efficient organism to fulfill that function. Thus, to alter the composition of the soil microbial community through inoculation, the alien organism must be able to survive in the ecosystem sufficiently long to become established and either a) to fill a niche that is not occupied in the soil ecosystem or b) to be more efficient than the indigenous microbes so as to be capable of dislodging preexisting populations.

These prerequisites underscore the situations where soil inoculation appears or has been documented to be successful. In most cases, a new niche is created in the soil for the inoculated organism to occupy. This is seen in inoculation of root tissue with *Rhizobium* strains (the new niche is provided by the invasion of the nascent root tissue), addition of nitrogen-fixing blue-green algae to rice paddy fields (the new niche is provided by flooding of a previously drained soil), or addition of a specialized degrader population to a site to be renovated (the new niche is provided by the necessity of decomposing the biodegradable pollutant). Future expansion of the lists of inoculation success will be based on better prediction of the nature of the unoccupied niche in soil and an improved understanding of phenotypic microbial traits that allow for augmented capacity to compete with indigenous microbes thereby usurping their position in the soil community.

6.6. CONCLUDING COMMENTS: IMPLICATIONS OF SOIL MICROBIAL INTERACTIONS

This analysis definitely defines a role for microbe–microbe interactions as well as the necessity of interactions with higher organisms for the development of a stable, active soil community. Furthermore, the myopic nature of past scientific endeavors in this realm is revealed. The vast majority of soil microbiologically oriented projects have involved evaluation of the native soil procaryotic populations and, to some degree, fungal activity in isolation. This failure to consider higher organisms is understandable because the primary organisms which are instrumental in decomposition of native biomass and xenobiotic compounds are the bacteria, fungi, and, to some degree, actinomycetes. Yet, it is mandatory to realize that the rate of decomposition processes in soil is controlled by the interactions of these primary feeders with a vast array of secondary feeders. The indirect activities of mixing organic debris with soil, inoculating this biomass with degrader populations, and mincing of large plant structures have a significant and obvious effect on decomposition rates. This rate is further controlled by the feeding activity of parasites and predators. Hence, any evaluation of ecosystem dynamics must not only include a consideration of the metabolic capacity of the soil microbial community and their physical and chemical needs but also be conducted with an appreciation of the impact of higher organisms on the vitality and survival of these front-line organic matter mineralizers.

REFERENCES

Alexander, M. 1981. Why microbial predators and parasites do not eliminate their prey and hosts. Annu. Rev. Microbiol. 35:113–153.

Anderson, J. M. 1988. Invertebrate-mediated transport processes in soil. Agric. Ecosystems Environ. 24:5–19.

Anderson, J. M., and P. W. Flanagan. 1989. Biological processes regulating organic matter dynamics in tropical soils. In D. C. Coleman, J. M. Oades, and G. Uehara (eds.), Dynamics of Soil Organic Matter in Tropical Ecosystems. University of Hawaii Press, Honolulu.

Anderson, J. M., S. A. Huish, P. Ineson, M. A. Leonard, and P. R. Splatt. 1985. Interactions of invertebrates, micro-organisms and tree roots in nitrogen and mineral element fluxes in deciduous woodland soils. In A. H. Fitter, D. Atkinson, D. J. Reed, and M. B. Usher (eds.), Ecological Interactions in Soil: Plants, Microbes and Animals, pp. 377–392. Blackwell Scientific Publications, Oxford.

Anderson, J. M., and T. G. Wood. 1984. Mound composition and soil modification by two soil-feeding termites (Termitinae, Termitidae). Pedobiologia 26:77–82.

Anderson, R. V., D. C. Coleman, C. V. Cole, and E. T. Elliott. 1981. Effect of the nematodes *Acrobeloides* sp., and *Mesodiplogaster iheritieri* on substrate utilization and nitrogen and phosphorus mineralization in soil. Ecology 62:549–555.

Bamforth, S. S. 1988. Interactions between protozoa and other organisms. Aric. Ecosystems Environ. 24:229–234.

Casida, L. E., Jr. 1980. Death of *Micrococcus luteus* in soil. Appl. Environ. Microbiol. 39:1031–1034.

Casida, L. E., Jr. 1983. Interaction of *Agromyces ramosus* with other bacteria in soil. Appl. Environ. Microbiol. 46:881–888.

Casida, L. E., Jr. 1989. Protozoan response to the addition of bacterial predators and other bacteria to soil. Appl. Environ. Microbiol. 55:1857–1859.

Conn, H. J., and J. W. Bright. 1919. Ammonification of manure in soil. J. Agr. Res. 16:313–350.

Danso, S. K. A., and M. Alexander. 1975. Regulation of predation by prey density: The protozoan–*Rhizobium* relationship. Appl. Microbiol. 29:515–521.

De Freitas, J. R., and J. J. Germida. 1991. *Pseudomonas cepacia* and *Pseudomonas putida* as winter wheat inoculants for biocontrol of *Rhizoctonia solani*. Can. J. Mirobiol. 37:780–784.

Elliott, E. T., H. W. Hunt, and D. E. Walter. 1988. Detrital foodweb interactions in North American grassland ecosystems. Agric. Ecosystems Environ. 24:41–56.

Freckman, D. W. 1988. Bacteriovorous nematodes and organic-matter decomposition. Agric. Ecosystems Environ. 24:195–217.

Fredrickson, A. G., and G. Stephanopoulos. 1981. Microbial competition. Science (Washington, DC) 213:972–979.

Grant, I. F., and M. Alexander. 1981. Grazing of blue-green algae (cyanobacteria) in flooded soils by *Cypris* sp. (ostracoda). Soil Sci. Soc. Am. J. 45:773–777.

Grant, I. F., A. C. Tirol, T. Aziz, and I. Watanabe. 1983. Regulation of invertebrate grazers as a means to enhance biomass and nitrogen fixation of cyanophyceae in wetland rice fields. Soil Sci. Soc. Am. J. 47:669–675.

Gupta, S. R., R. Rajvanshi, and J. S. Singh. 1981. The role of the termite *Odontotermes gudaspurensis* (Isoptera, Termitidae) in plant litter decomposition in a tropical grassland. Pedobiologia 22:254–261.

Hamdan, H., D. M. Weller, and L. S. Thomashow. 1991. Relative importance of fluorescent siderophores and other factors in biological control of *Gaeumannomyces graminis* var. *tritici* by *Pseudomonas fluroescens* 2-79 and M4-80R. Appl. Environ. Microbiol. 57:3270–3277.

Hornby, D. 1983. Suppressive soils. Annu. Rev. Phytopathol. 21:65–85.

Hossain, A. K. M., and M. Alexander. 1984. Enhancing soybean rhizosphere colonization by *Rhizobium japonicum*. Appl. Environ. Microbiol. 48:468–472.

Ingham, R. E., J. A. Trofymow, E. R. Ingham, and D. C. Coleman. 1985. Interactions of bacteria, fungi, and their nematode grazers: Effects on nutrient cycling and plant growth. Ecol. Monogr. 55:119–140.

Katznelson, H. 1940. Survival of microorganisms inoculated into sterilized soil. Soil Sci. 49:211–217.

Kloepper, J. W., J. Leong, M. Teintze, and M. N. Schroth. 1980. *Pseudomonas* siderophores: A mechanism explaining disease-suppressive soils. Curr. Microbiol. 4:317–320.

Lee, K. E. 1974. The significance of soil animals in organic matter decomposition and mineral cycling in tropical forest and savanna ecosystems. Trans. Int. Contr. Soil Sci. 10th. 3:43–51.

Leong, J. 1986. Siderophores: Their biochemistry and possible role in the biocontrol of plant pathogens. Annu. Rev. Phytopathol. 24:187–209.

Li, D.-M., and M. Alexander. 1986. Bacterial growth rates and competition affect nodulation and root colonization by *Rhizobium meliloti*. Appl. Environ. Microbiol. 52:807–811.

Li, D.-M., and M. Alexander. 1988. Co-inoculation with antibiotic-producing bacteria to increase colonization and nodulation by rhizobia. Plant Soil 108:211–219.

Lochhead, A. G. 1957. Qualitative studies of soil microorganisms: XV. Capability of the predominant bacterial flora for synthesis of various growth factors. Soil Sci. 84:395–403.

Lochhead, A. G., and M. O. Burton. 1957. Quantitative studies of soil microorganisms. XIV. Specific vitamin requirements of the predominant bacterial flora. Can. J. Microbiol. 3:35–42.

Mallory, L. M., C.-S. Yuk, L.-N. Liang, and M. Alexander. 1983. Alternative prey: A mechanism for elimination of bacterial species by protozoa. Appl.Environ. Microbiol. 46:1073–1079.

Ofer, J., R. Ikan, and O. Haber. 1982. Nitrogenous constituents in nest soils of harvester ants *Messor-Ebeninus* and their influence on plant growth. Commun. Soil Sci. Plant Anal. 13:737–748.

Polyanskaya, L. M., P. A. Kozhevin, and D. G. Zvyagintsev. 1983. The dynamics of populations of an antagonist and an antibiotic sensitive microorganism in nonsterile soil. Mikrobiologiya 52:145–148 (Russian).

Pomeroy, D. E. 1983. Some effect of mound-building termites on the soils of a semi-arid area of Kenya. J. Soil Sci. 34:555–570.

Reddy, K. R., R. Khaleel, and M. R. Overcash. 1981. Behavior and transport of microbial pathogens and indicator organisms in soils treated with organic wastes. J. Environ. Qual. 10:255–266.

Rosenzweig, W. D., and G. Stotzky. 1979. Influence of environmental factors on antagonism of fungi by bacteria in soil: Clay minerals and pH. Appl. Environ. Microbiol. 38:1120–1126.

Rosenzweig, W. D., and G. Stotzky. 1980. Influence of environmental factors on antagonism of fungi by bacteria in soil: Nutrient levels. Appl. Environ. Microbiol. 39:354–360.

Santos, P. F., J. Phillips, and W. G. Whitford. 1981a. The role of mites and nematodes in early stages of buried litter decomposition in a desert. Ecology 62:664–669.

Santos, P. F., and W. G. Whitford. 1981b. The effect of microarthropods on litter decomposition in a Chihuahan desert ecosystem. Ecology 62:654–663.

Soma, K., and T. Saito. 1983. Ecological studies of soil organisms with reference to the decomposition of pine needles. II. Litter feeding and breakdown by the woodlouse. *Procellio scaber.* Plant Soil 75:139–151.

Szabo, I., A. Benedek, and G. Barabas. 1985. Possible role of streptomycin released from spore cell wall of *Streptomyces griseus*. Appl. Environ. Microbiol. 50:438–440.

Tate, R. L., III. 1980. Microbial oxidation of organic matter in histosols. In M. A. Alexander (ed.), Advances in Microbial Ecology, Vol. 4, pp. 169–201. Plenum, New York.

Tate, R. L., III, and R. E. Terry. 1980. Effect of sewage effluent on microbial activities and coliform populations in Pahokee muck. J. Environ. Qual. 673–677.

Thomashow, L. S., and D. M. Weller. 1990. Role of antibiotics and siderophores in biocontrol of take-all disease of wheat. Plant Soil 129:93–99.

Thomashow, L. S., D. M. Weller, R. F. Bonsall, and L. S. Pierson, III. 1990. Production of the antibiotic phenazine-1-carboxylic acid by fluorescent *Pseudomonas* sp. in the rhizosphere of wheat. Appl. Environ. Microbiol. 56:908–912.

Usher, M. B. 1989. Population and community dynamics in the soil ecosystem. *In* A. E. Fitter (ed.), Ecological Interactions in Soil: Plants, Microbes, and Animals, pp. 243–265. Blackwell Scientific Publications, Boston.

Utkhede, R. S., and J. E. Rahe.1980. Biological control of onion white rot. Soil Biol. Biochem. 12:101–104.

Varon, M., and B. P. Zeigler. 1978. Bacterial predator-prey interaction at low prey density. Appl. Environ. Microbiol. 36:11–17.

Visser, S. 1985. Role of the soil invertebrates in determining the composition of soil microbial communities. *In* A. E. Fitter (ed.), Ecological Interactions in Soil: Plants, Microbes, and Animals, pp. 297–317. Blackwell Scientific Publications, Boston.

Wiggins, B. A., and M. Alexander. 1985. Minimum bacterial density for bacteriophage replication: Implications for significance of bacteriophages in natural environments. Appl. Environ. Microbiol. 49:19–23.

Wood, T. G. 1988. Termites and the soil environment. Biol. Fert. Soils. 6:228–236.

Wood, T. G., R. A. Johnson, and J. M. Anderson. 1983. Modification of soils in Nigerian savanna by soil-feeding *Cubitermes* (Isoptera, Termitidae). Soil Biol. Biochem. 15:575–579.

Woods, L. E., C. V. Cole, E. T. Elliott, R. V. Anderson, and D. C. Coleman. 1982. Nitrogen transformations in soil as affected by bacterial-microfaunal interactions. Soil Biol. Biochem. 14:93–98.

Zachmann, J. E., and D. R. Linden. 1989. Earthworm effects on corn residue breakdown and infiltration. Soil Sci. Soc. Am. J. 53:1846–1849.

7 The Rhizosphere/Mycorrhizosphere

The soil microbial community is driven by solar power. In other words, their primary energy supply is photosynthetically fixed carbon that enters the soil ecosystem primarily as plant biomass and exudates. Hence, a site of maximized biological activity is the portion of the ecosystem encompassed by the soil and root interface—the rhizosphere. The provision of energy to the microbial community by root exudates, dead roots, and sloughed cells results in intense microbial metabolic activity plus an enhancement of microbial interactions (e.g., competition, symbiosis, amensulism, and predation) over levels occurring in nonrhizosphere soil. Furthermore, because plants benefit from the mineralization capacity of the soil microbial community, the rhizosphere microbial community represents that portion of the soil biota with maximal effect on the aboveground community.

During the growing season, the rhizosphere is an expanding focus of biological energy. New microbial habitats are produced continually through root growth. Thus, scientific interest in explaining rhizosphere processes is derived from the desire to understand the basic biological activities occurring therein as well as the fact that the rhizosphere interaction provides a model for successful modification of native microbial community dynamics (e.g., root inoculation with symbiotic nitrogen-fixing populations) as well as for study of the potential uses of genetically modified microbes. Opportunities for using genetically modified microorganism in the rhizosphere and adjacent soils include population manipulation to reduce problems with plant pathogenic microbes as well as to achieve soil renovation or other bioremediation objectives.

Therefore, the overall objective of this analysis of rhizosphere/mycorrhizosphere properties is to present the properties of the microbial processes occurring within the rhizosphere and those factors that make it a unique portion of the soil ecosystem. Traits of the total rhizosphere are examined along with the microbe–plant interactions affecting plant growth and microbial community development. This analysis includes an evaluation of mycorrhizal symbioses and the impact of these associations on properties of the rhizosphere.

7.1. THE RHIZOSPHERE

By definition, the **rhizosphere is that portion of the soil under the direct influence of the roots of higher plants, whereas the rhizoplane encompasses the root**

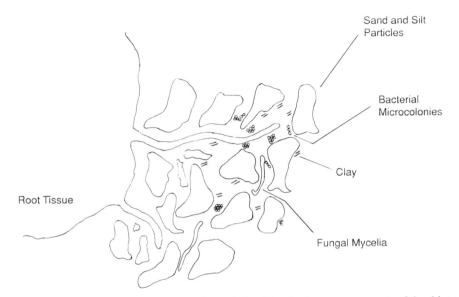

Sand and Silt
Particles

Bacterial
Microcolonies

Clay

Root Tissue

Fungal Mycelia

Fig. 7.1. Schematic diagram of the physical relationship of primary components of the rhizosphere (not drawn to scale).

surface and its adhering soil. As shown diagramatically in Fig. 7.1., the soil microbes in the vicinity of the growing roots are stimulated by the provision of a surface upon which to grow as well as the nutrients contained in leachates, sloughed cells, and decaying roots. Microbial colonies develop unevenly along the root surface with maximal development occurring where all prerequisites for growth are met. Therefore, occurrence of microcolonies is a good indicator of carbon and energy source leakage from the root.

The rhizosphere is a unique environment in the soil system largely because of the following properties:

- The properties of the plant are the predominant controls of the ecosystem rather than those of the soil. This situation results from the fact that the plant supplies the nutrient and energy source, provides space for the microbes to grow, as well as alters the gaseous regime of the soil.
- The microorganisms prevailing in the rhizosphere originate in the soil and are characteristic of the soil microflora.
- Because of the combination of increased microbial respiration occurring in the root zone as compared to the adjacent nonrhizosphere soil and the root respiration, the gas regime around the root differs from that of the surrounding soil. Carbon dioxide levels are generally elevated and molecular oxygen tensions are reduced in comparison with the gaseous content of nonrhizosphere-impacted soil.
- The roots provide a reasonably constant supply of nutrients to the microbial community. This influx of carbon and energy selects for those microbes capa-

ble of using these nutrients most efficiently—the fast growing microbes. In comparison, nonrhizosphere microbial communities typically exist in a "feast-or-famine" type of existence. Nutrients provided to nonrhizosphere populations primarily result from the influx of nutrients contained in infiltrated water (including fixed carbon mobilized by the decay of surface litter) or are carried into the site through movement of animal life (including the nutrients contained in the animal body when the organism expire). It is tempting to consider the rhizosphere community to be continuously bathed in nutrients. This concept may be of value in developing an overall model of the community, but it is far from the situation at the microsite. As is discussed in greater detail later in this chapter, quantities of root exudates as well as their chemical composition vary along the root, change throughout the growing season, and are dependent upon root density and plant–plant interactions.

- Since many of the products of root metabolism are acids (including carbon dioxide, which forms a weak acid when dissolved in soil water), soil minerals are solubolized in the rhizosphere. Thus, osmotic impacts on microbial growth are accentuated in the rhizosphere, mineral nutrient availability is modified, and clay mineral transformation may occur. For example, linkage of potassium availability and loss of potassium ion from mica flakes plus transformation of phylogopite into vermiculite in inoculated pine root (*Pinus sylvestris* L.) rhizosphere soils was recently shown by Leyval and Berthelin (1991).

7.1.1. The Microbial Community

The rhizosphere effect or the degree of stimulation of an activity of population by the energy input of the plant root has classically been described by evaluating the ratio of activity per unit weight of rhizosphere soil and the activity per unit weight of nonrhizosphere soil (the R/S value). Values greater than 1 indicate selective stimulation in the rhizosphere, equivalence of the two activities suggests no rhizosphere effect, whereas a R/S ratio of less than 1 reveals inhibition of the activity in the rhizosphere.

Rhizosphere Bacterial Populations: Bacterial populations are stimulated by root exudates. Populations greater than $10^9 g^{-1}$ rhizosphere soil are commonly detected. Selective stimulation of rapidly growing, gram negative, rod-shaped bacteria (*Pseudomonas* sp., *Flavobacterium* sp., *Alcaligenes* sp. especially, and *Agrobacterium* sp. occasionally) is observed. These bacterial species are found as microcolonies covering approximately 4–10 percent of the root surface. Competition, the primary selective, intermicrobial interaction, is based on growth rate, metabolic versatility, and growth factor requirements.

- *Growth Rate:* The conditions of the rhizosphere favor those bacteria that have short generation times. These organisms can compete for the limited nutrients when they are released by the root.
- *Nutrients:* A variety of fixed-carbon compounds are available to the microorganism within the rhizosphere. Stress of the competition may result in the

exhaustion of a particular growth substrate, whereas abundant supplies of others remain. Thus, successful bacterial populations are capable of using a variety of biochemical growth substrates.

- *Growth Factors:* Root exudates commonly contain a variety of amino acids and vitamins. These compounds may be used as microbial nutrients or growth factors. Determination of rhizosphere bacteria capable of synthesizing growth factors leads to the conclusion that some advantage is conferred on organisms requiring amino acids and no other growth factors. Interestingly, the proportion of bacteria in the rhizosphere that require complex nutritional growth factors declines, but their actual numbers increase. This somewhat confusing observation results from the fact that the number of these organisms in the ecosystem increases, but the total population increases at an even faster rate. Therefore, although there are more growth-factor-requiring bacteria in the rhizosphere than are found in nonrhizosphere soil, they comprise a small portion of the total population.

Two other factors affecting microbial growth in the rhizosphere are surface area available for colonial development and antibiotic production. Because adding nutrients to soil can also stimulate the same group of bacteria that are stimulated in the rhizosphere, nutrient supply has to be a primary selective factor in the rhizosphere; but, as nutrient needs are met more fully, the point of competition may be shifted to factors impacted by the minute distances between individual cells and physical obstructions to growth (e.g., surface area and toxicant production). Competition for surface area affects population development primarily at the microsite where cell–cell interactions are intensified. As indicated in chapter 6, antibiotic production appears to play a role in amensal interactions in the rhizosphere. With the close proximity of bacteria, the intense competitions occurring, and the positive impact of inoculation of plant roots with antibiotic producing microbial strains, antibiotics could be of major importance in this ecosystem. but a complete understanding of antibiotic production and its importance in the root ecosystem has yet to be developed.

Fungi in the Rhizosphere: Quantification of the rhizosphere effect on soil fungal populations is equivocal at best. This difficulty is derived, to a large degree, from inability to separate active from inactive fungal propagules. Fungal population densities may be quantified through enumeration of colonial development on defined media, or mycelial strands may be counted directly in soil samples. Colonies growing on defined media may result from outgrowth of spores contained in the soil sample or from mycelial strands. Direct observation of hyphal strands in soil provides a measure of total fungal biomass, but not species or genus identity. Colonies growing on the defined media can readily be classified to genus and species, but they may have developed from a quiescent spore serendipitously existing in the environs of the root or from an actively respiring fungal colony. Thus, data collected for the evaluation of rhizosphere effect on fungal populations is difficult to interpret. Total fungal counts appear not to be impacted by association with the root, but some individual populations may be stimulated. Plant roots do produce compounds that stimulate outgrowth of fungal resting structures and hyphal development. Similarly, zoospores of phycomycetes are chemotactically attracted to the growing root tip. These observations suggests a positive rhizosphere effect on fungal populations.

Other Rhizosphere Inhabitants: Organisms whose populations tend not to react to influxes of decomposible organic matter are usually not affected by root growth. This grouping includes actinomycetes, protozoa, and algal populations. The actinomycetes generally derive their energy supply from decomposition of less readily decomposible soil organic matter components, whereas algal populations use solar energy. Protozoan populations are limited by the distribution and density of prey populations. Apparently, the augmentation of the number of prey cells in the rhizosphere is insufficient to result in significant increases in numbers of preditors.

7.1.2. Sampling Rhizosphere Soil

The basic definition of the rhizosphere (that portion of the soil profile under the direct influence of the plant root) provides an excellent conceptual resolution of the rhizosphere and nonrhizosphere domains, but from a practical viewpoint it provides little upon which to base a soil sampling procedure. Questions that may arise in preparation of rhizosphere soil samples pertain to the distance from the root that the plant-effect extends and the degree of the impact (minute changes in soil properties at considerable distance from the growing root are caused by the presence of the root). Hence, most studies of the rhizosphere have involved an arbitrary soil collection procedure and a number of ways to physically separate roots and soil in laboratory and greenhouse studies.

The arbitrary nature of collecting such soil is demonstrated by examining methods of collecting rhizosphere soil from field sites. The more time-consuming procedure is to dig a soil pit around a plant and aseptically remove soil within a predetermined distance of the plant root—commonly a 1 cm distance. For example, Ruark et al. (1991) compared the interaction of acid rain and ozone treatments on hydrogen, aluminum, calcium, and magnesium ion concentrations in root-adjacent (rhizosphere) and bulk (nonrhizosphere) soils for loblolly pine (*Pinus taeda* L.) seedlings growing in the field. The procedure was adequate to demonstrate gross differences in anion concentration due to atmosphere problems and plant growth. Alternatively, the plant may be carefully removed from the soil. Nonadhering soil is allowed to fall to the ground whereas that soil adhering to the root system is gently shaken into a collection container. The latter soil is defined to be the rhizosphere soil. Soil in direct contact with the root, which cannot be gently shaken from the root surface but must be washed from the root to be collected, is defined as **rhizoplane soil.** The limitation associated with either of these procedures is that root-impacted and unaffected soils are inevitably mixed at unknown ratios. Thus, rhizosphere effects are at least underestimated and potentially may be totally occluded should sufficient dilution with unaffected soil occur.

Techniques similar to those used to separate nonrhizosphere and rhizosphere soils in the field may be applied to microcosm studies incubated in the laboratory or in the soil greenhouse. For examples, see Norton and Firestone (1991) where soil was sampled at small intervals from coarse and fine young roots of Ponderosa pine (*Pinus ponderosa* Laws.) seedlings for analysis of the metabolic status of the bacterial and fungal communities. Alternatively, rhizosphere and nonrhizosphere soils may be physically isolated in the microcosm (e.g., Hartel et al., 1989; Martens, 1982). Tate et al. (1991) isolated pitch pine (*Pinus rigida* Miller) seedling rhizosphere soils from nonrhizosphere soils with a 1-μm mesh fabric. Soluble exudates could pass through the membrane, but root tissue was retained. With either procedure, limitations of dilution of plant-impacted soil by non-plant-affected soil compromises data interpretation as was indicated for field soil-sampling procedures.

Results from the study of bulked soils probably underestimate the true stimulatory (or even inhibitory) effect of the plant root on soil microorganisms. This diminishing effect is further accentuated when we consider the distribution of the active microbes along the root (recall that root exudates are not released continuously along the root; hence, microbial colonies are found in "hot spots" in the rhizosphere). Thus, the bulked rhizosphere soil sample not only is diluted by nonrhizosphere soil but also is a composite of microsites containing highly active microbial communities plus microsites in which little or no stimulation of normal soil microbial activity has occurred.

7.1.3. Plant Contributions to the Rhizosphere Ecosystem

The role of the plant in the rhizosphere is to provide a surface for microcolony development and to produce the fixed-carbon compounds used by the soil microbes for carbon and energy and as growth factors. Major questions regarding this process are associated with quantifying the quantities of photosynthate entering the rhizosphere, identifying the organic components of root exudates, and explaining the variation in these parameters throughout the growing season and between plant species. A related concern in native systems and with dual cropping systems is determining the impact of the interaction of root systems from different plant species on the quantities and composition of root exudates.

Because, in native soils, photosynthate produced by higher plants (as compared to more limited quantities of carbon fixed by algal activity) generally provides the primary energy source for the soil microbial community, there has been a long-standing interest in quantifying the proportion of carbon fixed by the plant released into the rhizosphere community and the ultimate deposition of the fixed carbon (i.e., incorporation into humic substances, microbial biomass, or oxidation to carbon dioxide by the soil biota). Microbiologists see this carbon/energy source emanating from the plant root structure as the major driving force for the soil ecosystem, whereas plant biologists may regard production of root exudates as both a benefit to the plant and a potential growth limitation. Benefits are accrued from the stimulation of soil microbial populations whose functions encourage plant development (e.g., through nitrogen mineralization, reduction of plant pathogenic interactions, and synthesis of growth-controlling plant hormones such as indol acetic acid and giberellic acid). Negative implications are associated with the potential for excessive loss of photosynthate and, therefore, carbon and energy available for plant biomass production.

A variety of means of quantifying and tracing the movement of root exudates have been developed. A useful procedure for assessing the nature and quantity of fixed-carbon products exuded from growing roots would seem to be to quantify the biochemicals accumulated in the growth medium of plants grown under axenic conditions in liquid culture either in a greenhouse or a growth chamber culture (e.g., see study conducted by Griffin et al., 1977). The primary difficulty encountered with such studies is that the microbial community has a stimulatory effect on root exudate production and may alter the nature of the biochemicals contained in root exudates (Barber and Martin, 1976; Klein et al., 1988). Barber and Martin (1976) found that between 5 and 10 percent of the photosynthetically fixed carbon in wheat and barley plants was lost from the roots grown under axenic conditions, whereas 12 to 18 percent of the photosynthate was lost from roots growing in unsterilized soils. Thus, studies of root exudation should be conducted in situations that reflect native soil conditions, as closely as possible.

Using radio-labeled carbon with plants growing in soils appears to provide an improved estimate of photosynthate incorporation into soil fractions. For this, root exudates are evaluated in greenhouse or growth chamber experiments where photosynthate can be labeled with ^{14}C-labeled carbon dioxide. Aboveground and belowground portions of the plant must be physically separated so that access of the soil microbial community to $^{14}CO_2$ or ^{14}C-labeled photosynthetically produced organic compounds can occur only through the root tissue. Movement of the ^{14}C-labeled photosynthate into the rhizosphere is determined by quantifying the $^{14}CO_2$ evolved from rhizosphere soil and the incorporation of ^{14}C into humic substances (e.g., using sodium hydroxide fractionation of soil organic matter) and microbial biomass (e.g., with the chloroform fumigation technique). An alternative to using ^{14}C-labeled substances for quantifying root exudate production was exploited by Barber and Lynch (1977). These workers used the relationship between the quantity of nutrients available for microbial growth and the mass of cells synthesized (i.e., growth yields). By calibrating rhizosphere microbial biomass produced per unit of growth substrate, a growth yield constant was developed. This constant was used to convert the quantities of microbial biomass produced in the presence of roots growing in liquid culture in growth chambers into mass of photosynthate available for their growth.

A wide range of values for root exudate production are reported in the literature. Nutrients available to rhizosphere microbes vary with plant species, plant developmental stage, soil temperature, as well as the species distribution of the plant community. Newman (1985) concluded that soluble root exudate ranges from 1 to 10 g/100 g dry roots. Values found in the literature vary depending upon the inclusiveness of the data collected. Some reports of plant carbon incorporated into soil fractions include total plant carbon retained in soil organic fractions, whereas in other studies, root exudate production is evaluated exclusive of carbon contributions by death of sloughed root cells and root death. Martin (1977) recorded a range of rhizodeposition of 14.3 to 44.4 percent of photosynthetically fixed carbon (for wheat plants), whereas Davenport and Thomas (1988) found 10 percent of $^{14}CO_2$ fixed by corn (*Zea mays* L.) in belowground plant components as compared to 40 percent for bromegrass (*Bromus inermis* Leyss). This differential separation of fixed carbon between aboveground and belowground plant biomass resulted in a rhizo-deposition rate in the bromegrass twice that of the corn. During the last 31 days of the 55-day period from germination to seed set of blue grama (*Bouteloua gracilis* [H.B.K.] Lag.), 33 percent of the fixed carbon was found in the root biomass, 23 percent was found in the root-derived organic matter contained in the soil, and 22 percent was released as carbon dioxide (Dormaar and Sauerbeck, 1983). These values contrast to early studies such as that by Martin (1977) where 0.8–1.3 percent of total organic carbon in root-free soils had been derived from plant photosynthate.

The rhizosphere microbial community can increase the rhizodeposition rate of photosynthetically fixed carbon. Microbes significantly increased root exudation of *Agropyron cristatum* and *Agropyron smithii* but had no effect on *Bouteloua gracilis* (Biondini et al., 1988). For the former grass strains, root exudates were reduced 40 and 83 percent, respectively, from native soil values when grown in axenic culture. The root biomass of *B. gracilis* was increased by the presence of an active rhizo-sphere community (Klein et al., 1988). This microbial stimulation of root exudation may result directly from synthesis of plant hormones by the microbes that stimulate exudate production. An indirect stimulation of exudation could result from micro-bial use of the exudates, thereby preventing their accumulation. This stimulation would increase outward diffusion of exudate components from the root.

Merckx et al. (1986) compared the release of photosynthate by maize and wheat roots growing in a growth chamber. After 6 weeks growth, 1.5 and 2.0 percent of the corn and wheat photosynthate, respectively, was found as soil organic carbon resi-dues. Photosynthetically fixed carbon dioxide was also detected as soil microbial biomass and as carbon dioxide respired by the soil microbial community. For the wheat plants, approximately 20 percent of the fixed carbon was detected in each of these carbon pools. Distribution of photosynthetically fixed carbon within these various pools can vary with soil temperature (Meharg and Killham, 1989). Increas-ing the soil temperature from 5 to 25°C increased root-soil respiration from 5.7 to 24.15 percent. The proportion of the photosynthate retained in the plant root and soil was greater at 5 and 25°C, with a minimum at 15°C. The total fixed carbon released as root exudates by *A. cristatum, A. smithii,* and *B. Gracilis* was estimated to be 8, 17, and 15 percent of photosynthetically fixed carbon.

Microbial growth in the rhizosphere can also be affected indirectly by the quan-tities of plant nutrients available for plant biomass production. Merckx et al. (1987) grew maize plants at high and low nutrient levels. Plant biomass production was limited at the low nutrient levels after 35 days growth. In soils with the lower nutrient level, 2 percent of the total $^{14}CO_2$ fixed by the plant was retained in the soil at all harvest times (38, 35, and 42 days), whereas with the nongrowth limiting nutrient levels, 4 percent of the plant photosynthate was retained in the soil. Maxi-mal soil incorporation was found in the lower nutrient soils at 35 days growth, whereas the quantities retained in the high nutrient soils increased throughout the study period. Microbial biomass contained between 28 and 41 percent of the total soil ^{14}C in the lower nutrient soil, whereas in the higher nutrient soil, this value ranged from 20 to 30 percent of soil residual photosynthetically fixed carbon.

When evaluating plant contributions to soil microbial energy transformation, it is important to consider annual turnover of the root biomass itself. This fixed carbon contribution to microbial nutrient cycling may be considerable. For example, Whitford et al. (1988) found that in the Chihuahuan Desert, 85 to 90 percent of the mass of herbaceous annual roots and 40 percent of the mass of woody shrub roots was decomposed annually. These rates are comparable or greater than similar processes reported for mesic ecosystems. For comparison, Dahlman and Kucera (1965) found an annual increment in the surface 25 cm of soil (the A horizon) of a prairie grassland in central Missouri (USA) of 429 g/m^2. This represented about 25 percent of the total root biomass in this horizon suggesting an annual turnover rate of the root biomass of about 25 percent. For comparison, in a study of live root decomposition in four forests (a mixed deciduous forest in Virginia, a mixed deciduous forest and a *Pinus resinosa* plantation in Massachusetts, and an *Acer saccharum-Quercus borealis* forest in Wisconsin, USA) losses of between 20 and 60 percent of their initial mass over a 4-year period was detected for each forest (McClaugherty et al., 1984).

The composition of root exudate is a primary parameter in selecting for individual species active in the rhizosphere community and the diversity of that community. The composition of root exudates varies with growth condition of the plant and its developmental stage. The variety of compounds that may be contained in root exudates is indicated by a list of candidates compiled by Alexander (1977). These soluble, organic carbon sources accumulated by plants grown aseptically include:

- *Amino acids:* All naturally occurring amino acids.
- *Organic acids:* Acetic, butyric, citric, fumaric, glycolic, lactic, malic, oxalic, propionic, succinic, tartaric, and valeric acids.
- *Pentoses and hexoses:* Arabinose, deoxyribose, fructose, galactose, glucose, maltose, mannose, raffinose, rhamnose, ribose, sucrose, and xylose.
- *Pyrimidines and puridines:* Adenine, cytidine, guanine, and uridine.
- *Vitamins: p*-Aminobenzoate, biotin, choline, inositol, nicotinic acid, pantothenate, pyridoxine, and thiamine.
- *Enzymes:* Amylase, invertase, phosphate, and protease.

This smorgasbord of readily catabolized organic substances is supplemented by all components contained in sloughed root cells and decaying root biomass. As was indicated previously, as much as one-fifth to one-third of the fine root biomass may turn over annually.

The activity of the rhizosphere population is determined not only by the composition of root exudate, sloughed cells, and decaying root tissue but also by the total quantities of fixed carbon produced, as was suggested previously. From considering the relationship of these three fixed carbon pools to total plant productivity, it is clear that rhizosphere activity is dependent upon plant nutrient status and reaction of the plant to environmental stresses. Data that reflect this microbe–plant growth state interaction are provided by studies of ryegrass (*Lolium perenne*) under a variety of fertilization strategies (Turner and Newman, 1984; Turner et al., 1985) and in studies of grassland management (Ingram et al., 1985). The ryegrass studies demonstrated increased root exudation under phosphorus deprivation. Evaluation of range plants showed an interaction between herbage removal and irrigation with irrigation causing significant declines in general rhizosphere activity that was mitigated by herbage removal. Active fungal hyphae were decreased by both treatments, whereas overall microbial activity was stimulated by herbage removal, suggesting a shift in the relationship of active fungal and bacterial populations in the stressed-plant rhizosphere.

A further question regarding the rhizosphere effect relates to the longevity of the plant impact on the soil community following the demise of the plant. Death of the plant results in a decline in the rhizosphere community. Little, if any, microbiological effect is carried over in the soil until the next invasion by root tissue. Although there is an impact of adjacent plants on the nature of the rhizosphere community, preceding plant species existent in the ecosystem have little effect on rhizosphere community composition of succeeding plants. The new vegetation determines, to a large degree, its own rhizosphere community composition. Exceptions to this statement of none carry-over of rhizosphere effects relate to soil properties that are altered by the presence of root surface or root-nodulating organisms. For example, fixed nitrogen accumulated in a soil environment due to the symbiotic association of *Rhizobium* sp. and legumes may provide a nitrogen nutrient source for succeeding plant communities. Similarly, physical or chemical modifications of the soil mineral environment necessarily persist.

7.1.4. Benefits to Plants Resulting from Rhizosphere Populations

Development of an active rhizosphere community has a variety of indirect and direct impacts on plant biomass production. Many of these benefits are proposed because it is difficult to quantify minute modifications of the plant root environment and the effect of these changes on overall plant development; however, logic combined with sound data document the fact that an active, dynamic rhizosphere microbial population is beneficial, if not essential, for optimal plant productivity. Rhizosphere property modifications beneficial to plant growth and development range from the more difficult to prove production of growth-promoting substances to the clearly documented gains of symbiotic root associations involved in nitrogen fixation.

A wide variety of rhizosphere microorganisms have been demonstrated to produce plant growth hormones, including indolacetic acid (IAA), gibberellins, and cytokinins (e.g., see Nietko and Frankenberg, 1989; Tien et al., 1979). These substances could be proposed to stimulate root tissue development, thereby increasing the capacity of the root system to provide nutrients and water required for above-ground biomass function.

A more indirect, but clearly beneficial, result of stimulation of soil community by root invasion is the contribution of rhizosphere microbial communities to development of a stable soil structure conducive to plant community development. (See Tate, 1987, for a further discussion of microbial involvement in soil structural development.) The rhizosphere community produces polysaccharide material such as capsules and slimes that cements soil mineral particulates into microaggregates. Furthermore, soil structural improvements are gained from fungal mycelial production as well as by association of soil particles with root tissue. This improvement in soil structure through increased soil aggregation results in improvements in soil aeration, water infiltration, and root penetration.

Perhaps, the primary benefits of an active rhizosphere community to the plant community result from mineralization of plant biomass. Organic nitrogen, phosphorus, and sulfur compounds are oxidized, thereby liberating ammonium, phosphorus, and sulfate. Thus, the nonbioavailable organic forms of these essential nutrients are converted to plant-available mineral forms.

Two root-based symbiotic associations are instrumental in producing a thriving, stable plant community, nitrogen fixation associations (*Rhizobium*-legume and actinorrhizal associations) (chapter 11), and mycorrhizal symbiosis (section 7.2). The intimate association of the bacteria or actinomycete partner (nitrogen fixation) or fungi with the plant root (mycorrhizae) increase fixed-nitrogen resources or facilitate nutrient transfer to the higher plants.

Additionally, plant growth stimulation may result from solubilization of inorganic plant nutrients by producing organic or inorganic acids by rhizosphere organisms. Microbially produced organic acids may solubolize essential minerals simply through acidification of rhizosphere soil or by chelation (e.g., siderophore production) of the metal. The capacity of organic acids excreted by soil bacteria to solubolize soil minerals is exemplified by studies of 2-ketogluconic acid production by gram-negative bacteria growing in glucose liquid media (Duff et al., 1963). Additionally, the acidification of the rhizosphere environs through metabolic production of hydrogen ions alters the pH sufficiently to mobilize soil minerals (e.g., see Gillespie and Pope, 1990a, 1990b).

Solubolization of a variety of metals contained in natural and synthetic insoluble salts and minerals has been demonstrated. Siderophores (see section 6.2.3) and other metal chelators are synthesized by a variety of bacteria, including common soil organisms (e.g., see Carrilo et al., 1992; Fekete et al., 1983; Jurkevitch et al., 1992; Hoefte et al., 1991; Kloepper et al., 1980a; Meyer et al., 1989; Pesmark et al., 1990) as well as mycorrhizal fungal associates (e.g., Cline et al., 1982; Federspiel et al., 1991). A number of examples of stimulation of plant biomass production by these microbially synthesized metal chelators have been published (e.g., Bar-Ness et al., 1991; Crowley et al., 1991; Derylo and Skorupska 1992; Kloepper et al., 1980b). Bosser and Vestraete (1986) have presented a model in which the rhizosphere is the most impacted soil microsite for siderophore synthesis.

7.1.5. Plant Pathogens in the Rhizosphere

A common maxim in plant pathology is that a soil-borne pathogen is more destructive to a plant growing in sterile medium than it would be to a plant developing in native soil where an active rhizosphere community develops. This observation suggests that rhizosphere microflora directly or indirectly inhibit the invasion of the plant tissue by the pathogen. This disease reduction could result from direct competition between the pathogen and the root microbes, antagonism of the pathogen by root microbes, or alteration of root exudate diffusion into the soil environment in a manner that interferes with chemotaxis of the pathogen to the root.

Root-inhabiting microbes and plant pathogens may compete for space, nutrients, or even binding sites on the root surface. Space and nutrient competition could result in failure of the pathogen to develop critical population densities for disease initiation, whereas competition for specific binding sites would reduce the capability of the plant pathogen to initiate the infection process.

As indicated previously (chapter 6), the rhizosphere is one soil site wherein antibiotic production may confer survival advantage upon microbial populations. Both antibiotics and siderophores produced by a variety of *Pseudomonas* strains have been shown to reduce or preclude infection of wheat roots by take-all disease (Hamdan et al., 1991; Kloepper et al., 1980b; Thomashow et al., 1990; Thomashow and Weller, 1990). Other suggested interactions of antibiotic microbes in biocontrol of root disease are associated with rhizoctonia infections (De Freitas and Germida, 1991) and onion white rot (Utkhede and Rahe, 1980). Similarly, rhizobacteria may also be used to suppress growth of sensitive plants. For example, *Pseudomonas* spp. have been isolated from roots of winter wheat (*Triticum aestivum* L.) and downy brome (*Bromus tectorum* L.), which suppress growth of the downy brome weed (Kennedy et al., 1991).

An additional role of rhizosphere microbes in reducing root disease incidence is in interfering with chemotactic attraction of the pathogen to root receptor sites. Chemotactic interactions are more difficult to document in the field but are readily deduced to be of importance in plant disease production. A variety of compounds listed previously as components of root exudates may serve as attractants for plant pathogens. Growth of root inhabitants (including mycorrhizal fungi) necessarily reduces both the quantity and diversity of organic compounds diffusing from the root, thereby diminishing the probability of encounter of a plant pathogen.

7.1.6. Manipulation of Rhizosphere Populations

Modification of the composition of any soil community structure through inoculation with exogenous microbes requires a thorough knowledge of microbial interactions as well as a little bit of luck if a reasonably predictable outcome of the procedure is desired. Historically, intervention into rhizosphere community dynamics has been a premier example of the microbiologists expertise—some might say utility. Biological nitrogen fixation has been encouraged through inoculation of legumes with *Rhizobium* sp. to encourage symbiotic nitrogen fixation for many decades (chapter 11). Long before the microbiology of the symbiosis was understood, legumes have been exploited in crop rotations for their capacity to increase soil fertility. This well-established soil community modification procedure has been supplemented with methods to supply fungal inocula to stimulate mycorrhizal development plus use of a variety of plant-growth stimulating bacteria, including the free living nitrogen-fixing bacteria of the genus *Azospirillum* (e.g., see examples listed in Table 7.1).

TABLE 7.1. Examples of Reports of Modification Root Communities by Amendment with Alien Microbial Populations.

Microbe	Recipient	Reference
Rhizobium spp.	Legume seeds	See chapter 11
Frankia spp.	Higher plants	See chapter 11
Mycorrhizal fungi	Higher plants	See text
Pseudomonas putida	Wheat	Vrany et al., 1981
Bacillus polymyxa	Lodgepole pine	Holl and Chanway, 1992
Pseudomonas fluorescens	Wheat	Parke et al., 1986
Azospirillium spp.	Spring and winter wheat	Bashan, 1986a,1986b
		Harris et al., 1989
	Maize	Fallik et al., 1988
Bacillus sp.	Spring wheat	Chanway et al., 1988
Azospirillum brasilience	Wheat	Bashan et al., 1987
Pseudomonas fluorescens	Potato	Bahme et al, 1988
Pseudomonas serrata	Wheat and peas	Astrom and Gerhardson, 1988

Considering that the potential for modifying soil microbial community structure through inoculation with alien strains has long been viewed with pessimism, the question of why root inoculation is frequently successful must be examined. Although as many failures in inoculation for population alteration are reported as are successes, a cursory review of the literature could lead to the conclusion that the only limitations to the procedure are selection of an appropriate microbial strain for inoculation, development of a carrier to sustain the microbial population, and selection of a delivery system. Indeed, a variety of strains, carriers, and inoculation procedures have been developed, but unanswered basic science questions still remain. Considerations that must be examined are why some inoculation studies are successful while others are not and why some microbial strains are established in the community more readily than others. Complete understanding of these queries is not available, but some basic information exists to direct our attention to means of improving the probability of successful intervention into rhizosphere community dynamics.

A central maxim in soil ecology is that if a niche exists in an ecosystem for a microbe to function, it will be occupied by a member of the community. Additionally, displacement of an established, functional member of the community is difficult. Yet, such displacement is frequently desirable in that an improved function of the ecosystem may be achieved by establishing a more efficient or active microbial strain.

Inoculation of the rhizosphere provides a good example of when this displacement or augmented activity is possible. The first consideration in inoculating a soil site with an exogenously developed microbial strain relates to the occurrence of a niche in which the microbe has the capacity to function. Note that a niche is a functional entity not a physical place in the ecosystem (a habitat). Nascent root tissue can be a provider of both a habitat and a niche. The external structure of the root is a surface that supports colonial development whereas a variety of functional opportunities (i.e., professions) are derived from the invasion of the root tip into the soil (e.g., decomposition of components of root exudates, nitrogen mineralization, nitrogen fixation), which at least initially have not been implemented by existing populations. Thus, the production of new root tissue overcomes the primary barrier to microbial community composition modification. Remaining obstacles are delivery of an appropriate inoculum to the habitat and the capacity of the alien organism to compete with the indigenous populations.

Clearly, inoculation of established root systems is difficult and results in a low probability of development of the inoculated populations. This problem of inoculating established roots has been overcome by drenching root systems (e.g., see Parke et al., 1986) or totally avoided by inoculating seeds before planting. Typically, seeds may be drenched with liquid inocula or pelleted solid carriers of the microbial inoculum prior to planting (see chapter 11 for a discussion of inoculation with rhizobia and Bashan, 1986c, for use with *Azospirillum* sp.). The assumption underlying this procedure is that the growing root will carry the inoculum with it as the root mass expands. Considering that rhizobial inocula are at least partially established in the field using these methods, seed inoculation may be of less than perfect efficiency, but it still occurs. *Azospirillum brasilense* is also mixed throughout the soil and from plant to plant by growth of wheat roots (see Bashan and Levanony, 1987, 1989). Alternatives to root drenches and soil inoculation include banding the microbial inoculum material into the soil adjacent to the expanding root mass. This procedure is based on the assumptions that the roots will grow through the band of inoculum, the inoculum will survive sufficiently long in the soil to encounter the root, and the inoculant will compete successfully with indigenous soil microbes for colonizable surfaces on the root.

Competition of the alien organisms with indigenous soil microbial populations is most likely the least understood aspect of any soil community inoculation procedure. The added organisms must be capable of overcoming the normal soil community defenses to invasion as well as of coping with soil physical and chemical barriers. Many of the growth requirements for the microbe are met by the provision of nutrients and growth surface by the root; hence, dealing with biological competitors becomes the primary difficulty to establishment.

Development of microbial strains that would be useful for field inoculation is generally initiated by isolation of an organism from the indigenous soil population which is amenable for selection of genetic variants. This isolate may then be modified either through selection of natural genetic variants (mutants) or by genetic engineering procedures where appropriate genes are inserted into the genome. In either case, the objective is to develop an organism that is more efficient in a given process (e.g., nitrogen fixation) or that possesses an improved metabolic capacity (e.g., ability to decompose an environmental contaminant).

By starting with a native member of the soil or root community, it may be assumed that most of the necessary traits for successful competition with the indigenous population preexisted in the parent strain. The question then becomes Did the genetic manipulation or laboratory culture reduce or eliminate this competitive capacity? The query is frequently answered negatively. For example, legumes nodules are commonly formed by indigenous rhizobial strains in spite of the inoculation of seeds with more efficient varieties. Development of *Rhizobium* strains with improved nitrogen fixation capabilities have been developed into a fine art; yet the art is commonly lost when the efficient strains do not enter the site of expression of their enhanced ability—the root nodule. Unfortunately, laboratory-selected rhizobia are not always the best competitors for nodule occupancy (e.g., Klubek et al., 1988; Materon and Hagedorn, 1982; Moawad and Bohlool, 1984; Robert and Schmidt, 1983; Smith and Wollum, 1989; van Rensberg and Strijdom, 1985). Variability in nodule occupancy by alien rhizobial strains is exemplified by data from van Rensberg and Strijdom (1985) where the range of nodule occupancy by inoculant strains was from 17.7 to 100 percent with the highest values associated with legumes growing in soils with limited native rhizobial populations of appropriate specificity for the legume. As discussed in chapter 11, successful occupancy of the nodule by the inoculant depends upon its capacity to deal with the general rhizosphere population as well as possession of appropriate biochemical traits to interact efficiently with the root to form an effective nodule.

Thus, the rhizosphere microbial community structure has been manipulated through inoculation of the plant with alien organisms to improve plant development (i.e., in most cases, improve crop yields). Great potential exists for further exploitation of these skills to alter the microbial community to accelerate plant community development and to reduce soil pollutant loads. The former objective may be achieved by establishing members of the rhizosphere community with the capacity to decompose plant toxicants or improve nutrient availability or augment soil aggregation whereas the rhizosphere community may also be used to couple the nutrient generation ability of higher plants (e.g., fixed-carbon supplies) with the ability of a laboratory-generated rhizosphere inoculant bacterial species to degrade a troublesome pollutant (perhaps, even cometabolically).

7.2. MYCORRHIZAL ASSOCIATIONS

Fungus-root or mycorrhizal associations have been known to exist since early descriptions by Frank in the early 1880s (cited in Allen, 1991). These associations are commonly studied and have been extensively described. The description, or definition, provided for these fungal and higher plant symbioses, to a large degree, is linked to the definer's bias toward the consideration of the importance of the plant or the microbe to ecosystem function. This is best seen by comparing statements by Marx and Allen. Marx (1972) describes mycorrhizal associations as follows:

> The infection of feeder roots of most flowering plants by symbiotic fungi and the transformation of these roots into unique morphological structures called mycorrhizae (fungus–roots) undoubtedly constitutes one of nature's most widespread, persistent, and interesting examples of **parasitism.** Most plants of economic importance to man are actually dual organisms—part plant and part symbiotic root-inhabiting fungi.

This plant-oriented view stresses the use or perhaps even the "theft" of plant photosynthate by the controlled parasites in this symbiotic association. In contrast Allen

(1991) described the fungal–plant interaction from a more neutral or microbially oriented aspect:

> A mycorrhizae is a mutualistic symbiosis between plant and fungus localized in a root or root-like structure in which energy moves primarily from plant to fungus and inorganic resources move from fungus to plant.

In this definition, the mutualistic benefits of the interaction are stressed.

From an environmental, or soil community consideration, the fungal–plant associations in mycorrhizal symbioses can be viewed on a gradient of increasing association of the plant and soil community (Table 7.2). At the most extreme position lie the general soil microbial populations. These organisms are involved in mineralization of native soil organic matter, soil incorporated organic components [plant and animal debris], microbial biomass, as well as soluble organic carbon components contained in infiltrated water. This community functions totally at the mercy of the nutrient supply, which may range from nearly total deprivation (i.e., desert soil) to absolute luxury (e.g., soil receiving nutrient-laden industrial effluent). Of more direct association with the discussion at hand is the fact that the nutrients produced by organic compound mineralization by these general soil microbes may not be

TABLE 7.2. Comparison of Properties of Nonrhizosphere, Rhizosphere, and Mycorrhizosphere Soils.

Nonrhizosphere	Rhizosphere	Mycorrhizosphere
All biogeochemical cycles—rates limited by organic matter (humus) availability	Maximal biogeochemical cycling	All biogeochemical cycles are supported—controlled by both plant inputs and soil humus contents
Nutrients may be leached to root tissue or groundwaters	Nutrients mineralized directly available for plant biomass as well as microbial biomass synthesis	Nutrients incorporated into fungal biomass and transported to plant tissue
Nutrients may be incorporated directly into microbial biomass		
Microbial biomass controlled by carbon and energy resources	Microbial biomass synthesis is controlled by plant productivity and rate of root exudate production	Biomass controlled by plant productivity and transfer of photosynthate to the fungus
Note that nonrhizosphere soil metabolic activity is not affected directly by rhizosphere interactions, but mycorrhizal fungi may enter this soil region and catalyze biogeochemical processes.		

directly available to the aboveground community. Clearly, nutrient transfer to higher plants is dependent upon proximity of the microbial colony to active roots. In this case, a locational displacement as small as a few millimeters could determine the rate of mineral nutrient recovery by the plant.

An enhancement of ecosystem three-dimensional associations is derived from the transition from general soil populations to rhizosphere-linked communities. Rhizosphere communities catalyze all the processes previously described for non-rhizosphere soil, but their value to ecosystem sustainability is augmented by the improved efficiency for transfer of mineral nutrients to the plant community.

The ultimate association of soil microbes with growing plants is symbiotic fungal–root associations. This mycorrhizal symbiosis could be said to bridge these community types—nonrhizosphere, rhizosphere, and fungus–root. Although there is an intimate association between the fungus and plant root cells (with some types of mycorrhizae, actual penetration of the root cells itself occurs), the fungal biomass extends well beyond the rhizosphere into non-root-impacted soil. Thus, nutrients may be accumulated from nonrhizosphere soil by the fungus and transported through the rhizosphere community into the plant root, which supports growth and development of the entire hyphal network by the photosynthate provided by the fungal-infected plant. Furthermore, this mycelial network may directly link roots of plants of the same or different species in that the same fungal body may infect a number of individual plants.

The physical association of mycorrhizal fungi with plant roots has been extensively described and serves as the primary basis for classification of mycorrhizae. These associations may be classed as ectomycorrhizae, endomycorrhizae, or ectendomycorrhizae. In this grouping of mycorrhizae, the commonly encountered vesicular–arbuscular mycorrhizae are considered endomycorrhizae. A number of reviews of the biology of mycorrhizae have been published (e.g., see Allen, 1991; Mukerji et al., 1991). Thus, only basic descriptions of these classes of mycorrhizae will be presented herein.

Ectomycorrhizal fungi penetrate intracellularly and partially replaces the middle lamellae cortical cells of feeder roots. These fungi form a dense mycelial net around and between the plant cells termed a **Hartig net.** Ectomycorrhizal associations are also characterized by a dense, generally continuous hyphal network over the feeder root surface called a **fungal mantle.** This fungal mantle varies from one to two hyphal diameters to as many as 30 or 40 depending upon the fungal associate, the host, and the environmental conditions. Examples of plant species forming ectomycorrhizal associations are species in the families *Pinaceae, Salicaceae, Betulaceae,* and *Fugaceae.* From the view of the fungus itself, most ectomycorrhizal fungi are basidomycetes (primarily of the families *Amanitaceae, Boletaceae, Cortinariaceae, Russulaceae, Tricholomataceae, Rhizopogonaceae,* and *Sclerodermataceae*).

Endomycorrhizae are distinguished by the fact that the fungus penetrates the cortical cells of feeder roots and may form large vesicles and arbuscles (hence the term **vesicular–arbuscular mycorrhizae**). These fungi do not form dense fungal mantles, but they do develop a loose, intermittent arrangement of mycelium on the root surface. Endomycorrhizae are formed by most agronomic, horticultural, and ornamental crops, as well as some forest tree species that do not form ectomycorrhizae. The fungal species are phycomyces—many of which are in the genus *Endogone*.

As with many, if not most, biological classification systems, organisms form associations that are not easily classified in any of the predominate groupings. For mycorrhizal associations, the **ectodomycorrhizae** constitute this grouping. These mycorrhizae resemble ectomycorrhizae in forming a Hartig net and a fungal mantle. A resemblance with the endomycorrhizae is associated with their penetration of cortical cells. This mycorrhizal grouping is the least studied, and the nature of the fungal symbionts has not been totally elucidated.

Although mycorrhizae provide intriguing subjects for biological and physiological analysis, the emphasis for this study is to delineate a) the role of mycorrhizal symbioses in ecosystem function, b) their importance in nutrient cycling in soil–plant interactions, and c) their use in bioremediation plans.

7.2.1. Mycorrhizae in the Soil Community

A variety of elegant studies of soil enzymes and respiratory activity in relationship to soil physical and chemical properties and biological interactions (both belowground and aboveground) have been conducted. Some of this research has been directed at quantifying fungal and bacterial contributions to these activities through incorporation of population-specific antibiotics (e.g., cycloheximide to inhibit fungal activity and streptomycin to control bacterial activity) into the laboratory-incubated soil–plant samples, but evaluation of the role of mycorrhizal fungi in these soil functions is rarely considered. Standard methods for soil collection rarely preserve the soil–root–fungal associations. Hence, even direct counts of fungal mycelia in the soil would rarely, if ever, be linked to plant roots. This omission of a consideration of these important trophic interactions occurs in spite of the knowledge that mycorrhizal fungi constitute a major portion of soil biomass and that these fungi extend throughout the soil profile—far beyond the regions classified as rhizosphere soil—or even that portion of the soil profile considered to be root-influenced. An example of the association of soil enzymatic activity with root and fungal development was provided by Spalding et al. (1975). In their study, cellulase, invertase, polygalacturonase, and peroxidase activities were found to increase in soil under lycopodium fairy rings, suggesting a mycorrhizal fungal association with organic matter metabolism.

The potential for mycorrhizal contribution to overall soil metabolic activities can be appreciated by examining examples of the quantities of biomass contained in these structures. Fogel and Hunt (1983) in a study of a young, second-growth Douglas-fir (*Pseudotsuga menziesii* [Mirb.] Franco) stand found that mycorrhizae constituted 6 percent of the total tree-standing crop. Furthermore, they found that roots and mycorrhizae contained larger reserves of nitrogen, phosphorus, potassium, and magnesium than did the forest floor or soil fungi. Fine roots and mycorrhizae contributed between 84 and 78 percent of the total tree organic matter to the soil.

This study suggests that although the mycorrhizal fungal biomass constitutes only a small portion of the standing crop, it may account for 50 percent of the organic matter throughput in the *Pseudotsuga menziesii* stand. Read (1984) suggests that mycorrhizal fungal biomass may be the largest microbial biomass component of many forest soils. Allen (1991) states that "mycorrhizal fungi may be the single largest consumer group of net primary production in many, if not most, terrestrial biomes." Vasicular–arbuscular fungal hyphae have been quantified at densities of up to 38 m cm^{-3} (Allen and Allen, 1986). Remember that, although a portion of the fungal biomass is intimately associated with the root, the fungal net extends throughout the soil surface (See Allen, 1991, and Bagyaraj, 1991 for a review of this topic.)

From these observations, a number of benefits of mycorrhizal development to the total soil microbial community can be delineated. Foremost among the contributions to the soil community is the capacity to cycle nutrients (i.e., mineralize accumulated biomass). Although the bulk of the research efforts have been directed at quantifying mineral nutrient transfer from fungal biomass or soil to plant tissue, it is reasonable to assume that a portion of the metabolic products of this fungal activity will be released to the soil microbial community. For example, mycorrhizal fungi have been shown to mineralize soil organic phosphate through synthesis of phosphatase (e.g., see Dighton, 1983, and Pasqualini et al., 1992) and to solubolize mineral phosphorus through the acidification of the soil habitat through carbon dioxide production (see Knight et al., 1989).

Indirect benefits to the soil community by mycorrhizal fungal growth results from a generalized improvement of soil structure. Fungal hyphal development increases the association of soil particulates into aggregates. (See Tisdall and Oades, 1982, and for a review of this topic, Tate, 1987.) This improved soil structure has a direct impact on the indigenous microbial community through improved aeration and moisture infiltration and an indirect effect via stimulation of plant-root growth. Clearly, any augmentation of root development increases the quantities of fixed carbon reaching the soil microbial community.

7.2.2. Symbiont Benefits from Mycorrhizal Development

Mycorrhizal fungi are instrumental in augmenting plant nutrient availability in nutrient-stressed ecosystems. A wide range of data (at times contradictory data) have been collected demonstrating nutritional benefits to plant communities in stressed communities through mycorrhizal symbiotic associations. Most research has shown improved nutrient transfer to the plant tissue through the augmentation of the adsorbing surface of roots through extension of the fungal mycelium into non-rhizosphere soil. Gains in phosphate, nitrogen, and water transfers are most commonly reported. The expansive literature documenting these benefits has been reviewed by Allen (1991), Baguarak (1991), Gupta (1991), and Mukerji et al. (1991).

In a more intriguing vein, mycorrhizal fungi may not only enhance soil–plant transfer of nutrients, but they may also be instrumental in movement of nutrients between plants (e.g., see Camel et al., 1991; Eason et al., 1991; Hamel et al., 1991;

and Read et al., 1989). Read et al. (1989) demonstrated through the use of $^{14}CO_2$ that carbon moves freely between plants connected by mycorrhizal mycelium. This plant–plant bridging by the fungal hyphae occurs between host plants of the same or differing species. Similarly, Chiariello et al. (1982) found that ^{32}P-phosphate sprayed on leaves of *Plantago erect* in a grassland was transferred to the shoots of about 20 percent of the close neighbors. Vesicular–arbuscular mycorrhizae were noted to connect the root systems of neighbors of different species and considered to have mediated the nutrient transfer. Camel et al. (1991) found that the hyphal front may advance at a rate of 2.3 cm week^{-1} in soil–sand mixtures and that bridges of at least 90 mm may be formed. These workers suggest a competitive advantage of this bridging in plant groupings where mycorrhizal inoculum is limited. Plants developing in a less desirable portion (e.g., a shaded site) of the ecosystem may benefit from the photosynthetic activity of the less-stressed members of the community as well as from nutrient production by the mycorrhizae of the interconnected root system.

A further stimulus to plant biomass productivity is associated with the capacity of mycorrhizal associations to reduce or prevent plant disease development. (See Jalali and Jalali, 1991, for a review of this topic.) The association of the fungus with plant-disease-susceptible root tissue may reduce pathogen vulnerability by modifying exudate composition or concentration, stimulating plant-disease-protective response through its own infection of the root tissue, inhibiting competing microbial populations directly through synthesis of antibiotics and limiting access of the plant pathogen to root tissue by physically occupying the root surface.

Enhancement of fungal biomass by the symbiotic association with root tissue is generally attributed to augmented availability of photosynthate in the form of simple sugars for fungal energy and carbon needs and the reduced competition with the general rhizosphere microflora. Kucey and Paul (1982) found that within mycorrhizal, nodulated faba beans (*Vicia faba* L.) about 4 percent of the carbon fixed by the host was transferred to the fungal associate. (For a more detailed examination of this topic, see Allen, 1991.) A competitive advantage for the fungal associate in dealing with soil microbial populations in general and more specifically rhizosphere microbes is derived from its close physical association with the host root. Intrusion into the interlameller space between cortical cells (as well as cell penetration by the endomycorrhizae) and intimate contact with the root surface facilitate nutrient recovery from the host tissue.

Of interest from the impact of the host plant on the behavior of mycorrhizal fungi in soil is the potential for selective stimulation of spore outgrowth. The published results regarding selective stimulation of fungal spore outgrowth by plant root product is mixed; yet a real microsite stimulation most probably occurs. In their study of carrots and the vesicular–arbuscular mycorrhizal fungus *Gigaspora margarita,* Bechard and Piche (1989) found that root volatiles provided little stimulation of spore germination and that the exudates alone had no effect. Considerable stimulation was detected from the synergistic interaction between the volatile and exudate root products. From their study of the effect of root exudates of *Trifolium repans* on *Glomus fasiculatus* germination, Elias and Safir (1987) conclude that the primary stimulation of spore outgrowth results from exudates from phosphorus-deficient clover seedlings when compared to the effect of exudates from phosphorus-sufficient seedlings, suggesting that it is the quality of the exudates altering the spore germination rates. Further studies (Nair et al., 1991) suggest that two iso-flavonoids (7-hydroxy, 4″-methoxy isoflavone and 5,7,-dihydroxy, 4″methoxy iso-flavone) may be signal molecules in vesicular–arbuscular mycorrhizal symbioses. Other workers (see Bowen, 1969) conclude that root exudates had no effect on spore outgrowth. From evaluation of these disparate observations, we can deduce that mycorrhizal fungal spores are sensitive to root exudates and that variability in composition (quality) of these plant products with host nutritional status greatly affects this interaction.

A further observation of the infection process in mycorrhizal development is a potential interaction of plant-growth-promoting bacteria (e.g., Bowen and Theodorou, 1979; Garbaue and Bowen, 1989; Meyer and Linderman, 1986a). Bowen and Theodorou (1979) found that a variety of bacteria could either depress, have no effect, or stimulate mycorrhizal development in the rhizoplane of *Pinus radiata,* suggesting the necessity of fungal compatibility with the rhizosphere bacterial populations. Garbaue and Bowen (1989) actually found that "helper" microorganisms which improve the efficiency of ectomycorrhizal infection of *Pinus radiata* D. Don growing in a sandy podzol, exist in the rhizosphere. Meyer and Linderman (1986a) reported that a combination of plant-growth-stimulating bacteria and mycorrhizal development of subterranean clover (*Trifolium subterraneum* L.) resulted in considerably greater plant biomass synthesis than with either microbial component taken singly.

7.2.3. Environmental Considerations

The statement by Allen (1991) that "Mycorrhizae represent one of the least understood, most wide spread, and most important biological symbioses on earth" is no more greatly appreciated then when application of our understanding of this fungal–plant interaction to ecosystem problems is considered. Understanding how to successfully exploit mycorrhizal associations is gained by considering the current applications of mycorrhizal interaction management (development of functional plant communities on disturbed sites—land reclamation, and inoculation of nursery stock for field planting) and an interesting, but as yet little explored, option (development of genetically engineered fungal associates for protection of host plants from ecosystem contamination).

The concepts underlying successful management of abused soils (i.e., land reclamation) have conventionally been associated with highly mismanaged sites, such as those receiving industrial waste materials or accidental spills of petroleum products as well as more traditional mineland sites. With the growing concern for extensively cropped agricultural soils (especially situations where, due to soil erosion, the subsoils are providing the basis for agricultural production) and clear-cut forest sites, applicability of the scientific undergirding for the development of proper soil management plans for a wide variety of "tired" soil systems emerges. Traditionally, this soil management includes procedures that encourage the development of a soil structure conducive to aboveground plant community development (see Tate, 1987, for a review of this topic) and of establishment of stable populations of organisms involved in biogeochemical cycling (see Tate, 1985). Due to their ability to improve longevity and productivity of aboveground plant communities, mycorrhizal associations are a critical component in soil reclamation management.

Plant community gains from management of mycorrhizal associations for soil-quality improvement are derived from both the soil structural enhancements resulting from the fungal contributions to soil aggregation as well as the improved availability of essential plant nutrients. Rapid and somewhat long-lasting benefits are accrued from amendment of degraded soil with a variety of organic matter sources and fertilizers (e.g., see Fresquez et al., 1990; Stroo and Jencks, 1985; Tester, 1990; Wong and Ho, 1991), but for minimization of anthropogenic intervention, a stable soil microbe–plant interactive community must be developed. As was indicated previously, mycorrhizal associations are developed best under stressed conditions—especially phosphate stress. Examples of additional limitations common to degraded soils that encourage mycorrhizal symbioses development include fertilizer application practice and cropping system (Ellis et al., 1992), soil erosion (i.e., loss of fungal propagules with surface soil losses, Day et al., 1987), vegetation (Newman et al., 1981), and soil organic matter (Harvey et al., 1981). Hence, for rapid and enduring development of a reclaimed ecosystem, site management should include insurance of inclusion of fungal propagules in the reclamation of target soil, implementation of plant nutrient management procedures that will not prevent mycorrhizal development, and, where possible, use of inoculated transplants for the establishment of aboveground plant communities.

Questions associated with inoculation of seedlings with mycorrhizal fungal propagules relate not to the benefit of such associations to plant survival and productivity—these have long been assumed to be major pluses—but current concerns are more properly directed at optimization of the inoculation processes and selection of the best fungal strains for the symbioses. For example, Marx et al. (1978) documented stimulation of loblolly pine seedlings by a variety of ectomycorrhizal fungi, whereas more recently Theodorou and Bowen (1987) evaluated the selectivity of basidiospore germination and selection by pine seedlings and development of an effective method for inoculum conservation in the absence of host plants. Duponnois and Garbaye (1991) investigated the use of mycorrhization helper bacteria in establishing ectomycorrhizal fungal associations in forest nurseries. The unquestionable utility of such inoculation is exemplified by studies of Lindemann et al. (1984) where mycorrhizal interactions among other soil microbial activities were evaluated in mine spoil reclamation and of Wilson et al. (1991) where mycorrhizae were used to reclaim portions of a dewatered flue-gas desulfurization sludge pond.

Future considerations of the utility of exploiting mycorrhizal associations for soil reclamation may also be based on extension of genetic engineering procedures to mycorrhizal fungi and modification of the carbon substrate metabolism capacity of these organisms. As discussed previously, mycorrhizal fungi extend from the growing root into the surrounding soil. The soil-particle-associated mycorrhizal fungal hyphae catabolize soil organic carbon components. A potential exists to develop mycorrhizal fungi with the capacity to decompose pollutants that may occur in the vicinity of the roots of both plants selected for soil reclamation (e.g., pine trees and associated shrubs) or even crop plants. The objective for the first situation would be to develop fungal associates that provide a benefit to the plant beyond those generally associated with mycorrhizal interactions—i.e., removal of potential plant toxicants. In the case of the crop plant, the fungi could serve as a "first line of defense" per se, thereby preventing toxicants that may contaminate food plants from encountering and being accumulated in sensitive root tissue.

7.3. MYCORRHIZOSPHERE

Traditionally, rhizosphere activity and mycorrhizal symbioses have been studied as if they represented separate ecosystems, but a brief consideration of the preceding discussion of these ecosystem phenomena leads at least to the suspicion that they occur concurrently on many plant roots rather than as distinct entities. Indeed, the latter conditions is most likely the more common situation. Clearly, in all soil systems a growing plant will develop some sort of microbial–plant interaction termed a rhizosphere. As our understanding of the occurrence of mycorrhizal symbioses expands, it is becoming more obvious that the vast majority of higher plants are mycorrhizal. Thus we cannot conclude that all rhizospheres are mycorrhizospheres, but it is also evident that most of the data collected to date from field studies of rhizosphere populations must have resulted from evaluation of mycorrhizospheres—with the experimenter being guilty of ignorance of the manifestations of fungal mycelium in association with the roots or not possessing the requisite training to differentiate mycorrhizal root structures. This conceptual oversight is gradually being corrected.

In evaluating the quantities of exudates and their chemical substituents, evaluation of the intensity of mycorrhizal association is necessary. With symbiotic fungal–plant associations, essentially the entire root surface may be covered by several layers of fungal mycelium (ectomycorrhizal associations), or a major portion of the root surface may be free of obvious fungal hyphal development with the fungus deriving its nutrient through intracellular penetration (endomycorrhizal associations). In either case, the heterotrophic fungi consume root exudates and organic carbon contained in sloughed cells, thereby altering both the quality and quantity of root exudates. Thus, some components of the exudate may be decreased or totally depleted. Photosynthetically produced fixed carbon may also become available to the soil and rhizosphere microbial community through death and decay of the fungal mycelia. (See Fogel, 1988, for further discussion of this topic.) Direct and indirect impacts of this interaction between the fungal symbiont and root exudates on the rhizosphere and rhizoplane microbial populations include determination of the species that are active and alteration of the total microbial biomass produced (e.g., Meyer and Linderman, 1986b), as well as the control of plant pathogen populations discussed previously. Furthermore, interaction of the rhizosphere bacteria and mycorrhizal fungi may improve plant growth (Azcon, 1989; Will and Sylvia, 1990).

These limited evaluations of the impact of both rhizosphere populations and mycorrhizal associations with plant growth emphasize the need to expand future root community studies to include all levels of trophic interactions—be they the result symbiotic fungi, nitrogen-fixation symbioses (rhizobium–legume or actinor-rhizal), or general rhizosphere bacterial, fungal, protozoan, or nematode population growth. Each of these entities withdraws nutrients from the common source—plant photosynthate—either directly or indirectly, produces biomass, and alters the productivity or lack thereof of their neighbors. Ultimately, total ecosystem dynamics are controlled by the complex interactions of each member of the soil community.

7.4. CONCLUSION

The rhizosphere is most certainly an oasis of biological activity within the soil ecosystem. This ecosystem is represented by a diversity of microbial populations, the complete range of plant–microbe and microbe–microbe interactions, and the relative inclusiveness of all essential biogeochemical processes for total ecosystem development. An overall view of the processes occurring in the rhizo-sphere/mycorrhizosphere may be summarized by three words—interactions, productivity, and protection. That is, the rhizosphere is populated by a vast variety of life forms each of which is impacted both positively and negatively by all other creatures existing in the system, the productivity of the microbial community as well as the aboveground community is interlocked with the viability and stability of the microbial community (i.e., neither the plant community nor the microbial components can function optimally singly), and lastly the microbial community confers a resistance to the plant component of the system by competing with and in many cases destroying potentially destructive plant pathogens.

REFERENCES

Alexander, M. 1977. Introduction to Soil Microbiology. John Wiley & Sons, New York. 467 pp.

Allen, E. B., and M. F. Allen. 1986. Water relations of xeric grasses in the field: Interactions of mycorrhizae and competition. New Phytol. 104:559–571.

Allen, M. F. 1991. The Ecology of Mycorrhizae. Cambridge University Press, New York. 184 pp.

Astrom, B., and B. Gerhardson. 1988. Differential reactions of wheat and pea genotypes to root inoculation with growth-affecting rhizosphere bacteria. Plant Soil 109:263–270.

Azcon, R. 1989. Selective interaction between free-living rhizosphere bacteria and vesicular-arbuscular mycorrhizal fungi. Soil Biol. Biochem. 21:639–344.

Bahme, J. B., M. N. Schroth, S. D. Van Gundy, A. R. Weinhold, and D. M. Tolentino. 1988. Effect of inocula delivery systems on rhizobacterial colonization of underground organs of potato. Phytopathology 78:534–542.

Bagyaraj, D. J. 1991. Ecology of vesicular–arbuscular mycorrhizae. In D. K. Arora, B. Rai,

K. G. Mukerji, and G. R. Knudsen (eds.), Handbook of Applied Mycology, Vol. 1. Soil and Plants, pp. 3–34. Dekker, New York.

Barber, D. A., and J. M. Lynch. 1977. Microbial growth in the rhizosphere. Soil Biol. Biochem. 9:305–308.

Barber, D. A., and J. K. Martin. 1976. The release of organic substances by cereal roots into soil. New Phytol. 76:69–80.

Bar-Ness, E., Y. Chen, Y. Hadar, H. Marschner, and V. Roemheld. 1991. Siderophores of *Pseudomonas putida* as an iron source for dicot and monocot plant. Plant Soil 130: 231–241.

Bashan, Y. 1986a. Enhancement of wheat root colonization and plant development by *Azospirillum brasilense* following temporary depression of rhizosphere microflora. Appl. Environ. Microbiol. 51:1067–1071.

Bashan, Y. 1986b. Significance of timing and level of inoculation with rhizosphere bacteria of wheat plants. Soil Biol. Biochem. 18:297–302.

Bashan, Y. 1986c. Alginate beads as synthetic inoculant carriers for slow release of bacteria that affect plant growth. Appl. Environ. Microbiol. 51:1089–1098.

Bashan, Y., and H. Levanony. 1987. Horizontal and vertical movement of *Azospirillum brasilense* CD in the soil and along the rhizosphere of wheat and weeds in controlled and field environments. J. Gen. Microbiol. 133:3473–4380.

Bashan, Y., and H. Levanony. 1989. Wheat root tips as a vector for passive vertical transfer of *Azospirillum brasilense* CD. J. Gen. Microbiol. 135:2899–2908.

Bashan, Y., H. Levanony, and O. Ziv-Vecht. 1987. The fate of field-inoculated *Azospirillum brasilense* CD in wheat rhizosphere during the growing season. Can. J. Microbiol. 33:1074–1079.

Bechard, G., and Y. Piche. 1989. Fungal growth stimulation by CO_2 and root exudates in vesicular–arbuscular mycorrhizal symbiosis. Appl. Environ. Microbiol. 55:2320–2325.

Biondini, M., D. A. Klein, and E. F. Redente. 1988. Carbon and nitrogen losses through root exudation by *Agropyron cristatum, A. Smithii*, and *Bouteloua gracilis*. soil Biol. Biochem. 20:477–482.

Bossier, P., and W. Verstraete. 1986. Ecology of *Arthrobacter* JG-9-detectable hydoxamate siderophores in soils. Soil Biol. Biochem. 18:487–492.

Bowen, G. D. 1969. Nutrient status effects on loss of amides and amino acids from pine roots. Plant Soil 30:139–142.

Bowen, G. D., and G. Theodorou. 1979. Interactions between bacteria and ectomycorrhizal fungi. Soil Biol. Biochem. 11:119–126.

Camel, S. B., M. G. Reyes-Solis, R. Ferrera-Cerrato, R. L. Franson, M. S. Brown, and G. J. Bethlenfalvay. 1991. Growth of vesicular–arbuscular mycorrhizal mycelium through bulk soil. Soil Sci. Soc. Am. J. 55:389–393.

Carrillo, G. C., G. Del Rosario, and M. Vazquez. 1992. Comparative study of siderophore-like activity of *Rhizobium phaseoli* and *Rhizobium fluorescens*. J. Plant Nutr. 15:579–590.

Chanway, C. P., L. M. Nelson, and F. B. Holl. 1988. Cultivar-specific growth promotion of spring wheat by coexistent *Bacillus* spp. Can. J. Microbiol. 34:925–929.

Chiariello, N., J. C. Hickman, and H. A. Mooney. 1982. Endomycorrhizal role of interspecific transfer of phosphorus in a community of annual plants. Science (Washington, DC) 217:941–943.

Cline, G. R., P. E. Powell, P. J. Szaniszlo, and C. P. P. Reid. 1982. Comparison of the abilities of hydroxamic, synthetic, and other natural organic acids to chelate iron and other ions in nutrient solution. Soil Sci. Soc. Am. J. 46:1158–1164.

Crowley, D. E., Y. C. Want, C. P. P. Reid, and P. J. Szaniszlo. 1991. Mechanisms of iron acquisition from siderophores by microorganisms and plants. Plant Soil 130:179–198.

Dahlman, R. C., and C. L. Kucera. 1965. Root productivity and turnover in native prairie. Ecology 46:84–89.

Davenport, J. R., and R. L. Thomas. 1988. Carbon partitioning and rhizodeposition in corn and bromegrass. Can. J. Soil Sci. 68:693–701.

Day, L. D., D. M. Sylvia, and M. E. Collins. 1987. Interactions among vesicular–arbuscular mycorrhizae, soil, and landscape position. Soil Sci. Soc. Am. J. 51:636–639.

De Freitas, J. R., and J. J. Germida. 1991. *Pseudomonas cepacia* and *Pseudomonas putida* as winter wheat inoculants for biocontrol of *Rhizoctonia solani*. Can. J. Microbiol. 37:780–784.

Derylo, M., and A. Skorupska. 1992. Rhizobial siderophore as an iron source for clover. Physiol. Plant. 85:549–553.

Dighton, J. 1983. Phosphatase production by mycorrhizal fungi. Plant Soil 71:455–462.

Dormaar, J. R., and D. R. Sauerbeck. 1983. Seasonal effects of photoassimilated carbon-14 in the root system of blue grama and associated soil organic matter. Soil Biol. Biochem. 15:475–479.

Duff, R. B., D. M. Webley, and R. O. Scott. 1963. Solubolization of minerals and related minerals by 2-ketogluconic-acid producing bacteria. Soil Sci. 95:105–114.

Duponnois, R., and J. Garbaye. 1991. Effect of dual inoculation of douglas fir with the ectomycorrhizal fungus *Laccaria laccata* and mycorrhization helper bacteria (MHB) in two bare-root forest nurseries. Plant Soil 138:169–176.

Eason, W. R., E. I. Newman, and P. N. Chuba. 1991. Specificity of interplant cycling of phosphorus: The role of mycorrhizas. Plant Soil 137:267–274.

Elias, K. S., and G. R. Safir. 1987. Hyphal elongation of *Glomus fasciculatus* in response to root exudates. Appl. Environ. Microbiol. 53:1928–1933.

Ellis, J. R., W. Roder, and S. C. Mason. 1992. Grain–Sorghum–soybean rotation and fertilization influence on vesicular–arbuscular mycorrhizal fungi. Soil Sci. Soc. Am. J. 56:789–794.

Fallik, E., Y. Okon, and M. Fisher. 1988. Growth response of maize roots to *Azospirillum* inoculation: Effect of soil organic matter content, number of rhizosphere bacteria and timing of inoculation. Soil Biol. Biochem. 10:45–50.

Federspiel, A., R. Schuler, and K. Haselwandter. 1991. Effect of pH, L-ornithine and L-proline on the hydroxamate siderophore production by *Hymenoscyphus ericae,* a typical ericoid mycorrhizal fungus. Plant Soil 130:259–261.

Fekete, F. A., J. T. Spence, and T. Emery. 1983. Siderophores produced by nitrogen-fixing *Azotobacter vinelandii* OP in iron-limited continuous culture. Appl. Environ. Microbiol. 46:1297–1300.

Fogel, R. 1988. Interactions among soil biota in coniferous ecosystems. Agric. Ecosystems Environ. 24:68–85.

Fogel, R., and G. Hunt. 1983. Contribution of mycorrhizae and soil fungi to nutrient cycling in a Douglas-fir ecosystem. Can. J. Forest Res. 13:219–232.

Fresquez, P. R., R. E. Francis, and G. L. Dennis. 1990. Sewage sludge effects on soil and plant quality in a degraded, semiarid grassland. J. Environ. Qual. 19:324–329.

Garbaue, J., and G. D. Bowen. 1989. Stimulation of ectomycorrhizal infection of *Pinus radiata* by some microorganisms associated with the mantle of ectomycorrhizas. New Phytol. 112:383–388.

Gillespie, A. R., and P. E. Pope. 1990a. Rhizosphere acidification increases phosphorus recovery of black locust: I. Induced acidification and soil response. Soil Sci. Soc. Am. J. 54:533–537.

Gillespie, A. R., and P. E. Pope. 1990b. Rhizosphere acidification increases phosphorus recovery of black locust: II. Model predictions and measured recovery. Soil Sci. Soc. Am. J. 54:538–541.

Griffin, G. J., M. G. Hale, and F. J. Shay, 1977. Nature and quantity of sloughed organic matter produced by roots and of axenic peanut plants. Soil Biol. Biochem. 8:29–32.

Gupta, R. K. 1991. Drought response in fungi and mycorrhizal plants. *In* D. K. Arora, B. Rai, K. G. Mukerji, and G. R. Knudsen (eds.), Handbook of Applied Mycology, Vol. 1. Soil and Plants, pp. 55–75. Dekker, New York.

Hamdan, H., D. M. Weller, and L. S. Thomashow. 1991. Relative importance of fluorescent siderophores and other factors in biological control of *Gaeumannomyces graminis* var. *tritici* by *Pseudomonas fluorescens* 2-79 and M4-80R. Appl. Environ. Microbiol. 57:3270–3277.

Hamel, C., C. Nesser, U. Barrantes-Cartín, and D. L. Smith. 1991. Endomycorrhizal fungal species mediate ^{15}N transfer from soybean to maize in non-fumigated soil. Plant Soil 138:41–47.

Harris, J. M.. J. A. Lucas, M. R. Davey, G. Lethbridge, and K. A. Powell. 1989. Establishment of *Azospirillum* inoculant in the rhizosphere of winter wheat. Soil Biol. Biochem. 21:59–64.

Hartel, P. G., J. W. Billingsley, and J. W. Williamson. 1989. Styrofoam cup-membrane assembly for studying microorganism-root interactions. Appl. Environ. Microbiol. 55:1291–1294.

Harvey, A. E., M. F. Jurgensen, and M. J. Larsen. 1981. Organic reserves: Importance to ectomycorrhizae in forest soils of Western Montana. Forest Sci. 27:442–445.

Hoefte, M., K. Y. Seong, E. Jurkevitch, and W. Verstraete. 1991. Pyoverdin production by the plant growth beneficial *Pseudomonas* strain 7NSK2: Ecological significance in soil. Plant Soil 130:249–257.

Holl, F. B., and C.P. Chanway. 1992. Rhizosphere colonization and seedling growth promotion of lodgepole pine by *Bacillus polymyxa*. Can. J. Microbiol. 38:303–308.

Holl, F. B., C. P. Chanway, R. Turkington, and R. A. Radley. 1988. Response of crested wheatgrass (*Agropyron cristatum* L.), perennial ryegrass (*Lolium perenne*) and white clover (*Trifolium repens* L.) to inoculation with *Bacillus polymyxa*. Soil Biol. Biochem. 20:19–24.

Ingram, R. R., D. A. Klein, and M. J. Trlica. 1985. Responses of microbial components of the rhizosphere to plant management strategies in a semiarid rangeland. Plant Soil 85:65–76.

Jalali, B. L., and I. Jalali. 1991. Mycorrhizae in Plant Disease Control. *In* D. K. Arora, B. Rai, K. G. Mukerji, and G. R. Knudsen (eds.), Handbook of Applied Mycology, Vol. 1. Soil and Plants, pp. 131–154. Dekker, New York.

Jurkevitch, E., Y. Haear, and Y. Chen. 1992. Differential siderophore utilization and iron uptake by soil and rhizosphere bacteria. Appl. Environ. Microbiol. 58:119–124.

Kennedy, A. C., L. F. Elliott, F. L. Young, and C. L. Douglas. 1991. Rhizobacteria suppressive to the weed downy brome. Soil Sci. Soc. Am. J. 55:722–727.

Klein, D. A., B. A. Frederick, M. Biondini, and M. J. Trlica. 1988. Rhizosphere microorganism effects on soluble amino acids, sugars and organic acids in the root zone of *Agropyron cristatum, A. Smithii* and *Bouteloua gracilis.* Plant Soil 110:19–25.

Kloepper, J. W., J. Leong, M. Teintze, and M. N. Schroth. 1980a. *Pseudomonas* siderophores: A mechanism explaining disease-suppressive soils. Curr. Microbiol. 4:317–320.

Kloepper, J. W., J. Leong, M. Teintze, and M. N. Schroth. 1980b. Enhanced plant growth by siderophores produced by plant growth-promoting rhizobacteria. Nature (London) 286:885–886.

Klubek, B. P., L. L. Hendrickson, R. M. Zablotowicz, J. E. Skwara, E. C. Varsa, S. Smith, T. G. Islieb, J. Maya, M. Valdes, F. B. Dazzo, R. L. Todd, and D. D. Walgenback. 1988. Competitiveness of selected *Bradyrhizobium japonicum* strains in midwestern USA soils. Soil Sci. Soc. Am. J. 52:662–666.

Knight, W. G., M. F. Allen, J. J. Jurinak, and L. M. Dudley. 1989. Elevated carbon dioxide and solution phophorus in soil with vesicular–arbuscular mycorrhizal western wheatgrass. Soil Sci. Soc. Am. J. 53:1075–1082.

Kucey, R. M. N., and E. A. Paul. 1982. Carbon flow, photosynthesis, and N_2 fixation in mycorrhizal and nodulated faba beans (*Vicie faba* L.). Sol Biol. Biochem. 14:407–412.

Leyval, C., and J. Verthelin. 1991. Weathering of a mica by roots and rhizospheric microorganisms of pine. Soil Sci. Soc. Am. J. 55:1009–1016.

Lindemann, W. C., D. L. Lindsey, and P. R. Fresquez. 1984. Amendment of mine spoil to increase the number and activity of microorganisms. Soil Sci. Soc. Am. J. 48:574–578.

Martens, R. 1982. Apparatus to study the quantitative relationships between root exudates and microbial populations in the rhizosphere. Soil Biol. Biochem. 14:315–317.

Martin, J. K. 1977. Factors influencing the loss of organic carbon from wheat roots. Soil Biol. Biochem. 9:1–7.

Marx, D. H. 1972. Ectomycorrhizae as biological deterrents to pathogenic root infection. Ann. Rev. Phytopathol. 10:429–454.

Marx, D. W., W. G. Morris, and J. G. Meral. 1978. Growth and ectomycorrhizal development of loblolly pine seedlings on fumigated and nonfumigated nursery soils infested with different fungal symbionts. Forest Sci. 24:193–203.

Materon, L. A., and C. Hagedorn. 1982. Competitiveness of *Rhizobium trifolii* strains associated with red clover (*Trifolium pratense* L.) in Mississippi soils. Appl. Environ. Microbiol. 44:1096–1101.

McClaugherty, C. A., J. D. Aber, and J. M. Melillo. 1984. Decomposition dynamics of fine roots in forested ecosystems. Oikos 42:378–386.

Meharg, A. A., and K. Killham. 1989. Distribution of assimilated carbon within the plant and rhizosphere of *Lolium perenne:* Influence of temperature. Soil Biol. Biochem. 21:487–490.

Merckx R., A. Dijkstra, A. den Hertog, and J. A. van Veen. 1987. Production of root-derived material and associated microbial growth in soil at different nutrient levels. Biol. Fert. Soils 5:126–132.

Merckx, R., J. A. van Ginkel, J. Sinnaeve, and A. Cremers. 1986. Plant induced changes in the rhizosphere of maize and wheat. I. Production and turnover of root-derived material in the rhizosphere of maize and wheat. Plant Soil 96:85–94.

Meyer, J.-M., D. Hohnadel, and F. Hallé. 1989. Cepabactin from *Pseudomonas cepacia.* a new type of siderophore. J. Gen. Microbiol. 135:1479–1487.

Meyer, J. R., and R. G. Linderman. 1986a. Response of subterranean clover to dual inoculation with vesicular–arbuscular mycorrhizal fungi and a plant growth-promoting bacteium. Soil Biol. Biochem. 18:185–190.

Meyer, J. R., and R. G. Linderman. 1986b. Selective influence on populations of rhizoplane bacteria and actinomycetes by mycorrhizas formed by *Glomus fasiculatum*. Soil Biol. Biochem. 18:191–196.

Moawad, H., and B. B. Bohlool. 1984. Competition among *Rhizobium* spp. for nodulation of *Leucaena leucocephala* in two tropical soils. Appl. Environ. Microbiol. 48:5–9.

Mukerji, S., K. G. Mukerji, and D. K. Arora. 1991. Ectomycorrhizae. *In* D. K. Arora, B. Rai, K. G. Mukerji, and G. R. Knudsen (eds.), Handbook of Applied Mycology, Vol. 1. Soil and Plants, pp. 187–215. Dekker, New York.

Nair, M. G., G. R. Safir, and J. O. Siqueira. 1991. Isolation and identification of vesicular–arbuscular mycorrhiza-stimulatory compounds from clover (*Trifolium repens*) roots. Appl. Environ. Microbiol. 57:434–439.

Newman, E. I. 1985. The rhizosphere: Carbon sources and microbial populations. *In* A. E. Fitter (ed.), Ecological Interactions in Soil: Plants, Microbes, and Animals, pp. 107–121. Blackwell Scientific Publications. Boston.

Newman, E. I., A. J. Heap, and R. A. Lawley. 1981. Abundance of mycorrhizas and root surface microorganisms of *Plantago lancelata* in relation to soil and vegetation. A multivariate approach. New Phytol. 89:95–108.

Nietko, K. F., and W. T. Frankenberg, Jr. 1989. Biosynthesis of cytokinins in soil. Soil Sci. Soc. Am. J. 53:735–740.

Norton, J. M., and M. Firestone. 1991. Metabolic status of bacteria and fungi in the rhizosphere of ponderosa pine seedlings. Appl. Environ. Microbiol. 57:1161–1167.

Parke, J. L., R. Moen, A. D. Rovira, and G. D. Bowen. 1986. Soil water flow affects the rhizosphere distribution of seed-borne biological control agent. Soil Biol. Biochem. 18:583–588.

Pasqualini, S., F. Panara, and M. Antonielli. 1992. Acid phosphatase activity in *Pinus pinea—Tuber albidum* ectomycorrhizal association. Can. J. Bot. 70:1377–1383.

Pesmark, M., T. Frejd, and B. Mattiasson. 1990. Purification, characterization, and structure of pseudobactin 589 A, a siderophore from a plant growth promoting *Pseudomonas*. Biochemistry 29:7348–7356.

Read, D. J. 1984. The structure and function of the vegetative mycelium of mycorrhizal roots. *In* D. H. Jennings and A. D. M. Rayner (eds.), The Ecology and Physiology of the Fungal Mycelium, pp. 215–240. Cambridge University Press, New York.

Read, D. J., R. Francis, and R. D. Finlay. 1989. Mycorrhizal mycelia and nutrient cycling in plant communities. *In* A. E. Fitter (ed.), Ecological Interactions in Soil: Plants, Microbe, and Animals, pp. 193–217. Blackwell Scientific Publications. Boston.

Robert, F. M., and E. L. Schmidt. 1983. Population changes and persistence of *Rhizobium phaseoli* in soil and rhizospheres. Appl. Environ. Microbiol. 45:550–556.

Ruark, G. A., F. C. Thornton, A. E. Tiarks, B. G. Lockaby, A. H. Chappelka, and R. S. Meldahl. 1991. Exposing loblolly pine seedlings to acid precipitation and ozone: Effects on soil rhizosphere chemistry. J. Environ. Qual. 20:828–832.

Smith, G. B., and A. G. Wollum, II. 1989. Nodulation of *Glycine max* by six *Bradyrhizobium japonicum* strains with different competitive abilities. Appl. Environ. Microbiol. 55:1957–1962.

Spalding, B., J. M. Duxbury, and E. L. Stone. 1975. Lycopodium fairy rings: Effect on soil respiration and enzymatic activities. Soil Sci. Soc. Am. J. 39:65–70.

Stroo, H. F., and E. M. Jencks. 1985. Effect of sewage sludge on microbial activity in an old, abandoned minesoil. J. Environ. Qual. 14:301–304.

Tate, R. L., III. 1985. Microorganisms, ecosystem disturbance and soil formation processes. *In* R. L. Tate III and D. A. Klein (eds.), Soil Reclamation Processes. Microbiological Analyses and Application, pp. 1–33. Dekker, New York.

Tate, R. L., III. 1987. Soil Organic Matter. Biological and Ecological Effects. John Wiley & Sons, New York. 291 pp.

Tate, R. L., III, R. W. Parmelee, J. G. Ehrenfeld, and L. O'Reilly. 1991. Enzymatic and microbial interactions in response to pitch pine root growth. Soil Sci. Soc. Am. J. 55:998–1440.

Tester, C. F. 1990. Organic amendment effects on physical and chemical properties of a sandy soil. Soil Sci. Soc. Am. J. 54:827–831.

Theodorou, C., and G. D. Bowen. 1987. Germination of basidiospores of mycorrhizal fungi in the rhizosphere of *Pinus radiata* D.Don. new Phytol. 106:217–224.

Thomashow, L. S., and D. M. Weller. 1990. Role of antibiotics and siderophores in biocontrol of take-all disease of wheat. Plant Soil 129:93–99.

Thomashow, L. S., D. M. Weller, R. B. Bonsall, and L. S.Pierson, III. 1990. Production of the antibiotic phenazine-1-carboxylic acid by flurosecent *Pseudomonas* species in the rhizosphere of wheat. Appl. Environ. Microbiol. 56:908–912.

Tien, T. M., M. H. Gaskins, and D. H. Hubbell. 1979. Plant growth substances produced by *Azospirillum brasilense* and their effect on the growth of pearl millet (*Pennisetum americanum* L.) Appl. Environ. Microbiol. 37:1016–1024.

Tisdall, J. M., and J. M. Oades. 1982. Organic matter and water-stable aggregates in soils. J. Soil Sci. 33:141–163.

Turner, S. M., and E. I. Newman. 1984. Growth of bacteria on roots of grasses: Influence of mineral nutrient supply and interactions between species. J. Gen. Microbiol. 130:505–512.

Turner, S. M., E. I. Newman, and R. Campbell. 1985. Microbial population of ryegrass (*Lolium perenne*) root surfaces: Influence of nitrogen and phosphorus supply. Soil Biol. Biochem. 17:711–716.

Utkhede, R. S., and J. E. Rahe. 1980. Biological control of onion white rot. Soil Biol. Biochem. 12:101–104.

van Rensburg, H. J., and B. W. Strijdom. 1985. Effectiveness of *Rhizobium* strains used in inoculants after their introduction into soil. Appl. Environ. Microbiol. 49:127–131.

Vrany, J., V. Vancura, and M. Stanek. 1981. Control of microorganisms in the rhizosphere of wheat by inoculation of seeds with *Pseudomonas putida* and by foliar application of urea. Folia Microbiol. 26:45–51.

Whitford, W. G., K. Stinnett, and J. Anderson. 1988. Decomposition of roots in a Chihuahuan desert ecosystem. Oecologia (Berlin) 75:8–11.

Will, M. E., and D. M. Silvia. 1990. Interaction of rhizosphere bacteria, fertilizer, and vesicular–arbuscula mycorrhizal fungi with sea oats. Appl. Environ. Microbiol. 56:2073–2079.

Wilson, G. W. T., B. A. D. Hetrick, and A. P. Schwab. 1991. Reclamation effects on mycorrhizae and productive capacity of flue gas desulfurization sludge. J. Environ. Qual. 20:777–783.

Wong, J. W. C., and G. E. Ho. 1991. Effects of gypsum and sewage sludge amendment on physical properties of fine bauxite refining residue. Soil Sci. 152:326–332.

8 Introduction to the Biogeochemical Cycles

Those elements and compounds essential for the development of living cells occur in our terrestrial system in a variety of chemical forms—some of which are directly available to cells for energy and biomass synthesis whereas others must be chemically or biologically modified before they can be incorporated into cellular structures. Furthermore, although levels of these compounds existent in any given ecosystem may be variable, total quantities available on our plant to support biotic systems are conserved. Thus **the chemically and biologically catalyzed cycling of these nutrient elements (e.g., nitrogen, sulfur, carbon and phosphorus) between cellular, water-soluble, gaseous, and soil mineral forms in the ecosystem comprises the various biogeochemical cycles.**

The transformation of the central element of any biogeochemical cycle may be considered at a variety of levels of complexity. Depending upon the objective of the exercise, only general nutrient pools may be considered or the complexity of the model may be increased until every organic and inorganic component of the cycle is included in the study. For example, a model of carbon mineralization may be as simple as designating the rates of cycling of carbon atoms between organic carbon (e.g., plant biomass) and the primary mineral pool of this element, carbon dioxide. The complexity may be increased to include various classes of plant biochemicals (e.g., polysaccharides, lignin, proteins), and even these groupings may be broken down further so that the subgroupings of these biochemicals and their components are considered (e.g., cellulose, hemicelluloses, starch, glucose, and a variety of disaccharides and oligosaccharides). Similarly, the complexity of a biogeochemical cycles may vary in relationship to chemical form plus solubility in a soil system (e.g., movement of sulfur-containing compounds between water-soluble forms, soil minerals, and the organic components of living cells). Each of these levels of inclusiveness can be used to define important elemental transformations in our terrestrial system.

The variability in the degree complexity associated with consideration of biogeochemical cycles occurring in soil reveals the diversity of chemical and physical states of the elements (e.g., carbon, nitrogen, sulfur, phosphorus) being studied. Furthermore, the many chemical states present in a microbial habitat varies between ecosystems. The central nutrients of biogeochemical cycles may pass through a variety of transitions within a selected ecosystem of limited dimensions (e.g., a particular soil horizon of interest) or integrity of the nutrient cycles may rely upon inputs from adjacent portions of a more extensively defined ecosystem (e.g., a soil community including aboveground populations plus atmospheric inputs).

In summary, the chemical components of biogeochemical cycle are distinguished by variation in the oxidation state of the central element or nutrient (e.g., for the sulfur cycle, the reduction of S^0 to S^{-2}), in their chemical form (i.e., mineral vs. organic), the degree of oxidation of the compound itself (e.g., the transition from ammonium through nitrite to nitrate), and the cyclical movement of the elements of interest between these chemical pools. A central component for all cycles, as they have been traditionally presented, is the biological community, which is generally viewed from the aspect of the aboveground plant populations.

It must be stressed that the realm of biogeochemical cycles is much larger than just the **transitions of chemical forms necessary for sustenance of plant, animal, and microbial communities and the biological or chemical reactions involved in these processes (nutrient cycles).** Nutrient cycles entail a limited portion of the subject area encompassed by biogeochemical cycles. A consideration of biogeochemical cycles should involve more of a total ecosystem outlook than is included in evaluating nutrient cycles; that is, the definition of biogeochemical cycles includes a consideration of the totality of biological, chemical, and physical interactions associated with the movement of nutrients between organic and inorganic pools and the variety of chemical reservoirs existent within an ecosystem.

This definition of the complete realm of the biogeochemical cycles emphasizes the point that studying these chemical transformations has real implications beyond the world of plant science and general biology; that is, studying these cyclic transformations of life-sustaining nutrients is also essential for such broad-based disciplines as general ecology or environmental science. For example, consideration of biogeochemical cycles in disturbed environments is essential in optimizing anthropogenic activities for ecosystem sustenance as well as in developing management plans for restoring or reclaiming damaged systems, that is, developing appropriate ecosystem stewardship procedures. Therefore, this chapter provides a basic understanding of biogeochemical cycles by evaluating the general properties of biogeochemical cycles and explaining the interconnectedness of these processes with the goal of maintaining an appreciation of the importance of these processes to overall ecosystem function and maintenance.

8.1. MODELING OF BIOGEOCHEMICAL CYCLES

As suggested by the foregoing analysis, a study of biogeochemical cycles requires a means of assessing a wide array of complex interactions within a constantly varying environment. Some details of the processes involved may be studied in isolation, but a complete picture of the cycles requires development of an overall framework. A means of accomplishing the objective of developing an overarching framework for studying biogeochemical cycles is to assemble the data into models. These models may be used to elucidate gaps in knowledge, provide hypotheses for future study, as well as provide a means of quantifying nutrient movement within the cycles and the impact of environmental modification on these processes. Initial modeling activity —or perhaps modeling activity at its simplest—involves developing conceptual models. These models provide a basis for designing experiments that yield the data necessary for developing the more complex and predictive mathematical models.

8.1.1. Development and Utility of Conceptual Models

A primary objective of studying nutrient flow in an ecosystem is to elucidate a) the chemical forms of a nutrient element (e.g., carbon as carbon dioxide and organic matter; nitrogen as mineral and organic forms; phosphorus in soil minerals, water-soluble, and organic forms) in an ecosystem of interest, b) the quantity of nutrients retained in these compounds (i.e., the size of the nutrient pool), and c) the rate of flux between these various nutrient pools. Once the identify of the various participants in biogeochemical cycles (even in lieu of determining all the pertinent intermediates) is known, conceptual models may be assembled. These models are most easily presented as a series of boxes representative of the primary nutrient pools with arrows connecting the boxes to indicate the directional transformation (i.e., nutrient flux) of the cycle components (e.g., see Fig. 8.1 for a depiction of the carbon cycle). Frequently, to augment the information contained in a conceptual model, the type of arrow connecting the various reservoirs (e.g., dotted, size differential, or boldface) may be varied to suggest differences in the magnitude of the process depicted, but generally a conceptual model is limited to indicating the nutrient reservoir and the direction of nutrient flow. Distinction between relatively stable reservoirs, such as soil minerals or peat deposits, and the more ephemeral cycle components exemplified by simple organic compounds or mineral nitrogen forms is minimal. A more quantitative representation of biogeochemical cycles or the individual reactions comprising them is entailed in extension of conceptual models into mathematical models as described in section 8.1.2.

Characteristically, conceptual models of biogeochemical cycles emphasize conservation of subject elements in a variety of chemical forms. The elements of interest (i.e., carbon, nitrogen, phosphorus, or sulfur as well as a variety of trace metals) are found to move between a variety of chemical (organic vs. inorganic) and/or oxidation states at predictable rates. Basic chemical components of plant and animal biomass are transformed from biomass-unavailable form to compounds that are readily incorporated into living cells. For example, the reasonably inert compound, dinitrogen N_2, is reduced by soil microbes to ammonium, which, once it is released by the microbial cell, can be incorporated into plant tissue. With the death of the plant tissue, the soil microbial community may return the nitrogen to the atmospheric pool of dinitrogen. This conceptualization of biogeochemical cycles is plant-centered (or perhaps life-centered, when microbial and animal biomass are also viewed as major nutrient recipients). Other models that could be centered on movement of elements and compounds to and from atmospheric pools, soil minerals, humus, or any other primary ecosystem component may be presented.

Conceptual models by their very design may support a biased view of biogeo-

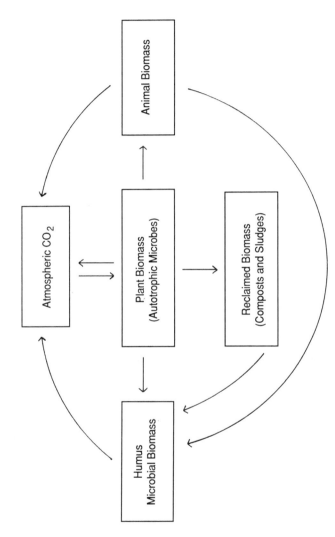

Fig. 8.1. A generalized carbon cycle conceptual model with biological and atmospheric aspects emphasized.

chemical cycles in that most representations are limited to designation of the basic chemical and biological forms of the cycle components. To provide a more complete model for soil microbiological or environmental science considerations, designation of the path for movement of the elements of interest through various ecosystem components (i.e., soil minerals, atmospheric gases, and biotic entities) is useful. A sound ecological approach to the study of biogeochemical cycles must include a) an appreciation of interactions between the compounds transformed, the biological community catalyzing the processes, and the physical environment in which the reactions occur and which is itself modified during the process, and b) the contribution of spontaneous chemical reactions to the overall nutrient-generation rates.

In biogeochemical cycling, the physical, chemical, and biological worlds interact in a manner that leads to an enhancement of ecosystem sustainability and controls successional development of the ecosystem. Thus, although conceptual models of biogeochemical cycles are usually presented as "boxes connected by arrows," it must be remembered that the boxes represent real-world nutrient reservoirs, such as plant biomass, and that the magnitude of the vectors designated as arrows is controlled not only by life-generated processes but also by the properties of the environment in which the plant, microbe, or animal is located.

Thus, when collecting data for the development of conceptual models, we should consider biotic as well as abiotic interactions. For example, mineral nitrogen availability determines productivity and nature of the aboveground community. This nitrogen can be derived through nitrogen fixation (i.e., the total amount of nitrogen available to the plants may increase through input from external sources), or the nitrogen contained in biomass may be recycled into nascent biomass through mineralization of deceased biomass and soil organic matter pools (i.e., the total amount of nitrogen contained within the soil system remains reasonably constant). In either of these situations, the productivity of the aboveground community is limited by the capacity of the soil microbial community (nitrogen fixers or nitrogen mineralizers) to provide a pool of nitrate and ammonium ions for incorporation into plant biomass. The activity of this soil microbial community is in turn controlled by the quantities of energy and nutrients provided by the aboveground community in the form of photosynthate, as well as by the physical and chemical environment. Anoxic conditions may favor loss of fixed nitrogen from the system through biological denitrification reactions, whereas chemical denitrification is accentuated in acidic soils. Furthermore, these life-limiting environmental properties are continually being modified by the physical presence and the biological products of the plant, animal, and microbial communities (see Tate, 1987). For example, improvements in the soil physical environment through enhancement of soil aggregate structure by growth of the microbial community augments plant biomass production, which may further stimulate microbial growth through augmented inputs of fixed carbon. This example could be expanded further by altering soil pH through growth of the plant biomass mineralizers and the resultant impact on solubility of other plant nutrients (sulfur and a variety of mineral nutrients).

Therefore, the essential steps in developing and refining a conceptual model of a particular biogeochemical process that occurs in soil involve elucidation of the forms of the nutrient element of interest in the system, perhaps tracing the movement of elements between these forms using such isotopes as ^{14}C or ^{15}N, and determination of the physical and chemical properties of the system controlling the relative magnitude of the various chemical pools. This process of model building is based primarily on characterization of the system and development of a reasonable understanding of the relationship between the various nutrient reservoirs occurring therein. These studies may entail determination of biological trophic interactions (e.g., Santos et al., 1981; Santos and Whitford, 1981; Santos et al., 1984), determination of the impact of management and climate changes on ecosystem function (e.g., Allen, 1990; Pastor and Post, 1988; Tabatabai and Chae, 1991; Walters et al., 1992), evaluation of impact of properties of ecosystem inputs in their mineralization (e.g., Meentemeyer, 1978; Fresquez et al., 1990), as well as examination of native ecosystem processes (e.g., Parker et al., 1983).

8.1.2. Mathematical Modeling of Biogeochemical Cycles

A conceptual model provides an indication of the "design" of the biogeochemical cycles in an ecosystem, but revelation of the quantitative relationships is minimal. Quantities of the various cycle components and their changes with time can be measured for each system studied, but the predictive qualities of such data are limited. More generally, experimental objectives should include development of concepts that allow us to apply the relationships derived for one environmental system to a variety of related situations. The more detailed the experimental design (i.e., inclusiveness of variables and relationships evaluated), the broader the applicability of the results. Conceptual models are commonly used to design experiments that incorporate the complexity of interactions involved. Extending the data to other ecosystem usually requires development of mathematical models.

A first attempt at understanding nutrient-cycling processes may involve analyzing the quantities of various compounds present within the soil site of interest, but it soon becomes clear that a detailed understanding of the intricacies of the reactions are obscured by site heterogeneity and the shear number of potential site properties that may alter the reaction rate and its final outcome. Thus, a mathematical approach at data summarization and interpretation becomes desirable.

The first step toward developing more generally applicable mathematical relationships may stem from statistical analysis procedures. A simple model may be yielded through a linear or nonlinear regression analysis of the data. Such models may describe the particular site under study, but, most commonly, their utility in deriving relationships and principles that may be extrapolated to other ecosystems is minimal. This conclusion stems from the fact that the equations derived are mathematical relationships fitted to specific data; that is, the relationship may be forced to fit a straight line, parabola, or hyperbola, for example.

Equations that apply more generally to biological, chemical, or physical phenomena may be adapted to environmental data to provide a more inclusive description of the phenomena of interest. Such models may be reasonably simple. For example, in an environment defined by a soil sample placed in a test tube, the transition of a reactant to a product as catalyzed by a soil enzymatic activity may be easily depicted by a first-order equation or by the Michaelis–Menten relationship (see chapter 4). As the complexity of the system increases (e.g., interest in the quantities of the reactant consumed within the profile of a conifer forest) and the variety of interactions with soil properties and climatic variations expands, inclusion of known relationships, such as the Q_{10} equation or Arrhenius relationship for temperature impact on reaction rates, may be necessary to increase model utility. Furthermore, in the complex, heterogeneous environment of soil, effective chemical concentrations for a reaction may be defined by properties other than total substrate present. For example, reaction rates may be dependent upon soil parameters such as the rate of dissolution of the substrate (Stucki and Alexander, 1987) or diffusion of the enzyme substrate into the microsite wherein the organisms reside (Scow and Hutson, 1992; Scow and Alexander, 1992). Also, in an ecosystem adapting to a perturbation, the rate of occurrence of a process may be proportional to the growth or death of the microbial populations producing the requisite enzymes (i.e., the reaction may be best described by a Monod equation; see chapter 3). Hence, a variety of mathematical relationships descriptive of not only biological but also chemical and physical interactions are combined to provide a mathematical relationship that, at least in part, describes the extent of a natural process and predicts rate changes with perturbation of the system. Because of the general applicability of the mathematical relationships (e.g., Michaelis–Menten equations provide a reasonable model of the kinetics of the vast majority of enzymatic processes), the biogeochemical models based on them provide reasonable descriptors of process rates in a diversity of situations similar to the individual soil site that may have been studied to develop the original mathematical model. See Tate (1987) for a description of model development and application.

The following studies exemplify the applicability of mathematical models of varying complexity to biogeochemical cycling:

- Soil properties controlling specific portions of a biogeochemical cycle as shown by study of the impact of distribution of aerobic and anaerobic microsites on denitrification rates (McConnaughey and Bouldin, 1985a, 1985b; McConnaughey et al., 1985) as well as evaluation of nitrogen mineralization (Deans et al., 1986), sulfur and nitrogen mineralization (Ellert and Bettany, 1988), and carbon mineralization (Stroo et al., 1989) in a variety of ecosystems.

- Soil process response to global changes as revealed by alteration of carbon dioxide emissions from soil due to global warming (Jenkinson et al., 1991).

- Nutrient cycling within a particular ecosystem as exemplified by studies of carbon, nitrogen, phosphorus, and sulfur cycling within a grassland as impacted by cultivation (Parton et al., 1988), variation in decomposer populations, and substrated in grasslands (Hunt, 1977).

- Movement of nutrients into or from a portion of a nutrient cycle such as microbial biomass in nutrient cycles (van Veen et al., 1984) or soil organic matter.

• From an ecosystem management view in evaluating the impact of soil steward-ship on nutrient cycling as exemplified by studying the erosion impact on plant phosphorus (Jones et al., 1984).

In each of these situations, varying combinations of mathematical relationships that describe biological transformations, pH variation, temperature effects on biological processes, adsorption processes, plant composition, and so on, were combined to provide a mathematical representation of nutrient movement between various bio-geochemical nutrient pools.

8.2. SPECIFIC MODELS OF BIOGEOCHEMICAL CYCLES AND THEIR APPLICATION

Basic conceptual models of the primary nutrient cycles (carbon, nitrogen, phospho-rus, and sulfur) are depicted in Fig. 8.1 through 8.4, respectively. These models are designed to emphasize the basic properties of biogeochemical cycles. The elements are transferred between organic and inorganic forms (all cycles) as well as through a variety of oxidation states (nitrogen and sulfur cycles). For example, note that the oxidation state of nitrogen varies from -3 in the ammonium ion to $+5$ in the nitrate ion, whereas in the sulfur cycle the oxidation state of the sulfur atom varies from -2 (in H_2S) to $+6$ (in H_2SO_4). Similarly, a variety of soil and atmospheric inorganic compounds (e.g., nitrate, ammonium, dinitrogen, elemental sulfur, phosphate rock, and sulfur-bearing minerals) and plant, animal, and microbial biomass organic components constitute primary nutrient pools in these models of the biogeochemical cycles.

Biogeochemical cycles are usually examined from the view of their contribution to higher plant productivity, but it must also be observed that a number of intermedi-ates in the various cycles may have a considerable impact on the integrity of the terrestrial ecosystem. Positive as well as negative implications can be discerned. Perturbations of nutrient cycling could result in overproduction of substances, which results in large-scale declines in system quality, altered biomass productivity, or modified terrestrial ecosystem sustainability. For example, variations in carbon dioxide evolution due to oxidation of soil organic matter resulting from changing land use may affect atmospheric carbon dioxide levels and plant biomass productivi-ty (e.g., see Adams et al., 1990; Allen, 1990; Dahlman et al., 1985; Reddy et al., 1989). Nitrous oxide, a key intermediate in denitrification, is also a greenhouse gas (e.g., see Firestone, 1982; Mosier et al., 1991). Nitrate and phosphate are often significant ground- and surface water contaminants, whereas sulfate and sulfur diox-ide are major components of acid rain and acid mine drainage problems.

In contrast to these associations of biogeochemical cycles with negative environ-mental situations, a number of benefits may also result from soil management to optimize biogeochemical cycling. For example, soil organic nitrogen and carbon levels are generally enhanced when conservative tillage practices are implemented on soils previously managed with conventional tillage methods (e.g., see Blevins et al., 1984; Cambardella and Elliott, 1992; Doran, 1980). Increases in soil organic

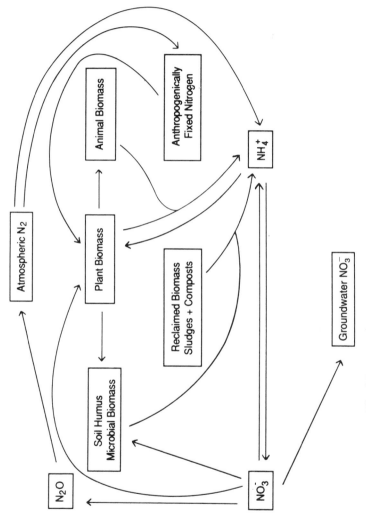

Fig. 8.2. A generalized conceptual model of the nitrogen cycle.

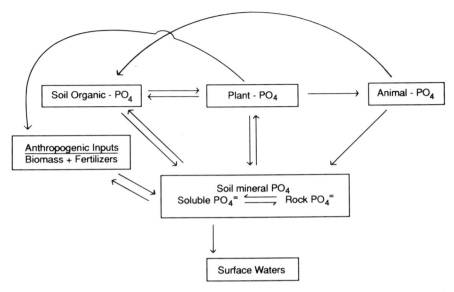

Fig. 8.3. A conceptual model of the soil phosphorus cycle with soil mineral reservoirs as the central component.

matter reserves could alter the rates of accumulation of atmospheric carbon dioxide. Furthermore, sulfate reduction to sulfide may be exploited to reduce the difficulties associated with acid mine drainage impacted soils (see chapter 14), and denitrification may be stimulated to reduce ground- and surface water contamination within nitrate (see chapter 13).

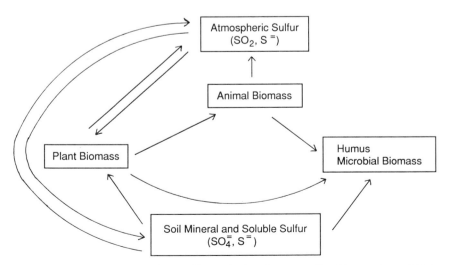

Fig. 8.4. A conceptual model of the sulfur cycle showing generalized interactions of biological resources with soil and atmospheric sources.

The penultimate impact of the soil microbial community on the total ecosystem is the product of their contributions to the orderly transition of the chemicals of life through their various oxidation states (i.e., their participation in the biogeochemical cycles). Optimizing ecosystem conditions for nutrient cycling results in a sustained system in which nutrients are cycled with minimal losses of essential cycle components from the system or movement of potentially troublesome intermediates into ecosystems where their presence may be less appropriate (e.g., in lake eutrophication or groundwater contamination). That is, microorganisms are instrumental in such essential ecosystem-sustaining activities as the conversion of plant-unavailable forms (e.g., dinitrogen N_2) to chemical entities that may be incorporated into plant biomass (nitrate NO_3^- or ammonium NH_4^+) as well as the return of organic components to inorganic forms that may be recycled into plant biomass (carbon or nitrogen mineralization). Loss of plant biomass nutrients phosphate or nitrate negatively impacts the system by reducing its productivity and could damage adjacent surface waters. Balance between mineral nitrogen production and consumption minimizes surface- and groundwater quality degradation.

8.2.1. The Environmental Connection

To appreciate the ramifications of any biogeochemical process, we must understand the implications of the synergism between the soil biota and their environment. Both the physical environment and the biological entities growing therein are modified by the products of biological growth. This synergism is absolutely essential for development of a sustainable ecosystem. For example, in a managed site, such as would be involved in a reclamation process, the sustainability of the developing ecosystem relies upon the capability of the microbial community to modify an initially hostile habitat to one more conducive for their own growth as well as that of the higher plant community. In highly damaged sites or in situations where rapid establishment of an aesthetic aboveground community is desirable or required, anthropogenic intervention, such as pH modification by liming or amendment with organic material to improve the initial conditions, may be desirable. In the absence of such site management, the longevity of the biological populations essential for establishing a viable aboveground and belowground community could easily be truncated. A typical scenario follows: To provide a rapid start to the soil microbial community and to sustain aboveground populations until the soil physical conditions are modified to those more conducive to plant developments, the soils may be amended with organic material (e.g., composts and sludges), fertilizers and limestone to stimulate indigenous microbial community development. Initially, this ecosystem development would be supported by external nutrient inputs through a high level of anthropogenic intervention. A stable situation requires the slow evolution of soil populations and modification of the soil physical properties to a point wherein nutrients required for continued biomass productions are produced internally. This transition from artificially sustained to a natural system requires such physical–biological synergistic interactions as the enhancement of soil physical structure through aggregation resulting from microbially synthesized polysaccharides (see chapter 1), as well as microbial amelioration of soil acidity in sulfate-impacted sites. See Tate (1985) for a more complete discussion of the importance of managing soil microbial populations for mineland reclamation. This system development can be termed quasi synergistic in that a feedback interaction develops between the microbial community and its physical environment. As the complexity of microbial community increases, the conditions of their habitat improves. This alteration in turn supports augmented microbial growth. This cycle is repeated until ultimately a steady state develops where an optimized balance between the microbes and their physical and chemical environment has been achieved.

These belowground processes not only stabilize the soil microbial populations, but also improve conditions for plant population productivity and longevity. For example, soil aggregate development results in augmented soil air diffusion and water infiltration rates. Not only is the soil microbial community stimulated by this habitat improvement but plant root growth and aboveground biomass synthesis are also increased. Because the microbial community is limited, to a large degree, by the fixed-carbon supply and this fixed carbon in nonmanaged ecosystems is the direct product of plant biomass production, this soil structure augmentation improves the conditions necessary for microbial community growth. Thus, the biological aspect of that portion of the carbon cycle represented by microbial mineralization has modified the physical environment in such a manner as to increase carbon fixation by the plants, which in turn allows for an increase in the soil microbial populations dependent on this carbon generation. This portion of the ecosystem development is a true synergistic interaction.

The interdependence of higher plants, microbes, and their physical environment (not simply from a fixed description, but more from the view that these interactions are resulting in a continuous evolution of the ecosystem) is demonstrated by this simple example of site management. Similar interactive effects can be explored by evaluating the biological and chemical interactions involved with acid mine drainage-impacted soil. For this analysis, the pH of a soil is considered to have been lowered through the input of sulfate contained in acid mine drainage to the point that the indigenous acid-sensitive microbial population has ceased to function. We can manipulate microbial activity to reduce the toxic conditions so that the concentrations of the toxic acid (in this case sulfate) may be reduced sufficiently for biological activity to be resumed. Such site management may include flooding of the soil to encourage development of sulfate-reducing bacteria populations (see chapter 13) and the subsequent removal of the sulfide from the system through precipitation with heavy metals, commonly iron, which are generally contained in the acid mine drainage. The ameliorative processes eventually result in reestablishing biogeochemical cycling of nutrients. (See Mills, 1985, for a further discussion of managing acid mine drainage.)

These examples demonstrate how optimizing microbial community development modifies the physical and chemical environment due to the action of the biological components. Traditionally, this "optimization" was thought to entail improvements in community structure, which translates, from an anthropogenic viewpoint, into an aesthetic aboveground community. But ecosystem evolution may also involve alterations in the chemical environment that cause declines in aboveground productivity.

Drainage of acid peats (either artificially or anthropogenically induced) stimulates the microbially catalyzed oxidation of the metallic sulfides contained therein into sulfate. The resultant declines in soil pH may limit aboveground plant productivity but stimulate the evolution of an acidophilic microbial community.

The biological cycling of nutrients in soils causes both permanent (long-term) and transient (seasonal) alterations of the physical environment. Solubilization of minerals and their subsequent movement into plant tissue or even leaching from the soil site results in an irreversible change in soil mineralogy. In contrast, more ephemeral changes in soil components are associated with the interaction between aboveground plant community and the soil microbes. For example, total system fixed-nitrogen loads may be increased through the activity of nitrogen-fixing bacteria. This nutrient augmentation results in a general improvement in soil fertility, but the effect may be limited chronologically should the newly incorporated fixed nitrogen be removed from the site (e.g., harvesting of crops—anthropogenically or by grazing animals), leached from the system in percolating waters, or denitrified.

In the preceding analyses, microbial and plant communities were shown to benefit from biogeochemical cycles and an added benefit resulted from modifying the physical environment by these organisms. It must be stressed that no volition on the part of the communities or individual organisms is involved in these adaptations. Beneficial alterations of the organism's surroundings are purely fortuitous. A developing situation is created wherein the microbial communities alter their habitat, which in turn results in the selection of more effective microbial populations. This interaction produces a continuously evolving ecosystem.

8.2.2. Interconnectedness of Biogeochemical Cycle Processes

Conceptually, the most aesthetic means of evaluating biogeochemical cycles is to view them as independent entities, as has been the historical practice. Yet, to some degree, this presents an impediment to developing an absolute appreciation of the intricacies of nutrient cycling in the soil system. Central to all nutrient cycles are the various ecosystem components composed of organic matter—a variety of biomass pools plus intermediates in the mineralization of these components and soil humic substances. This organic-matter-centered concept of biogeochemical cycling is depicted in Fig. 8.5, which shows the assembly of the components of biomass and in Fig. 8.6, which represents the disassembly or mineralization of organic substances to the primary mineral pools.

The pivotal role of organic matter in biogeochemical cycles is derived primarily from the fact that many of the organic compounds contain a variety of elements— carbon, oxygen, nitrogen, phosphorus, sulfur. A conceptual model depicting this interaction is shown in Fig. 8.7. Movement of the primary components of biomass can be envisioned to be traveling along a highway that converges on a "square" consisting of living cells. Some organic substances are lost from the biomass and exit the square as biomass products (e.g., root exudates, extracellular enzymes) but most enter "decomposition lane" upon demise of the cell. Some of the organic substances take a detour through the neighborhood of humic substances, but eventually

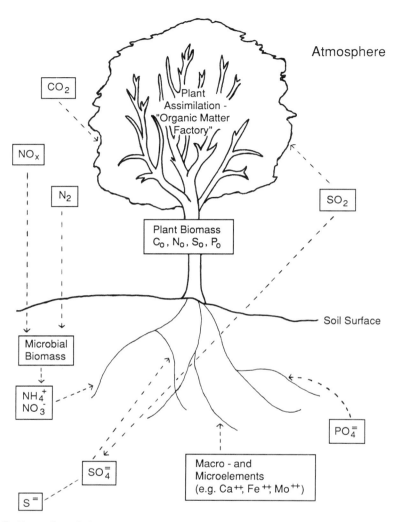

Fig. 8.5. Examples of the central role of plant biomass in sequestering various nutrient elements. (The subscript o refers to various organic pools of the individual element.)

all enter a square consisting of mixed composition organic compounds (e.g., amino acids, nucleotides, carbohydrates, phytin). Continued mineralization of these products yields the original mineral components.

A practical difficulty resulting from this interaction of the biogeochemical cycles is the designation of a particular mineralization process as being driven by the needs of an individual nutrient cycle. For example, amino acid mineralization in soil can be easily quantified. Major products of the process are cellular energy, carbon dioxide, and ammonium. The question is whether the ammonium is produced because of a requirement of the decomposer population for ammonium to synthesize new biomass or whether the mineral nitrogen was produced as a by-product of the

Fig. 8.6. Dissambly of plant organic matter into various primary inorganic nutrient pools. (The subscript o refers to various organic pools of the individual element.)

carbon and energy recovery processes. In fact, mineralization of carbon and nitrogen are sufficiently linked in soil that assessment of carbon dioxide evolution rates has been suggested as a means of estimating net nitrogen mineralization (Gilmour et al., 1985). It is easy to produce models that describe various nutrient cycles but difficult to designate particular processes as being primarily driven by needs of any specific nutrient cycle.

8.3. BIOGEOCHEMICAL CYCLES AS SOURCES OF PLANT NUTRIENTS FOR ECOSYSTEM MAINTENANCE

An ecosystem is generally appreciated and classified by the properties of the above-ground plant and perhaps, to some degree, the animal community, but that portion of

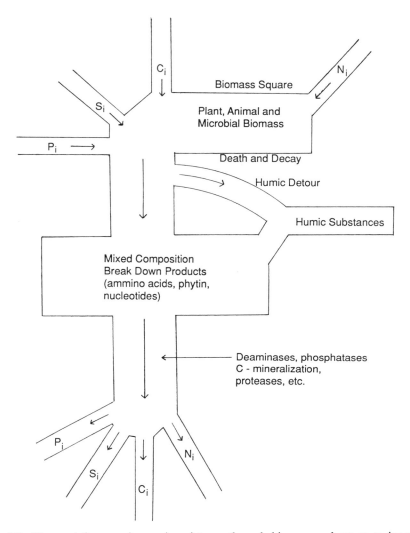

Fig. 8.7. The road from an inorganic existence through biomass and return to inorganic forms. (The subscript *i* refers to inorganic forms of the element.)

the system cannot function without an active soil biological community. A good way of appreciating the interactive nature of the above- and belowground portions of the ecosystem is to examine the nutrient-cycling interactions between the two regions. Biomass nutrients enter soil through rainfall, dry deposition (e.g., sulfate, nitrate, and minerals contained in dusts), atmospheric gases (e.g., sulfur dioxide), anthropogenic sources (e.g., soil amendments), and fixation (e.g., dinitrogen, carbon dioxide). Primary sources of the basic building-blocks of organic matter are divided between the atmosphere (carbon dioxide and dinitrogen) and soil (phosphate or sulfate-bearing soil minerals). Sulfur sources could be considered to be divided between soils (e.g., sulfur-bearing minerals) and atmospheric sources (e.g., sulfur dioxide generated by power generation and, to some extent, automobile exhausts). With the activities directed toward alleviating air pollution, especially from automobiles and power plants, the latter source of sulfur compounds is being reduced resulting in a need to consider soil mineral and organic-matter-derived sulfur inputs for plant biomass productivity.

As an ecosystem matures (approaches equilibrium), reliance on external nutrient inputs declines, and the majority of the materials necessary for biomass productivity are produced by mineralization of resident organic matter pools. This situation is most clearly seen with nitrogen cycling in a mature forest. In a climax forest, the vast majority of the nitrogen is contained in soil organic matter and plant biomass nitrogen pools (e.g., Tate, 1987). In these situations, nutrient availability to the plant community relies almost totally on the activity of the soil microbial community.

In these ecosystems receiving minimal inputs of nutrients, a truncated nutrient cycle can be said to exist. In these situations, the various nutrients are cycled between organic and mineral nutrient pools. Organic nutrient resources in such systems are composed of plant biomass and photosynthate entering the soil and native soil organic matter. These organic matter pools can be separated into a) those substances separated on basis of solubility (humic acids, humin, and fulvic acids) and b) recognizable biochemical components of living cells (i.e., polysaccharides, amino acids, nucleotides). The latter substances are usually classed as readily available organic matter (See chapter 9 for a discussion of the importance of this reservoir of organic material in ecosystem productivity.) Readily available organic matter is that portion of the soil organic matter reservoir that is easily metabolized by the soil microbial community. It is distinguished by its short residence time (generally less than 1–2 years) and the fact that it provides essentially all the metabolic energy for the soil microbial community.

Ecosystems may also derive a major portion of their energy from externally supplied organic substances. These situations are usually the product of anthropogenic inputs—sites managed for biomass productivity waste disposal sites. Specific examples are intensively cultivated agricultural soils, where crop debris and anthropogenically produced fertilizers provide the bulk of plant nutrients, plus soils impacted with varying inputs of sludges and composted materials. Soil amendment with these societal residues may be significant plant nutrient sources (e.g., see Boyle and Paul, 1989; Geiger et al., 1992; Holland and Coleman, 1987).

8.4. GENERAL PROCESSES AND PARTICIPANTS IN BIOGEOCHEMICAL CYCLES

The soil biological community mobilizes minerals, mineralizes organic substances, and immobilizes nutrients directly through a combination of oxidation and reduction reactions or indirectly through environmental modification. Direct reactions may involve alteration of the valence state of the particular element of interest as well as oxidation or reduction of the compound containing the element. For example, energy recover and mineralization of carbon-based compounds involves the oxidation (electron removal) of the carbon compounds to carbon dioxide. Similarly, ferrous iron may provide energy for microbial growth by removal of electrons to produce ferric iron. Similar alterations of the valence state as well as increases in the oxidation state of the total compound are associated with oxidation of ammonium to nitrate and sulfide to sulfate. In each of these instances, the energy yielded by the electron transfer yields increases in microbial biomass.

Lest a false impression be created regarding the direction of the oxidation–reduction processes in soil, note that a variety of reductive transformations are essential for biogeochemical cycle integrity. For example, while organic compounds are oxidized by denitrifying bacteria, the electrons so generated are used to reduce nitrate to dinitrogen. Similarly, sulfate may also serve as an electron acceptor—producing hydrogen sulfide among the products—along with the oxidation of a variety of organic carbon compounds.

Minerals may also be solubolized within the ecosystem by less direct means than direct alteration of their redox state by actively growing microorganisms. For example, chelators solubolize metals, and modification of the microsite pH may result in solubolization of a variety of cations. Acidification results in solubilization of metals such as aluminum as well as dissolution of carbonate rocks, thereby freeing associated cations.

The primary mediators of these processes are the soil bacterial, actinomycete, and fungal communities. Most research has historically been associated with these groups of organisms because they are responsible for catalyzing the bulk of the reactions involved in biogeochemical cycling, but recall from chapter 7 that the activity of these organisms in soil is controlled to a large degree by their interactions with soil protozoa, nematodes, microarthropods, as well as a variety of soil animals (earthworms, ants, millipedes, and insects). Also, although the contribution of these latter biological groupings may be less than those of the primary decomposers, the higher members of the soil community also participate in mineralization of organic matter through the respiratory processes necessary to maintain their own biomass.

The interactions of protozoa and higher animals that live in soil are key to mineralization and cycling of nutrients through a variety of means.

- These organisms are instrumental in the mixing of organic matter within the soil profile, thereby augmenting the contact between the plant residues and the decomposer populations. Key participants in these processes include earthworms, ants, termites, and even, to some extent burrowing animals.
- Soil animals are involved in inoculation of plant litter with propagules of decomposers. Movement of ants, termites, and other insects between soil nests and surface litter ensures contact of active microbial cells with plant debris.
- The soil animals indirectly stimulate decomposition by modifying soil properties. For example, formation of earthworm casts and ant hills result in increased aeration.

- The soil animals may increase the surface area of the organic debris indirectly by moving through dry deposits of the plant biomass as well as directly by chewing and tearing the plant parts. Increased surface area augments the potential contact with decomposer populations (see chapter 5).
- As previously indicated, the organisms may mineralize the plant biomass directly. For example, the woodlouse (*Porcello scaber*) feeds on decaying pine needles (Soma and Saito, 1983).
- Trophic interactions described in chapter 7 also stimulate decomposition. Interactions between the primary decomposers (bacteria and fungi) with population-controlling nematodes, protozoa, and mites are key in maintaining active primary feeder populations.

These soil populations, primary as well as secondary decomposers, derive energy, reducing power, and the ability to grow in anaerobic and anoxic conditions from the transformations of organic and mineral intermediates of the various nutrient cycles. The incidental transfer of products of the decomposer populations to the aboveground populations, plants directly and animals indirectly, is a by-product of their activity, which produces a stable, developing ecosystem.

8.5. MEASUREMENT OF BIOGEOCHEMICAL PROCESSES: WHAT DATA ARE MEANINGFUL?

Design of field and laboratory experiments for biogeochemical process assessment is complicated by the generally prevalent desire to extrapolate the data from soil samples to an ecosystem-wide or even terrestrial view. Questions impinging on validity of data extrapolations include the inclusiveness (or appropriateness) of the data collected and the selection of representative soil samples. Furthermore, the manipulation of the soil sample (i.e., storage conditions, mixing, air drying) subsequent to collection may preserve or alter the relationship of process rates occurring in the sample from those occurring in the field.

8.5.1. Assessment of Biological Activities Associated with Biogeochemical Cycling

As indicated in chapter 2, a variety of methods are available for assessing biological activity in soil samples. Furthermore, a multitude of assays of enzymatic activities, which are useful in estimating transformation rates associated with various biogeochemical processes (e.g., dehydrogenase, phosphatase, as well as assessment of the oxidation rates of a variety of carbonaceous substrates, such as glucose and aromatic compounds), have been adapted for use with soil samples. Applicability of the data derived from these studies relates to inherent limitations of the procedures themselves plus difficulties associated with providing a meaningful extrapolation of the results beyond the individual soil sample studied.

Problems with quantifying biological activities in soil in order to determine nutrient-cycling potential and rates can be exemplified by evaluating the applicability of various measures of soil microbial biomass. Several methods have been described for estimating microbial biomass levels and activity (e.g., respiratory activity [oxygen uptake or carbon dioxide evolution plus dehydrogenase], chloroform fumigation, ATP measurement, pcr analysis, and direct examination). See chapter 2 for a discussion of the basis for each of the methods. Before using each procedure we need to consider the intrinsic limitations of the assay (i.e., what is and is not measured?) plus the variability of microbial biomass densities between soil samples and within a soil site.

For any scientific procedure, the researcher must be cognizant of what the assay is truly measuring. For example, it is easy to postulate that carbon dioxide evolution rates from soil are proportional to the activity of the decomposer population in that soil sample, without realizing that carbon dioxide is also produced by those organisms feeding upon the decomposers plus any plant root tissue in the sample. Thus, such data could yield an overestimation of carbon mineralization in an ecosystem. Similarly, we could easily think that the total active soil microbial biomass is related to carbon mineralization potential in a soil sample, but microbial biomass assessment techniques—chloroform fumigation, ATP measurement, and direct counts—all result in the quantification of both active and inactive microbial propagules.

8.5.2. Soil-Sampling Aspects of Assessment of Biogeochemical Cycling Rates

Data interpretation is further complicated by the fact that the assays are generally conducted in the laboratory. Assuming that a representative field soil sample may be collected, data yielded from analysis of the soil sample in the laboratory at some distant time may not reflect actual population densities or activities in the field. A variety of changes in the soil properties generated by removing the soil sample from its native site, its storage conditions, and preparation of the soil for study, may alter the biological activity occurring therein. For example, collecting the soil sample as well as any sieving or mixing of the soil that occurs in the laboratory releases native soil organic matter trapped within soil aggregates. At least a portion of this organic matter can be metabolized by the microbial community. Thus, the biomass and the enzymes requisite for catabolism of the organic matter may be augmented above levels existent in the field. This complication is increased should storage of the sample be necessary. Even at refrigerator temperatures, significant growth of the microbial populations can occur. Some enzymatic activities are preserved at reduced temperatures, whereas other activities are lost. (See Tate, 1987, for a more complete discussion of this topic.) Hence, for any measure of biological activity in a field sample, control experiments must be conducted to determine stability of the variable of interest.

Extrapolation of laboratory-generated data relies upon the initial soil sample being representative of the field situation. This prerequisite for data analysis is essential in studies of biogeochemical cycles. With few exceptions, laboratory assays are conducted with soil sample sizes ranging from less than 1 g to perhaps 100 g at most. As indicated in chapter 1, horizontal and vertical variation of microbial populations in soils as well as the containment of microbial populations within discrete microsites within the soil matrix makes collection of manageable soil samples that represent an entire field or ecosystem site highly difficult. Even if care is taken to produce a series of soil samples that contain a balance of soil mass between pertinent portions of the ecosystem, the samples are preserved in a manner to minimize or preclude alteration of the biological activities prior to their quantification. Portions of the biological community that represent the nutrient transformation of interest are quantified. We cannot guarantee that the collected data collected represent any area larger than that which was directly sampled.

8.5.3. Environmental Impact of Nutrient Cycles

As indicated previously, biogeochemical cycles contain a variety of chemical compounds that are linked in a cyclic manner and whose interconversion is accomplished by a combination of biological and chemical processes. Although a mass of data documenting the size of various nutrient pools within a number of ecosystems has been collected and published, considerable interest necessarily must be directed at the rate of movement between these pools and the environmental conditions altering these rates. This necessarily requires an accurate assessment of the distribution of nutrients within a given site. Difficulty in achieving this goal is exemplified by the pathways for nutrient loss from an undisturbed, native ecosystem. Complications arise from the failure to consider fully inputs or losses from the system. For example, nutrients may leave a system through leaching, denitrification, and volatilization (including ammonium volatilization), as well as being lost in harvested crop. The latter category is not limited to the biomass removed through agricultural processes. Wild and domestic animals may be prime movers of nutrients from an ecosystem through their grazing activities.

This movement of nutrients between various reservoirs has been exploited for agricultural crop production. "Mining" of fixed nitrogen contained in biomass has long been the basis for successful subsistence agriculture, but this enhancement of transfer of nutrients from organic to mineral reservoirs is not limited to low input, limited harvest cropping systems. Histosols, organic soils, are also so exploited. Peats and swamps represent sidetracks in nutrient cycles. Carbon, nitrogen, phosphorus, and sulfur compounds are incorporated into plant biomass, which due to water-logged conditions may be preserved for thousands of years. Removal of impediments to microbial mineralization of the organic matter results in rapid mineralization of the biomass, which produces sufficient nitrogen to support an intensive agricultural cropping system as is exemplified by the Everglades Agricultural Area (South Florida, USA). Decomposition of the organic matter comprising the muck soils of that region produces as much as 3.3×10^4 kg C ha^{-1} year^{-1}, which is accomplished by production of approximately 1400 kg N ha^{-1} year^{-1} (Tate, 1980). For comparison, soils of a tropical rain forest are estimated to produce $405-2117$ g C m^2 year (Schlesinger, 1977). The foregoing carbon evolution rate for the histosols in southern Florida is equivalent to 3300 g C m^2 year^{-1}.

In undisturbed ecosystems, such as forest sites, mineralization and biomass synthesis rates are nearly balanced. That is, recovery of the mineral nitrogen produced by the soil microbial community through plant roots is sufficiently efficient that little nitrogen leaves the site through such processes as leaching to groundwater. The slight losses occurring through transfer of mineral nutrients to ground- or surfacewaters are balanced by nitrogen inputs due to nitrogen fixation. Concerns arise from those systems, especially disturbed sites, where nitrogen leaching may reach ground or surface waters (e.g., the movement of nitrate and phosphates into Lake Okeechobee due to oxidation of this histosols of the Everglades Agricultural Area). Economic considerations arise when externally supplied nutrients (fertilizers) added to agricultural soils leach from the system rather than being retained in the crop.

8.5.4. Examples of Complications in Assessing Soil Nutrient Cycling: Nitrogen Mineralization

Nitrogen mineralization rate is a key indicator of the potential biomass production that can be supported by indigenous mineral nitrogen production. As indicated previously, the vast majority of the nitrogen contained in soils exists as organic nitrogen. Thus, baring any exogenous inputs, new biomass cannot be synthesized exclusive of mineralization of the resident organic nitrogen pools. The size of these nitrogen pools (organic and mineral) and the flux rate between them are, therefore, meaningful values in the characterization of an ecosystem.

Two procedures most commonly used to measure soil nitrogen mineralization potential are: a) the Stanford and Smith (1972) procedure, which was originally proposed for the study of soils contained in a column but has also been used for soils incubated as a batch sample in a beaker, and b) the buried bag method (Eno, 1960; Pastor et al., 1984). With the Stanford and Smith method, soil samples are collected and transported to the laboratory where they are sieved to increase homogeneity and to remove plant debris. The prepared soil samples are then dispensed into soil columns (leached method) or distributed into beakers (batch method). Mineral nitrogen is extracted from these samples with a dilute $CaCl_2$ solution. Nitrate plus nitrite and ammonium production are assayed over a time frame usually varying from 2 to 8 weeks. With the buried bag procedure, soils are collected in the field and placed in a polyethylene bag either with minimal manipulation or following sieving. The bag containing the field soil is returned to the soil horizon in a position as closely replicating the situation of the collected soil as much as possible. In theory, the conditions within the bag represent the field situation except that free exchange of soil water is prevented. That is, it is usually assumed that gaseous conditions within and without the bag are equivalent. Data from Bremner's laboratory (Bremner and Douglas, 1971) indicate that free oxygen concentrations decline inside the bag to levels that would reduce the activity of aerobic microbial populations. For an example of the use of these two procedures with forest soils, see Poovarodom and Tate (1988) and Poovarodom et al. (1988). With each procedure, mineral nitrogen production rates are determined by quantifying ammonium and nitrate plus nitrite concentrations in the soil samples before and after incubation.

The primary objective of conducting these assays is to measure the quantities of organic nitrogen that can be mineralized (size of the nutrient pool) and how fast it is being mineralized (the flux rate). Note that these two values are not directly measured by the previously mentioned procedures. Although direct assessment of the size and mineralization rate of labile soil organic nitrogen pool would be ideal, due to the complexity of the soil organic nitrogen pool, this is impossible. Soil organic nitrogen compounds are a highly diverse mixture of compounds, only a portion of which is easily metabolized by the soil microbial community. Furthermore, no soil organic nitrogen component has been found to be mineralized at a rate directly proportional to the overall nitrogen mineralization rate. Therefore, estimates of nitrogen mineralization must be derived indirectly. Both the quantities of mineralizable nitrogen contained within a soil sample and its mineralization rate are calculated from nitrogen mineralization data. Such data are predicted upon the assumption that a) soil organic nitrogen mineralization occurs in laboratory-incubated or buried bag samples at rates comparable to the field situation, b) preparation of the soil samples for analysis does not alter the availability of the organic nitrogen to the microbial community, and c) mineralization is a first-order process occurring at the same rate throughout the incubation period. (See section 10.4 for further discussion of the assumptions and the alteration of the mathematical models describing nitrogen mineralization to account for their failure in native soil samples.)

Derivation of nitrogen mineralization parameters from the Stanford and Smith procedure or the buried bag data may be calculated as follows: Determination of the nitrogen mineralization rate is usually predicated upon the observation that nitrogen mineralization follows first-order kinetics:

$$\frac{dN}{dt} = -kN \tag{8.1}$$

where N is the concentration of mineralizable organic nitrogen, k is a rate constant, and t is time. Integration of this equation produces the following:

$$N_t = N_0 e^{-kt} \tag{8.2}$$

where N_t is organic nitrogen at time t and N_0 is organic nitrogen at time zero. Since the nitrogen at time t is equivalent to the initial nitrogen minus the nitrogen mineralized N_m, the equation can be reduced to one containing a single unknown:

$$N_t = N_0 - N_m \tag{8.3}$$

$$N_0 - N_m = N_0 e^{-kt} \tag{8.4}$$

$$N_m = N_0(1 - e^{-kt}) \tag{8.5}$$

With these calculations we can estimate the original organic nitrogen and its mineralization rates under the conditions where the data was collected; that is, the derived parameters are not necessarily applicable to field situations (see section 10.4). For example, N_0 represents the initial nitrogen concentration in the disturbed soil sample, that is, the nitrogen available to the microbial population after the aggregates were disrupted by sampling and sieving of the soil sale. A significant proportion of that nitrogen could have been trapped in soil aggregate prior to sampling. Thus, the mineralization rate derived from the laboratory-incubated and the buried-bag-retained soils may be an overestimation of that occurring in the field.

Further inaccuracies in the data derived by this method are added by the potential variation in nitrogen production due to diminished oxygen availability in the buried bag procedure (and the potential for this to occur in batch incubated samples), which would result in an underestimation of N_m. Complications associated with the Stanford and Smith leaching procedure involve the removal of mineralizable organics during leaching of the columns (Smith et al., 1980). With the latter nitrogen mineralization assay method, the soil columns are leached with a salt solution prior to and at intervals during the incubation time. The resultant leachate contains the soil mineral nitrogen, which is the objective of the leaching procedure, but it also includes a portion of the water-soluble organic nitrogen. Removal of this latter soil nitrogen fraction from the pool of organic nitrogen may reduce the actual mineralization observed during the incubation procedure. Each of these difficulties in soil incubation—modification of soil atmosphere and removal of mineralizable organic nitrogen—may alter the rate of ammonium, nitrate, and nitrite production by the soil microbial community. Thus, with all these procedures, at best, an estimate of the field-available organic nitrogen pool and its mineralization rate is provided.

The model described previously is based on observations that nitrogen mineralization in soil generally follows first-order kinetics. Other mathematical models for deriving N_0 and k are based on alternative nitrogen mineralization kinetics (e.g., Bonde and Lindberg, 1988; Broadbent, 1986; Cabrera and Kissel, 1988; Deans et al., 1986; Juma et al., 1984), but the first-order model described here appears to be the most commonly used for derivation of these nitrogen mineralization constants.

These observations demonstrate the difficulties that are encountered in producing data representative of field-nutrient pool size and the flux between the various pools. Such considerations of cycles where large mineral pools exist are complicated by difficulties in assessing small changes in mineral nutrient against a large background. This problems is commonly encountered when fluxes of organic phosphate and sulfate are measured against backgrounds of soil-soluble and rock phosphate. These observations clearly reveal the basis for problems with ecosystem-wide and global extrapolation of such data.

8.6. CONCLUSIONS

Appreciating the properties of the reactions that occur in biogeochemical cycles and the impact of variation in environmental factors on their reaction rates is essential to understand a native ecosystem and develop soil management or reclamation plans. These cycles are commonly appreciated solely for their contributions to aboveground plant productivity and their impact on ecosystem development, but biogeochemical processes are much more complex than the simple biological- or chemical-catalyzed transformation of an essential plant macro- or micronutrient. A complete depiction of biogeochemical cycles entails an understanding of not only the biological entities involved but also their synergistic interactions with the soil physical and chemical properties as they contribute to native ecosystem development. For reclaimed soils, this synergism impacts the development of a self-sustainable ecosystem.

REFERENCES

Adams, R. M., C. Rosenzweig, R. M. Pert, J. T. Ritchie, B. A. McCarl, J. D. Glyer, R. B. Curry, J. W. Jones, K. J. Boote, and L. H. Allen, Jr. 1990. Global climate change and US agriculture. Nature (London) 345:219–224.

Allen, L. H., Jr. 1990. Plant responses to rising carbon dioxide and potential interactions with air pollutants. J. Environ. Qual. 19:15–34.

Blevins, R. L., M. S. Smith, and G. W. Thomas. 1984. Changes in soil properties under no-tillage. *In* R. E. Phillips and S. H. Phillips (eds.), No-Tillage Agriculture, pp. 190–230. Van Nostrand Reinhold Co. New York.

Bonde, T. A., and T. Lindberg. 1988. Nitrogen mineralization kinetics in soil during long-term aerobic laboratory incubations: A case study. J. Environ. Qual. 17:414–417.

Boyle, M., and E. A. Paul. 1989. Nitrogen transformation in soils previously amended with sewage sludge. Soil Sci. Soc. Am. J. 53:740–744.

Bremner, J. M., and L. A. Douglas. 1971. Use of plastic films for aeration in soil incubation experiments. Soil Biol. Biochem. 3:289–296.

Broadbent, F. E. 1986. Empirical modeling of soil nitogen mineralization. Soil Sci. 141:208–213.

Cabrera, M. L., and D. E. Kissel. 1988. Evaluation of a method to predict nitrogen mineralization from soil organic matter under field conditions. Soil Sci. Soc. Am. J. 52:1027–1031.

Cambardella, C. A., and E. T. Elliott. 1992. Particulate soil organic-matter changes across a grassland cultivation sequence. Soil Sci. Soc. Am. J. 56:777–783.

Dahlman, R. C., B. R. Strain, and H. H. Rogers. 1985. Research on the response of vegetation to elevated atmospheric carbon dioxide. J. Environ. Qual. 14:1–8.

Deans, J. R., J. A. E. Molina, and C. E. Clapp. 1986. Models for predicting potentially mineralizable N and decomposition rate constants. Soil Sci. Soc. Am. J. 50:323–326.

Doran, J. W. 1980. Soil microbial and biochemical changes associated with reduced tillage. Soil Sci. Soc. Am. J. 44:765–771.

Ellert, B. H., and J. R. Bettany. 1988. Comparison of kinetic models for describing net sulfur and nitrogen mineralization. Soil Sci. Soc. Am. J. 52:1692–1702.

Eno, C. F. 1960. Nitrate production in the field by incubation of soil in polyethylene bags. Soil Sci. Soc. Am. J. 24:277–279.

Firestone, M. K. 1982. Biological denitrification. *In* F. J. Stevenson (ed.), Nitrogen in Agricultural Soils, Agron. Monogr. 22, pp. 289–326. Agronomy Society of America. Madison, WI.

Fresquez, P. R., R. E. Francis, and G. L. Dennis. 1990. Sewage sludge effects on soil and plant quality in a degraded, semiarid grassland, J. Environ. Qual. 19:324–329.

Geiger, S. C., A. Manu, and A. Bationo. 1992. Changes in a sandy Sahelian soil following crop residue and fertilizer additions. Soil Sci. Soc. Am. J. 56:172–177.

Gilmour, J. T., M. D. Clark, and G. C. Sigua. 1985. Estimating net nitrogen mineralization from carbon dioxide evolution. Soil Sci. Soc. Am. J. 49:1398–1402.

Holland, E. A., and D. C. Coleman. 1987. Litter placement effects on microbial and organic matter dynamics in an agroecosystem. Ecology 68:425–433.

Hunt, H. W. 1977. A simulation model for decomposition in grasslands. Ecology 58:469–484.

Jenkinson, D. S., D. E. Adams, and A. Wild. 1991. Model estimates of CO_2 emissions from soil in response to global warming. Nature (London) 351:304–306.

Jones, C. A., C. V. Cole, A. N. Sharpley, and J. R. Williams. 1984. A simplified soil and plant phosphorus model: I. Documentation. Soil Sci. Soc. Am. J. 48:800–805.

Juma, N. G., E. A. Paul, and B. Mary. 1984. Kinetic analysis of net nitrogen mineralization in soil. Soil Sci. Soc. Am. J. 48:753–757.

McConnaughey, P. K., and D. R. Bouldin. 1985a. Transient microsite models of denitrification: I. Model development. Soil Sci. Soc. Am. J. 49:886–891.

McConnaughey, P. K., and D. R. Bouldin. 1985b. Transient microsite models of denitrification: II. Model results. Soil Sci. Soc. Am. J. 49:891–895.

McConnaughey, P. K., D. R. Bouldin, and J. M. Duxbury. 1985. Transient microsite models of denitrification: III. Comparison of experimental and model results. Soil Sci. Soc. Am. J. 49:896–901.

Meentemeyer, V. 1978. Macroclimate and lignin control of litter decomposition rates. Ecology 59:465–472.

Mills, A. 1985. Acid mine waste drainage: Microbial impact on the recovery of soil and water ecosystems. In R. L. Tate, III and D. A. Klein (eds.), Soil Reclamation Processes. Microbiological Analyses and Applications, pp. 35–81. Dekker. New York.

Mosier, A., D. Schimel, D. Valentine, K. Bronson, and W. Parton. 1991. Methane and nitrous oxide fluxes in native, fertilized and cultivated grasslands. Nature (London) 350:330–332.

Parker, L. W., J. Miller, Y. Steinberger, and W. G. Whitford. 1983. Soil respiration in a Chihuahuan desert rangeland. Soil Biol. Biochem. 15:303–309.

Parton, W. J., J. W. B. Stewart, and C. V. Cole. 1988. Dynamics of C, N, P and S in grassland soils: A model. Biogeochem. 5:109–131.

Pastor, J., J. D. Aber, and C. A. McClaugherty. 1984. Above-ground production and N and P cycling along a nitrogen mineralization gradient in Black Hawk Island, Wisconsin. Ecology 65:256–268.

Pastor, J., and W. M. Post. 1988. Response of northern forests to CO_2 induced climate change. Nature (London) 334:55–58.

Poovarodom, S., and R. L. Tate, III. 1988. Nitrogen mineralization rates of the acidic, xeric soils of the New Jersey Pinelands: Laboratory studies. Soil Sci. 145:337–344.

Poovarodom, S., R. L. Tate, III, and R. A. Bloom. 1988. Nitrogen mineralization rates of the acidic, xeric soils of the New Jersey Pinelands: Field rates. Soil Sci. 145:257–263.

Reddy, V. R., D. N. Baker, and J. M. McKinion. 1989. Analysis of effects of atmospheric carbon dioxide and ozone on cotton yield trends. J. Environ. Qual. 18:427–432.

Santos, P. F., N. Z. Elkins, Y. Steinberger, and W. G. Whitford. 1984. A comparison of surface and buried Larrea tridentata leaf litter decomposition in North American hot deserts. Ecology 65:278–284.

Santos, P. F., J. Phillips, and W. G. Whitford. 1981. The role of mites and nematodes in early stages of buried litter decomposition in a desert. Ecology 62:664–669.

Santos, P. F., and W. G. Whitford. 1981. The effects of microarthropods on litter decomposition in a Chihuahuan desert ecosystem. Ecology 62:654–663.

Schlesinger, W. H. 1977. Carbon balance in terrestrial detritas. Ann. Rev. Ecol. Syst. 8:51–81.

Scow, K. M., and M. Alexander. 1992. Effect of diffusion on kinetics of biodegradation: Experimental results with synthetic aggregates. Soil Sci. Soc. Am. J. 56:128–134.

Scow, K. M., and J. Hutson. 1992. Effect of diffusion and sorption on kinetics of biodegradation: Theoretical considerations. Soil Sci. Soc. Am. J. 56:119–127.

Soma, K., and T. Saito. 1983. Ecological studies of soil organisms with reference to the decomposition of pine needles. II. Litter feeding and breakdown by the woodlouse, *Porcellio scaber.* Plant Soil 75:139–151.

Smith, J. L., R. R. Schnabel, B. L. McNeal, and G. S. Campbell. 1980. Potential errors in the first-order model for estimating soil nitrogen mineralization potentials. Soil Sci. Soc. Am. J. 44:996–1000.

Stanford, G., and J. J. Smith. 1972. Nitrogen mineralization potentials in soils. Soil Sci. Soc. Am. J. 36:465–472.

Stroo, H. F., K. L. Bristow, L. F. Elliott, R. I. Papendick, and G. S. Campbell. 1989. Predicting rates of wheat residue decomposition. Soil Sci. Soc. Am. J. 53:91–99.

Stucki, G., and M. Alexander. 1987. Role of dissolution rate and solubility in Biodegradation of aromatic compounds. Appl. Environ. Microbiol. 53:292–297.

Tatabai, M. A., and Y. M. Chae. 1991. Mineralization of sulfur in soils amended with organic wastes. J. Environ. Qual. 20:684–690.

Tate, R. L. III. 1980. Microbial oxidation of organic matter of histosols. *In* M. Alexander (ed.), Advances in Microbial Ecology, Vol. 4, pp. 169–201. Plenum. New York.

Tate, R. L. III. 1985. Microorganisms, ecosystem disturbance and soil-formation processes. *In* R. L. Tate, III and D. A. Klein (eds.), Soil Reclamation Processes. Microbiologial Analyses and Applications, pp. 1–33. Dekker, New York.

Tate, R. L. III. 1987. Soil Organic Matter. Biological and Ecological Effects. John Wiley & Sons. New York. 291 pp.

van Veen, J. A., J. N. Ladd, and M. J. Frissel. 1984. Modelling C and N turnover through microbial biomass in soil. Plant Soil 76:257–274.

Walters, D. T., M. S. Aulakh, and J. W. Doran. 1992. Effects of soil aeration, legume residue, and soil texture on transformations of macro- and micronutrients in soils. Soil Sci. 153: 100–107.

9 The Carbon Cycle

Conceptually, models of the carbon cycle can easily be reduced to representations of the processes associated with the transformation of carbon dioxide into organic matter and return of this primarily photosynthetically fixed carbon to its mineral form via biological mineralization (e.g., Fig. 8.1). This simplistic depiction of the carbon cycle processes provides a means for developing generalized principles, but amassing the knowledge necessary for managing soil sites for ecosystem sustenance or reclamation relies upon separation of these two primary carbon reservoirs into pools designated by location in the ecosystem, residence time, and/or chemical form. Commonly studied soil organic carbon reservoirs include living and dead microbial biomass, animal and plant debris, plant root exudates, as well as humic and fulvic substances (Fig. 9.1). Even this grouping of soil organic components must be expanded conceptually before we can present a detailed understanding of the properties of both the individual compounds themselves and their interactions with ecosystem components that limit the rate of flux between these organic reservoirs and the mineral pools. In other words, we must be able to quantify biochemicals (e.g., protein, polysaccharide, lignin, fat, waxes, and resins) comprising each major soil organic reservoir that provide nutrients and energy to the soil microbial population, evaluate the structural associations of these components in cells, and explain the inclusion of these components in soil aggregates before we can develop a detailed model of the soil carbon cycle.

Some soil organic matter pools can be distinguished by their degradation resistance or vulnerability (e.g., humic substances vs. animal or plant biomass), but most of the organic matter pools designated in Fig. 9.1 are mixtures of susceptible and relatively stable organic components. Consequences of interactions within the soil environment are illustrated by an analysis of plant biomass decomposition in soil. Accessibility of plant biomass components to the soil microbial population is determined by the rate of breaching of the plant cell wall, which protects the more easily decomposed substances contained in the cytoplasm, as well as by the amount of biomass occluded by inclusion in soil aggregates. Thus, the kinetics of movement of carbon among easily decomposable carbon pools may be retarded by its inclusion within a soil aggregate or stabilization of plant cellular structure. Similarly, a normally decomposable xenobiotic compound, such as a pesticide, may persist as the result of sorption to soil mineral particles or of weak associations with soil humic substances.

In conclusion, to understand the complexity of the processes leading to increases or decreases in quantities of sequestered organic carbon in soil, the nature of the

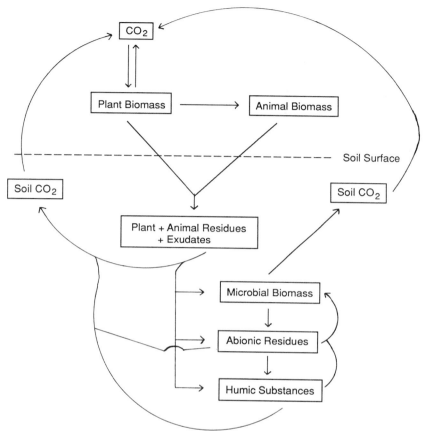

Fig. 9.1. Conceptual model of carbon cycle with transfers between major soil organic matter reservoirs emphasized.

components of this soil fraction and the processes leading to their decomposition must be elucidated. Each of the major organic matter reservoirs in soil is a mixture of readily metabolized organic compounds plus relatively biodegradation-resistant substances, such as lignin and humic acids. Accumulating decomposition-resistant compounds increases the portion of terrestrial carbon retained in soil, whereas the easily metabolized resources provide the energy necessary to drive the soil microbial population and "propel carbon through the carbon cycle." Hence, this chapter will a) evaluate the nature of organic compounds retained within major soil organic matter reservoirs, b) explain the mechanisms involved in the catabolism of these compounds, and c) determine the properties of the carbon molecules, the microbial community, and the soil ecosystem itself, which are all instrumental in controlling the return of soil organic matter to atmospheric carbon dioxide pools.

9.1. ENVIRONMENTAL IMPLICATIONS OF THE SOIL CARBON CYCLE

Although extensive anthropogenic intervention may allow development of a plant community on essentially any supportive matrix, a sustained, viable, productive ecosystem is dependent upon the integrity of all soil biogeochemical cycles. The carbon cycle occurring in soil serves as an excellent model of the localized impact of organic matter decomposition as well as the worldwide implications of modification of the quantities of fixed-carbon retained within soils.

Organic matter flux studies involving various soil reservoirs are frequently inspired by concern for such ecological and societal problems as demonstrating the potential for using soil systems in recycling organic-carbon-based wastes, reclaiming sites impacted by mismanagement—accidental or purposeful—and, from a less direct viewpoint, assessing the impact of soil carbon mineralization processes on atmospheric composition (e.g., the greenhouse effect). Furthermore, historical interest has also emphasized determining the best management procedures for enhancing soil structural quality and optimizing agricultural production. In summary, societal applications of studies of soil mineralization processes therefore include strong implications for decisions involving the management of greenhouse gas fluxes, maintenance of natural ecosystems, disposal of societal wastes, and reclamation of damaged soil systems as well as aspects of enhancement of agricultural production of food and fiber.

9.1.1. Soils as a Source or Sink for Carbon Dioxide

Soil ecosystems are both producers and consumers of greenhouse gases, especially the products of carbon mineralization, carbon dioxide, and methane. Increases in soil organic matter levels result in the sequestering of excess carbon dioxide into soil organic matter (soils are a net pool for carbon dioxide), whereas management of the soil ecosystem in a manner causing declines in colloidal soil organic matter or ecosystem productivity converts the ecosystem into a source of mineral carbon. For example, conservation tillage practices could have a substantial impact on terrestrial carbon dioxide production. Kern and Johnson (1993) estimate that maximizing the implementation of no-tillage soil management practices could result in sequestering of carbon in soil organic matter equivalent to 0.7 to 1.1 percent of the total projected US fossil fuel carbon emissions for a 30-year period. A less direct but significant interaction between soil organic matter reservoirs and atmospheric mineral carbon concentrations results from managing soil properties to optimize plant biomass production (i.e., ecosystem productivity). As is discussed later, plant biomass is a primary carbon reservoir. Among the soil properties controlling biomass synthesis is soil aggregation, which is determined by interactions of soil minerals with humus (chapter 1). Thus, improvements in soil structure resulting from increases in colloidal organic mater may indirectly reduce atmospheric carbon dioxide loads by increasing terrestrial biomass pools.

Evidence for the role of soil organic matter in modulating increases in atmospheric carbon dioxide is provided by observing the incongruity between known carbon dioxide sources and measurable atmospheric loads of this gas. That is, known sources of atmospheric carbon dioxide exceed the identifiable sinks. Soil organic matter and terrestrial biomass most likely account for at least a portion of this unknown fate of the carbon dioxide (e.g., Harden et al., 1992, and Sundquist, 1993). Since the terrestrial biosphere and associated soils contain far more carbon than the atmosphere, these carbon pools may serve as buffers or moderators of atmospheric carbon dioxide levels. A recent example of estimates of mass of carbon in various biogeochemical pools is provided by Sudquist (1993), where it is estimated that 750 Gt (1 Gt = 10^{15}g) of carbon was contained in the atmospheric pool in 1990 A.D. compared to 2160 Gt for the terrestrial biosphere, with the soils component accounting for approximately two-thirds of this total. Eswaran et al. (1993) estimated that 1576 Gt of carbon is stored in soils, with about 32 percent of this carbon retained by tropical soils.

Furthermore, to gain somewhat of a visual perspective on the ratio between living matter and dead organic matter in terrestrial ecosystems, this relationship can be considered in forest systems. It has been estimated that between two- and three-fold the carbon retained in forest biomass is found in dead organic matter on the forest floor (Kögel-Knaber, 1993).

Thus, with the nearly threefold concentration of carbon retained within the biosphere, minor changes biomass production and retention of fixed carbon within soil can result in significant increases or decreases in atmospheric carbon dioxide levels. Examples of such potentially world-impacting ecosystem modifications include cultivation of virgin soils (generally resulting in a decline in native soil organic matter pools; see Tate, 1987 for a review of this topic), drainage of peat soils (e.g., Armentano, 1980, Tate, 1980) as well as clear cutting forest biomass and its replacement with less productive plant species.

9.1.2. Diffusion of Soil Carbon Dioxide to the Atmosphere

A further consideration in modeling the exchange rate between soil organic carbon pools and atmospheric carbon dioxide is the diffusion rate of carbon dioxide between soil air and the atmosphere. Variation in these localized soil properties (i.e., moisture, structure, temperature) control not only the gaseous exchange rates but also delimit the mineralization rate (i.e., the mass of carbon dioxide available to move from soil to atmospheric pools) characteristic of the soil site (see Oades, 1988). The direction of movement is from soil air to the overlying atmosphere in that concentrations of gaseous carbon dioxide existent in soil pores generally exceeds atmospheric carbon dioxide loads. A variety of soil properties interact to facilitate or inhibit this gas exchange. Along with the increases in the mass of carbon dioxide resulting from ecosystem modifications listed previously, soil moisture, temperature, structure (aggregate formation and pore structure), biological activity (root development and earthworm activity), and mass movement of water (e.g., rainfall) also control carbon dioxide distribution. Less dramatic variability in ecosystem properties than those resulting from ecosystem-wide alterations, natural and anthropogenically induced, may determine the carbon dioxide exchange rate between soil and atmospheric reservoirs. For example, a minor alteration in soil moisture and temperature, such as those resulting from developing crop canopies, can alter the carbon dioxide flux rate. These interactions were presented in a model developed by Ouyang and Boersma (1992a, 1992b).

A less direct but significant interaction between the terrestrial carbon pool and the atmosphere is the effect of atmospheric carbon dioxide availability and plant biomass synthesis rate. The most obvious effects of varying biomass production rates are exemplified by total ecosystem modification that occur with forest establishment (e.g., Vitousek, 1991) or deforestation (e.g., Woodwell, 1983), but increases in atmospheric carbon dioxide can also lead to augmented plant biomass production in established ecosystems (e.g., see McMurtrie et al., 1992; Polglase and Wang, 1992, Rogers et al., 1983). In either situation, plant productivity enhancement can also alter the equilibrium concentration of organic carbon retained in the underlying soil. Most of the photosynthetically fixed carbon entering soil is mineralized to carbon dioxide by the soil microbial community, but some carbon will be retained in soil in more biodegradation-resistant fractions—microbial biomass and humic substances. In other words, increased fixed-carbon inputs may lead to increased soil microbial populations, their products (e.g., extracellular polysaccharides), as well as greater accumulation of humified substances than existed prior to stimulation of aboveground plant community growth. Thus, stimulation of biomass productivity by increased carbon dioxide loads could lead to a reduction (albeit of variable magnitude) in atmospheric carbon dioxide and a resultant sequestering of carbon in soil organic matter pools.

9.1.3. Managing Soils to Augment Organic Matter Contents

Because of an historical predisposition for studying soil processes that affect crop yields, most of the data useful for explaining processes that control soil humus levels are associated with research whose primary objective is to improve plant biomass productivity (crop yields). The results of decades of research regarding establishment of natural organic matter levels in soil could be succinctly summarized as follows (see chapter 1 and Tate, 1987, for a more detailed review of this topic):

- The amount of soil organic matter retained in a particular soil system is a product of the balance between organic matter input (primarily aboveground biomass production) and decomposition (predominantly microbial mineralization of plant biomass).
- An equilibrium level of soil organic matter is attained in an undisturbed soil ecosystem that is controlled by total ecosystem properties—that is, the nature of such soil ecosystem properties as the plant community, soil texture, climate, topography, and anthropogenic intervention. (See Amundson and Jenny, 1991; Jenny, 1980).
- Of ultimate importance in evaluating the impact of soil organic matter transformations on terrestrial carbon cycling is the fact that disturbance of a soil, as

exemplified by cultivation or clear-cutting of a forest, results in establishing a new equilibrium level of soil organic matter that is generally lower than that in the pristine or undisturbed ecosystem.

• Processes that reduce soil aggregation, aerate the soil, or reduce biomass incorporation rates result in reduced retention of organic matter.

Many recent examples of alterating soil organic matter resources because of an ecosystem disturbance (e.g., clear-cutting of forests, cultivation of forest or grassland soils) and of the relationship of subsequent soil management techniques on establishing new steady-state organic matter percentages in the soil are available. For example, see Aguilar et al. (1988). Alegre and Cassel (1986), Mann (1986), and Rasmussen and Rohde (1988).

The changes in soil organic matter levels in reaction to ecosystem disturbances tend to be proportional to the amount of soil structure disruption associated with the initial site clearing, the extent of cultivation or related management of the soil site subsequent to the initial disturbance (i.e., continued degradation of soil structure and mixing of the soil results in greatest losses of native soil humus), and the quantities of organic matter produced in or amended to the soil. For example, in the study by Alegre and Cassel (1986), it was shown that in clearing forest soils in the Amazon jungle of Peru, slash-and-burn clearing allowed greater retention of native soil organic matter than did mechanical clearing with a bulldozer. Similarly, maintenance of soil aggregate structure by bedding and lime application reduced organic matter losses compared to that observed in cleared soils, which were managed by more structurally damaging procedures. Alteration of inputs of organic matter can be through maintenance of plant productivity (aboveground as well as root mass) and return of *in situ* produced plant biomass to the soil ecosystem (Barber, 1979; Richter et al., 1990, Wood et al., 1992) as well as through soil amendment (e.g., for the long-term impact of manure amendments to cropped soils, see Jenkinson and Rayner, 1977).

Reduction in site intervention (e.g., conversion from intensive tillage to no-till management of soil) reduces susceptibility of organic matter to microbial decomposition through enhancement of soil aggregate development. See Havlin et al. (1990) and Scott and Wood (1989) for examples of studies of the impact of variation in tillage practices on soil organic matter.

The overall environmental impact of each of these localized soil management procedures is alteration in the quantities of carbon dioxide released to the atmosphere (potential greenhouse gas problems) or conversely augmentation of the proportion of total terrestrial carbon residing in the soil organic matter pool. The latter situation is of greater overall benefit because it results in a reduced level of greenhouse gases and improves the stability and productivity of the ecosystem, a factor that can also reduce greenhouse gas production.

9.1.4. Soils as a Carbon Recycling Processor

A major product of human societies is organic waste products such as industrial and residential garbage, sludges, as well as a variety of petroleum waste products.

Incineration of these substances converts essentially all the carbon to atmospheric carbon dioxide, whereas use of the substances, where appropriate, as soil amendments recycles plant nutrients contained therein into plant biomass and allows sequestering of portions of the organic matter into the various soil organic carbon pools. Thus, the fixed-carbon components and associated nutrients are recycled through the soil biomass and associated carbonaceous components.

In using soils to recycle organic waste substances, the natural biological decomposers resident in soil can be exploited as an economical means of returning bulky, waste vegetative material to less troublesome mineral carbon or stable soil organic carbon reservoirs. For example, see review by Boyle (1990) describing the land disposal of sludge or studies of landfarming of petroleum wastes (see Bossert et al., 1984; Dibble and Bartha, 1979a, 1979b; Norris, 1980).

Primary questions regarding soil system's use to recycle the organic waste products of society do not pertain to the capacity of soils to decompose such materials because the organic components contained therein differ little from organic substances that enter soils naturally as fresh plant biomass, litter, and thatch or chemically related substances that are normally decomposed by soil populations (i.e., petroleum products). Instead, they relate a) to the total quantities that can be added before the system is overloaded (i.e., either natural decomposition processes are reduced, ecosystem sustainability is threatened, or groundwater contamination is threatened), b) to difficulties with contaminants of the waste material that may impede soil biological activity or may be resistant to mineralization (e.g., heavy metal contaminants and synthetic organic chemicals), and c) to the potential for spreading of pathogenic microbes (e.g., fecal coliforms, protozoa, and viruses). For example, see review by He et al. (1992) regarding the land disposal of composted municipal solid waste. Societal benefit results not only from the disposal of troublesome waste substances, but if the process is properly conducted, soil properties can also be improved (e.g., Douglas and Magdoff, 1991; Epstein et al., 1976; Mays et al., 1973; Mitchel et al., 1978; Piccolo and Mbagwu, 1990).

These limited examples of the impact of soil carbon cycle processes on the total terrestrial ecosystem emphasize the necessity to develop a clear understanding of the types of organic carbon compounds mineralized in soil, their distribution in major carbon cycle pools (Fig. 9.1), and their decomposition kinetics.

9.2. BIOCHEMICAL ASPECTS OF THE SOIL CARBON CYCLE

The complexity of the biochemical processes associated with carbon cycling in soil is illustrated by the diversity of organic compounds mineralized and the array of microbes involved. The soil microbial community is adept at converting native biomass components plus a variety of xenobiotic compounds to atmospheric carbon dioxide, albeit with varying rates dependent upon the substrate structure and ecosystem properties. This seemingly limitless capability of the soil microbial community to mineralize organic carbon compounds could be considered to be the ultimate in recycling. Indeed, at one time in the development of the concepts of soil microbiology, the soil microbial community was considered to be infallible; that is, failure to discern measurable mineralization of an organic compound was thought to relate to the inability of the researcher to provide conditions conducive to the activity of the requisite microbes, rather than to be an inherent genotypic limitation of the microbes With the advent of the "chemical age," we know that the concept of microbial infallibility was at best optimistic from the view of microbial capacities and pessimistic with regard to the innovative abilities of chemists and the magnitude of the societal appetite for unique industrially synthesized materials.

As a result of the expanding capacity to synthesize organic carbon-based compounds that bear little resemblance to those produced naturally, this fallibility/infallibility concept of the soil microbial populations must be adjusted. More accurately, the soil microbial community can be considered to be infallible only at mineralizing biologically synthesized organic compounds. This conclusion is based on current concepts of genotypic diversity of indigenous soil microbes. Interestingly, the potential now exists to expand soil metabolic capabilities through bioengineering. With the increasing capability of developing "designer genes," the enzymatic array of the soil microbes may come to include a variety of "unnatural" enzymes. These enzymes would be characterized by the fact that they result from modifications of cellular DNA to allow the production of enzymes with a specificity that includes the capacity to cleave molecular bonds not normally encountered in the soil system, thereby resulting in mineralization of the chemically synthesized organic compound. In that situation, perhaps, our concept of microbial infallibility may return to something approaching the optimistic view of the prechemical age. The challenge is for soil microbiologists to become as adept at developing novel degraders of xenobiotic compounds as chemists are at producing such substances.

9.2.1. Individual Components of Soil Organic Carbon Pools

The organic components of soil amendments and native soil organic matter are easily separated by chemical class (e.g., polysaccharides, proteins, lignin, humic acids, and fulvic acids). Each of these groups of compounds is distinguished by specific chemical properties that delimit their utility for carbon and energy metabolism of the soil microbial community. Because all but the humic and fulvic acids are sufficiently defined chemically that they can be studied in the test tube, the biochemical processes associated with their mineralization appear to be reasonably predictable in soil. Thus, we can understand the biochemistry of the soil environment by analyzing the organic compounds available as carbon and energy resources.

Understanding the relative biodegradation susceptibility of each chemical group allows us to predict the relative activity of different soil ecosystems. For example, a site containing large quantities of polysaccharide supports a more active microbial community than one in which essentially all the organic carbon is associated with humic and fulvic acids (i.e., a site where external plant or animal biomass inputs are nonexistent or minimal). Quantification of biologically decomposed substrates in a soil sample only provide, at best, an estimate of the potential biological activity contained therein. This conclusion results from the fact that the organic compounds do not exist in isolation in soil but instead occur in complex mixtures of easily decomposed and more complex organic substances, not all of which are available to the soil microbial community. For example, proteins and starch are readily detected in soil. These substances may be decomposed within a matter of hours when amended as chemically pure preparations to soil, but both starch and proteins could have half-lives of days or longer in soil should they be retained within a reasonably biodegradation-resistant structure, such as a plant cell wall.

Polysaccharides are the primary energy source of the soil biological community. The predominant source of soil polysaccharides is plant tissue, although locally important contributions may be derived from influxes of animal biomass. Internally synthesized polysaccharides originate mainly from microbial synthesis. Between 5 and 25 percent of the carbon contained in native soil organic matter (i.e., that organic matter remaining in soil after all recognizable plant and animal debris has been removed) is composed of polysaccharides. Of this soil humus polysaccharide, that originating in newly synthesized biomass is most susceptible to biodegradation, whereas polysaccharides covalently bound to humic acids, and polysaccharides coating soil minerals are more resistant. A diverse mixture of polysaccharides and their decomposition products have been isolated from soil (for more detail on this subject, see Gupta, 1967, or Stevenson, 1982):

- Monosaccharides including hexoses (D-glucose, D-galactose, D-mannose) and pentoses (D-ribose, D-arabinose, D-xylose).
- Polysaccharides such as cellulose and a variety of hemicelluloses from plants as well as amino sugar containing polysaccharides including chitin as found in some fungi and in insect exoskeletons and the complex mucopeptides of bacterial cell walls.
- Uronic acids and polyuronic acids such as glucuronic acid, galacturonic acid, and polygalacturonic acid (i.e., pectin).
- Methylated sugars: 2-*O*-methyl-D-xylose, 2-*O*-methyl-D-arabinose, 2-*O*-methyl rhamnose, 4-*O*-methyl galactose.

These polysaccharides are distinguished by the diversity of linkages between the monosaccharide units (e.g., cellulose vs. amylose), inclusion of nonsugar moieties (e.g., chitin and microbial cell walls), as well as combination of a variety of monosaccharides within a single polysaccharide molecule (e.g., hemicelluloses). Variation in each of these molecular properties impacts decomposition susceptibility through alteration of the crystalline structure of the molecule and/or the number of enzymes necessary for its mineralization.

The effect of linking the sugar moieties on biological stability is best exemplified by comparing the structures of two glucopyranoses, amylose and cellulose (Fig. 9.2). Amylose, α-(1 $\rightarrow$ 4)-D glucose, is usually decomposed in soil in a matter of hours whereas cellulose, β-(1 $\rightarrow$ 4)-D-glucose, has a half-life in soil that may be measured in years. Amylose is hydrolyzed by the enzyme amylase, whereas a complex of enzymes is required for disruption of the crystalline structure and

Fig. 9.2. Structural formulae of cellulose and amylose.

cleavage of cellulose into monomers (e.g., Béquin, 1990). Chitin, β-(1 $\rightarrow$ 4)-N-acetylglucosamine, also forms a rigid structure that retards catabolism by most soil bacteria.

In contrast to these reasonably simple structures, the bacterial cell wall (Fig. 9.3) is composed of regular repeating units of N-acetylglucosamine and N-acetylmuramic acid interlinked by short peptide chains. The latter structures slowly contribute to soil carbon cycling through death and decay of the bacterial populations.

Hemicellulose is second in abundance in plant tissue only to cellulose. It is found in proximity to cellulose in primary and secondary cell walls. Structurally the only commonalty between cellulose and hemicellulose is that they are both polysaccharides. Two groupings of hemicellulose are introduced into soil—homoglucans and heteroglucans. Homoglucans, the less common grouping of plant hemicelluloses, consist of single monosaccharide units. Examples include xylan (polyxylose), mannan, and galactan. Heteroglycans are composed of more than one monosaccharide or uronic acid. Most contain two to four different monosaccharides, but examples of five to six components are detected. The most abundant moiety is listed last in the name (e.g., mannose for glucomannan and arabinomannan). Hemicelluloses as a group are structurally complex, containing 50 to 60 monosaccharide units with branching. A small group of monosaccharides (xylose, arabinose, mannose, glucose, galactose, glucuronic acid, and galacturonic acid) provide the bulk of the sugar moieties of hemicelluloses.

Lipids are a more heterogeneous group of soil organic compounds. This soil component is defined operationally. Soil lipids are those soil organic components that are soluble in organic solvents. They include fats, waxes, and resins. Chemical

Fig. 9.3. Structural formula for *Escherichia coli* peptidoglycan cell wall structure.

components include waxes (long chain fatty acids and higher chain aliphatic alcohols and some cyclic alcohols), organic acids (both short and longer chain), normal parafins (C_{16} to C_{32}), as well as polycyclic hydrocarbons. The latter groups include polycyclic aromatic hydrocarbons, sterols, terpenoids, and chlorophyl. (See Stevenson, 1982, for a more complete discussion of these compounds.)

Lignin is a major component of plant biomass. Stems of woody angiosperms contain between 18 and 25 percent lignin (on a dry weight basis). For comparison, gymnosperms are composed of 25 to 35 percent lignin, whereas 10 to 30 percent of the weight of monocotyledons is lignin (Crawford, 1981). Lignin is a random polymer of sinapyl, coniferyl, and coumaryl alcohols containing a variety of complex organic linkages, which are considerably less common in other plant substituents. Spruce (*Picea abies*), for example, contains 48 percent arylglycerol-β-aryl ethers, 6 to 8 percent noncyclic benzyl aryl ethers, 9.5 to 11 percent biphenyl, 7 percent 1,2-diarylpropane structures, 9 to 12 percent phenylcoumaran structures, and 3.5 to 4 percent diphenyl ethers (Crawford, 1981). Because of the complex aromatic structure of this plant component, lignin is decomposed slowly in soil (see review by Kirk and Farrell, 1987, for discussion of the biochemistry of lignin decomposition). The most labile components of this complex molecule are the side chains and methoxyl groups. For example, after a 6-month incubation period in a neutral sandy loam, about 23 percent of the ring carbons and 2-carbons and 39 percent of the methoxyl carbons of coniferyl alcohol units linked into model and cornstalk lignins were evolved as carbon dioxide (Martin and Haider, 1979). (See Tate, 1987, for a more detailed examination of the decomposition of this compound in soil.) A summary of the traits of lignin decomposition that are significant to soil microbiologists and that can be used to evaluate the impact of lignin on the rate of cycling plant carbon through soil follows:

- The complexity of the structure results in extended residence time for the plant carbon contained in lignin in soil.
- Although lignin contains a variety of chemical components that could yield energy to growing microbes, recovery of this energy by the soil microbial community is unlikely. The variety of components and their random assembly into the lignin molecule increases the quantities of energy that the microbes must expend to recover energy from the mineralization of lignin.
- The extended residence time of lignin results in incorporating significant portions of the lignin molecule into soil humic acids, which extends the longevity of these carbons in the soil ecosystem.

Humic and fulvic acids are the most biodegradation-resistant of the soil organic components. Although lignin presents a formidable barrier to microbes that are instrumental in cycling organic carbon that enters or is contained within the soil ecosystem back to mineral pools, an even more complex structure is encountered by soil microbial populations active in carbon mineralization in humic and fulvic acids. Humic substances are commonly extracted from soil with dilute alkaline solutions (e.g., 0.1 M NaOH). Humic acid is the substance that is soluble in the alkaline solution but precipitates at pH less than 1, whereas fulvic acid is soluble in both the acidic and alkaline conditions. Humin is not soluble in acidic or alkaline solutions.

Humic acids are random polymers of aromatic and aliphatic moieties. Molecular weights for these substances have been shown to range from a few thousand to over one million daltons (e.g., see Stevenson, 1982). Malcolm (1990) specifies a molecular weight range for soil humic acids of 50,000 to 500,000 daltons. Although traditional concepts of the generalized structure of humic acids are based on the assumption that these substances are predominantly aromatic, recent research has shown that humic substances are frequently primarily aliphatic (Malcolm, 1990). For example, Hatcher et al. (1981) calculated aromaticities for humic acids from a variety of climatic zones ranging from 35 to 92 percent. Thus, the commonly encountered hypothetical structure of humic acids must be interpreted to contain representative aromatic and aliphatic moieties, but the ratio between these individual components differs from that shown.

A variety of proteins and polysaccharides may be covalently bound to the humic acid molecules. Furthermore, organic compounds entering soil or produced therein may become associated with humic acids through a variety of weak associations (e.g., hydrogen bonding, van der Waals forces, ionic bonds). Individual components of the humic acid complex may be microbiologically synthesized aromatic compounds, xenobiotic compounds (e.g., pesticides), or even extracellular proteins and polysaccharides released into the environment by microbial and plant biomass either as a normal process of cellular metabolism or through death and decay of the biomass.

The soil fulvic acid fraction contains a mixture of organic compounds that includes monosaccharides, oligosaccharides, polysaccharides, amino acids, peptides, organic acids, plus fulvic acids. Removing the defined organic compounds through ion exchange chromatographic procedures (Malcolm, 1990), leaves the more ill-defined soil organic component, fulvic acid. Fulvic acid is composed of a light-brown, low-molecular-weight organic material of only a postulated composition. These materials have a molecular weight range of about 1000 to 5000 daltons (Malcolm, 1990) and are more oxidized than humic acids (Stevenson, 1982). Their origin in soil is more problematic. They may be decomposition products of the larger, more complex humic acids, precursors of these molecules, or, more likely, both.

Humic and fulvic acids present a formidable, but not insurmountable, barrier to the microbes involved in mineralization of soil organic matter. A long list of papers have been published in which it is concluded that specific microbial strains with the capability of mineralizing these complex compounds have been isolated (e.g., see Andreyuk and Gordienko, 1978; Andriiuk et al., 1973; Blondeau, 1989; Federov and Il'ina, 1963). The fact that an equilibrium level of these substances is reached in soil further indicates that these substances are mineralized, albeit at a slow rate, in the soil environment. The latter conclusion is supported by the fact that ages of humic acids in soils have been estimated to range from a few decades to several thousands of years (see review by Stout et al., 1981).

Examining the structure of soil humic acids and their distribution in soil indicates two sources of the resistance of these substances to microbial catabolism. Biochemical stability is derived from the diversity of the components of the molecule and their random polymerization. For a microbial species to use these complex molecules for a carbon and energy source, a large number enzymes must be synthesized. The energy expended by the cell in producing the enzymes would most likely exceed that which could be derived from catabolism of the complex aromatic/aliphatic structure. Further impediments to microbial decomposition of humic acids develop from the association of the large molecule with soil minerals, especially clay particles. This chemical and physical interaction inhibits approaching the molecule by either cells or enzymes (see review by Theng et al., 1989).

Because of the diversity of the biological compounds associated with humic acids either through covalent linkage or a weaker association, the potential exists for microbes to gain at least a limited amount of energy from catabolism of portions of the humic acid molecule. The most easily envisioned mechanism for energy production is the mineralization of humified proteins and polysaccharides. Even the oxidation of these normally easily decomposed substances is slow when they are associated with the large humic acid molecule. Not only are the proteins and polysaccharides anticipated to be widely spaced on the humic acid molecules, but the humic acids are dispersed within the soil mineral components. Furthermore, the approach of the protease of hydrolase is inhibited sterically by the obtrusive aliphatic/aromatic humic acid component. Thus, it can be proposed a priori that the organisms involved with mineralization of humified substances would grow slowly. This hypothesis appears to be true in that most reports of microbial growth on humic acids involve the activity of actinomycetes, nocardia, and related microbes (e.g., Andreyuk and Gordienko, 1978; Andriiuk et al., 1973; Blondeau, 1989; Federov and Il'ina, 1963). Linking the observations that a) humic acid molecules are decomposed biologically, b) oxidation of the random structure of the aliphatic/aromatic portion of the molecule would not yield sufficient energy to support microbial growth, and c) humified polysaccharides and proteins could be oxidized to yield energy suggests that the soil microbial community decomposes the aliphatic/aromatic humic acid core cometabolically in the vast majority of the situations. Energy for microbial growth could be supplied by oxidation of the humified proteins and polysaccharides as well as by nonhumified organic matter located in the same microsite as the humic acid molecule.

9.2.2. Analysis of Soil Organic Carbon Fractions

Difficulties in quantifying components of soil organic matter are not generally the result of a shortage of analytical methods. An immense and diverse body of literature exists in which the quantification of xenobiotic chemicals or specific biologically synthesized compounds has been described. Primary difficulties result from attempts to apply the methods developed for analyzing reasonably homogenous or mildly complex samples to chemically complex soil samples. Association of the target organic compound with humic substances or soil minerals reduces recovery efficiencies. Furthermore, the reactivity of soil organic components may be enhanced during the extraction period, thereby creating a diverse array of procedural artifacts. Thus, implementation of any analytical procedure for use with soil systems requires modification in order to maximize extraction efficiency and to minimize artifact generation.

Analysis of a defined organic component in a heterogeneous matrix is complicated by the degree of interaction of various substituents with soil organic and mineral components. Thus, the intensity of the extraction procedure employed necessarily varies depending upon the extent of association of the organic grouping of interest with the mineral and humic components, the complexity of the material to be extracted, as well as the solubility of the substance to be analyzed. The least complex procedure would be that applied to analyzing substances dissolved in soil interstitial water or easily dissolved from particulate substances (e.g., the extraction of free carbohydrates, water-soluble proteins, or simple organic compounds such as acetate and amino acids with water or ethanol). As the solubility of the organic component in water decreases or the association with soil minerals increases, use of more stringent extractants is necessitated. Differential solubility is commonly used to separate soil mineral and organic fractions. This is exemplified by using ether or methylene chloride to extract aliphatic compounds. Perhaps, the most extreme extraction procedure is the separation of humic substances from mineral matter with strongly alkaline solutions (usually 0.1 M NaOH).

A clear understanding of the limitations of each fraction procedure is necessary to determine its general applicability as well as to interpret the data yielded. Concerns range from the more mundane (how to remove the soil from the extraction suspension) to the highly critical determination of artifacts resulting from the extraction procedure. Some specific questions regarding soil organic matter fractionation follow:

- *What is the best method to reduce mineral contamination of the extract?* Soil extracts contain mixtures of the target compound plus dissolved soil minerals and suspended colloidal minerals such as clays. The degree of contamination with these substances is termed the **ash content** of the extract. Because these substances interfere with many of the analytical procedures used to quantify soil organic fractions subsequent to extraction, they must be removed or minimized. Ionic contaminants may be removed with ion exchange resins. Mineral inclusions are reduced by pretreatment of the soil with hydrofluoric acid prior to extraction.

- *Does the extraction procedure modify the organic carbon pool?* Strongly acidic or alkaline solutions are frequently used to dissociate soil organic components from soil minerals. This may result in oxidation, hydrolysis, or even total destruction of some compounds. Polysaccharides and proteins may be hydrolyzed under acidic conditions, the oxidation state of humic acids is increased under alkaline conditions in the presence of oxygen, and some amino acids are destroyed by heating in the acidic solutions generally used for extraction of proteins from soil.

- *Are the extractions quantitative?* Two questions are generally considered in evaluating a procedure for extracting organic matter from soil: Is the procedure efficient? Is the array of compounds contained in the extract representative of that existing *in situ*? An indication of the efficiency of extraction is required to determine *in situ* concentrations, but for many studies a representative sample, not total extraction, is acceptable. As long as a representative, reproducible extraction of the organic matter is conducted, 100 percent efficiency is not required. Greater problems are encountered if the extraction efficiencies of components of the fraction of interest varies. For example, should it be necessary to elucidate the diversity of polysaccharide monomers in an organic soil, all polysaccharide pools must be represented in the extract in proportion to their occurrence in soil. This mixture of sources includes the more easily extracted plant-associated materials as well as those less readily solubolized bacterially synthesized heteropolysaccharides coating soil minerals.

- *How does the extraction procedure affect post-extraction analysis?* Because soil is complex and the extraction procedure necessitated by the association of organic materials with soil mineral and humic components is intense, it is not uncommon to produce a soil extract containing high concentrations of substances that interfer with subsequent characterization of extract components. Probably the best example of this situation is the fulvic acid fraction. The acidity and salt content of the extract must be reduced before the individual organic components can be studied.

Specific extraction procedures for analyzing soil organic matter typically involve dissolution of the organic matter in highly alkaline or acidic solutions (e.g., Stevenson, 1982). All the questions listed previously pertain to these procedures. These methods are characterized by the intensity of the extractants (highly acidic or alkaline solutions), incomplete removal of the fraction from soil, modification of the substance of interest during the extraction, and production of an extraction solution that must be neutralized and desalted prior to analysis.

These limitations of extraction procedures create significant problems with data interpretation, particularly in comparing results attained with different soils or laboratories. Data must be interpreted with the understanding that extraction efficiencies may vary with soil type and that the degree of modification of the soil organic component during extraction may not be equivalent for different laboratories. The latter condition is particularly acute when comparing historical data. Many early experiments were conducted without an appreciation of the degree of artifacts created by the extraction procedure or of the means of reducing these artifacts (e.g., conducting alkaline extracts of soils under a nitrogen atmosphere to reduce oxidation of the organic matter). See Stevenson (1982) for a more detailed discussion of specific procedures used in isolating soil organic matter fractions, their applicability, and limitations.

9.2.3. Conceptual Transition from Evaluation of Chemistry to Determination of the Function of Organic Matter in Soil

As outlined in the previous section, procedures for analyzing organic components in soil are reasonably straightforward. For the soil microbiologist, the greater difficulty is derived from attempting to relate the chemical data to a specific role in the soil ecosystem. None of the extraction procedures commonly applied to the soil system differentiate organic matter pools on a functional basis. More commonly, substances are extracted and quantified based on their solubility (e.g., humic substances), chemical class (e.g., protein, polysaccharide, lipid), or even distribution in specific biomass components (microbial, plant, or animal). Thus, the resultant extracts contain a mixture of functionally active and inactive materials. For example, polysaccharide, which is actively contributing to microbial respiration, may be found in the fulvic acid, humic acid, and humin soil fractions. In contrast, protein, all of which can be quantified in an acid hydrolysate of soil, is a mixture of protein directly contributing to soil respiration and resistant or stabilized protein (i.e., that protein stabilized by covalent linkage into soil humic substances).

The capacity to analyze soil organic matter fractions based on their role in the ecosystem is limited. One approximate division of soil organic substances of some functional value—at least conceptually—is their separation based on decomposition susceptibility. Organic components may be classified as being easily (readily) decomposable or relatively biodegradation-resistant. This elementary division allows discernment of that portion of fixed carbon which provides energy for growth and development of the rapidly dividing microbial populations and that fuels those more ephemeral processes commonly associated with hourly, daily, or even seasonal changes in soil carbon as opposed to the more constant processes driven by the gradual decomposition of the more stable organic fractions.

Readily metabolized soil organic matter provides a carbon and energy source for the soil microbial community with a minimal energy expenditure by the microbe. The microbial investment in energy acquisition involves synthesis of catabolic enzymes and direct substrate activation (as involved with the substrate level phosphorulation of glucose in the hexose monophosphate pathway of glucose catabolism). Compounds that require the action of a few enzymes prior to producing energy for the microbe are more easily decomposed than complex substances that require an interaction with a variety of enzymes prior to imposition of an oxidative change that provides energy to the microbe. It could be stated that the greater the energy deficit created by the microbe in preparation for catabolism of a substance, the less likely that the substance will serve as a primary energy source for the soil microbial community. For example, glucose and other monosaccharides, disaccharides such as sucrose, amino acids, proteins, polypeptides, and simple aromatic compounds are all classified as easily metabolized compounds, whereas humic substances and lignin are extreme examples of more biodegradation-resistant substances.

A useful means of grouping compounds into readily decomposed and biodecomposition-resistant fractions is to evaluate the diversity of enzymes that must be synthesized to mineralize the substrate in relationship to the quantities of energy provided to the microbes. Simple organic compounds are easily decomposed and provide adequate energy to support viable microbial populations. Similarly, polysaccharides, although they may have extended longevity in soil as long as a few years still yield growth-supporting energy with minimal expenditures by the microbes for enzyme synthesis. In contrast, considerable energy for microbial growth is contained in lignin and humic acids, but the energy expenditure involved in enzyme synthesis that would be required for the microbes to recover this energy in growth supporting quantities is prohibitive.

Stabilization of Easily Decomposed Substances in Soil: A substance may be easily grouped in one of two metabolic categories by analyzing its decomposition rate by axenic microbial cultures or even when added to soil samples. Again, the situation is complicated by the heterogeneous distribution of organic compounds and the diversity of their chemical associations in soil. An unquestionably easily metabolized substance, such as carbohydrate or proteinaceous material, may exhibit an extreme longevity in a soil ecosystem. This may result from its association with residual cell structures or with soil humic or mineral matter.

Inherent properties of a chemical compound control its longevity in soil provided the substance exists in a free state and the microbial community (or their enzymes) enjoy unimpaired access to the material. Reductions in substrate concentrations decrease the probability of such access as does occlusion within soil mineral structures or retention within cell walls. All cytoplasmic components, irrespective of the mineralization potential, are occluded from microbial attack until the cell wall is breached. Similarly, access to these energy-yielding substances is also precluded or limited sterically by linkage to soil humic acids.

Carbon Dioxide Yields from Catabolism of Easily Metabolized Substances: The role of readily decomposable organic compounds in soil carbon cycling is unquestioned. These compounds are the primary sources of energy and nutrients in an active soil ecosystem. Acetate catabolism in soil provides an excellent example of the decomposition kinetics and fate of this chemical grouping. After 6 days of incubation in soil, typically greater than 70 percent of acetate carbon is usually evolved as carbon dioxide with the remainder incorporated into microbial biomass components, such as carbohydrates and amino acids (Sørensen and Paul, 1971). As much as 90 percent of the carbon of readily decomposable compounds has been detected as carbon dioxide (e.g., Kassim et al., 1982) with as much as 95 percent of the carbon contained in carbon dioxide plus microbial biomass pools. These decomposition efficiencies are sufficiently reproducible with this class of compounds that if less than 70 percent of the carbon contained in a readily decomposibile substance is evolved as carbon dioxide in a particular experiment then the procedure should be examined to determine whether some soil property is inhibiting microbial activity. For example, in a test of decomposition of a metabolizable pesticide in soil, reduced mineralization could be the result of moisture limitations or failure to account for total disposition of the amended carbon rather than of the inability of the microbial community to catabolize the amendment.

For comparison, a small portion of the carbon contained in a biodegradation-resistant compound is evolved as carbon dioxide, even after extended incubation periods. In a study of melanin decomposition, after a 1-year incubation period, between 5 and 13 percent of the carbon had been evolved as carbon dioxide with less than 0.7 percent of the carbon retained in microbial biomass (Stott et al., 1983). With this class of compounds, the bulk of the amended carbon is retained in unmodified or slightly changed parent molecules. The biodegradation-resistant compounds have an augmented probability of being incorporated into soil humic substances. (See Tate, 1987.)

9.2.4. Microbial Mediators of Soil Carbon Cycle Processes

The concept of microbial infallibility presented previously was soundly based on an appreciation of the diversity of microbial species that function as primary decomposers in soils. Biologically synthesized organic carbonaceous compounds are mineralized in essentially all soil systems where biological life is supported by the chemical and physical traits of the site. Organic carbon compounds are oxidized in soil with a variety of compounds, such as nitrogen oxides, molecular oxygen, organic substances, and sulfur compounds, serving as electron acceptors. Active populations include primary decomposers (bacteria, actinomycetes, and fungi) as well as secondary feeders.

The breadth of organisms involved in carbon mineralization is exemplified by listing a portion of the microbial world involved in cellulose catabolism. This primary soil community energy source is used as a carbon and energy source by fungi (including *Alternaria, Aspergillus, Chaetomium Coprinus, Penicillium, Phanerochaete, Polyporus, Rhizopus, Trichothecium,* and *Zygorhynchus*), yeasts (including *Kluyberomyces* and *Candida*), actinomycetes (including *Micromonospora, Microbispora, Nocardia, Streptomyces,* and *Streptosporangium*), as well as a wide variety of commonly occurring bacterial genera (including *Bacillus, Cellulomonas,*

Clostridium, Corynebacterium, Cytophaga, Pseudomonas, Sporocytophaga, and *Vibrio*) (Alexander, 1979; Béquin, 1990).

Considering the vast number of microbial genera that are involved in carbon-cycling processes, a functional division of cycle participants is commonly more useful in assessing metabolic potential and kinetic individual soil sites than is a detailed analysis of the microbial species present in the sample.

Perhaps the most commonly encountered grouping of soil microbes is based on the growth substrate and the rate of catabolism of the substrate. For example, soil bacteria may be grouped into such metabolic categories as amino acid oxidizers or cellulose degraders. The capacity of the carbon substrate to provide energy for growth and the capability of the microbial species to compete with other members of the microbial community result in their separation into two groups based on growth rate. Rapidly growing soil bacteria, which are primarily involved in catabolism of easily metabolized or fresh biomass, are termed **zymogenous** bacteria, whereas the slow growing organisms (i.e., those predominantly associated with the steady decomposition of native soil organic matter) are classed as **autochthonous** (Winogradsky, 1924).

This growth substrate/rate division of soil bacteria is useful to some degree; but, as the metabolic diversity of the various species (especially members of the genera *Arthrobacter* and *Pseudomonas*) and their survival capacity in soil are elucidated, it becomes clear that an unequivocal separation of soil microbes into these defined groups is difficult. This delineation of soil bacteria is further complicated by the observation that growth rate of an organism is the result of the expression of genetically determined capabilities within the physical and chemical framework of the soil microsite. Thus, an organism that grows rapidly with an easily metabolized substrate as a source of carbon and energy source in the test tube may have generation times of days or longer in soil. Autochthonous or zymogenous designation become cloudy at best.

Other large-scale groupings that are commonly encountered and are perhaps as tenuous as the autochthonous/zymogenous grouping include division of the soil population into native and alien species and, a more modern apparition of this separation, natural vs. bioengineered. As with the bipartite division of autochthonous and zymogenous bacteria, these designations are useful in clear-cut cases but become confusing or even convey an erroneous impression of the diversity of the soil system with the more cosmopolitan or even little-studied microbes. For example, organisms alien to the soil system can be cynically defined as those organisms that have not, to date, been isolated from soil. In such situations, future studies may change "alien" to "native." Further confusion may be derived from varying degrees of inclusiveness associated with native. Are native soil organisms related to a specific soil site, ecosystem, or all terrestrial soils? Difficulties in determining whether organisms should be designated bioengineered or natural will likely emerge as nonengineered organisms with the same or nearly identical genotypes as the laboratory-produced microbes are isolated from soil.

A caveat associated with analyzing data in which soil microbes are divided into any of these general groupings is that we must evaluate the subjectivity of the decision process to ensure that we have not introduced a bias into the conclusions of the experimental study as a result of this artificial grouping of soil populations.

9.3. KINETICS OF SOIL CARBON TRANSFORMATIONS

Identifying and quantifying the proportion of the soil carbon distributed in various organic matter pools is both interesting and useful. However, from the viewpoint of determining the impact of this soil component on our ecosystem, it is the movement of carbon between these various pools and its rate of return to the atmosphere that better describe the dynamics of carbon cycle processes occurring in any specific soil site. As discussed previously, questions that depend upon carbon flux measurements (kinetics of the process) include the evaluation of such terrestrially important processes as the capacity of the soil community to reduce or to modulate atmospheric carbon dioxide concentrations and its utility in recycling organic wastes.

Several mathematical models that describe carbon mineralization processes in soil and the effects of soil environmental properties have been developed (see Tate, 1987). Based on the observation that most carbon mineralization results from biological decomposition, we can reasonably assume that the carbon transformations are best described by Michaelis–Menten kinetics (see chapter 5). When evaluating the decomposition of simple biochemicals, their decomposition can easily be characterized by this simple descriptor of enzyme kinetics or similar equations describing microbial growth kinetics (the Monod equation).

More commonly, organic-carbon-based substances decomposed in soil are a complex mixture of biochemicals that are decomposed by a variety of soil microbes as described previously in a heterogeneous ecosystem. In this situation, carbon dioxide flux rates are a summation of the kinetics of individual biodegradation processes. Thus, any Michaelis–Menten type relationships can be obscured by the variety of active microbes present, the diversity of carbon-based substrates catabolized, the differences in individual growth rates and enzyme efficiencies, and the variation in the impact of the physical environment on the activity of the organisms and the enzymes. That is, kinetics of carbon dioxide evolution become the sum of all sources of this gas. Therefore, the rate of decay of these substances is frequently more easily described by a simple first-order equation with a series of terms for each biochemical group present in the complex substrate. A generic representation of the equation follows:

$$Y_t \text{ (Biomass residue)} = C_1 e^{-k_1 t} + C_2 e^{-k_2 t} \ldots \qquad (9.1)$$

where C_1 and C_2 are concentrations of the individual organic carbon fraction, t is time, and K_1 and K_2 are decomposition rate constants. An individual term can be added to this equation for each biochemical component of the complex organic substance studied, but experience demonstrates that the decomposition rates can be grouped into labile and more resistant fractions. That is, the decomposition of complex organic substances in soil is biphasic (Fig. 9.4) with an initially rapid

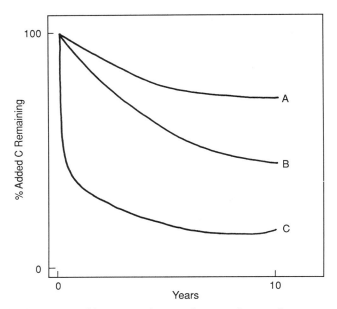

Fig. 9.4. Biphasic decomposition curves for complex organic amendments to soil. The percent of easily decomposable organic matter in A ≪ B ≪ C.

decomposition period (readily metabolized or labile organic carbon) followed by a slow decomposition phase (resistant and occluded organic matter). The magnitude of the rapid decomposition phase is proportional to the quantity of readily decomposed organic matter in the tissue and its accessibility. This phase could represent the majority of the carbon contained in the substrate (e.g., succulent green grass tissue mixed into soil) or a minority of the carbon (e.g., woody tissue).

Decomposition kinetics of rice and barley straw (Murayama, 1984), rye straw and maize (Jenkinson and Ayanaba, 1977), and water hyacinth (*Eichornia crissipes* [Mart.] Solmes) (Moorhead et al., 1986) provide examples of the applicability of this two-term equation in evaluating transformations of complex organic substances in soil. These materials are all constituted primarily of labile organic carbon; hence, most of their biomass is mineralized to carbon dioxide during the initial year following incorporation into soil. With these types of material, we can reasonably anticipate that approximately two-thirds of the carbon will be evolved as carbon dioxide during the first year of incubation in soil. In contrast, the first year weight loss of leaf litter, which contains more biodegradation-resistant structures than do the succulent grass tissues and water hyacinths discussed previously, has been shown to be considerably smaller (e.g., *Fagus grandifolia,* 21 percent; *Acer saccharum,* 32 percent; *Quercus alba,* 39 percent) (Shanks and Olson, 1961).

These data were collected in aerobically incubated soil samples, but similar results can be derived from anaerobically incubated soil samples, if all decomposition products are considered. Gale and Gilmour (1988) derived rate constants for decomposition of alfalfa (*Medicago sativa* L.) in aerobically and anaerobically incubated soil samples. The aerobically incubated samples were incubated at optimum soil moisture under a carbon-dioxide-free atmosphere, whereas the anaerobic soils were flooded under a nitrogen atmosphere. Carbon dioxide, methane, and water-soluble organic carbon were assayed. Three decomposition phases were distinguished (rapid, intermediate, and slow). Aerobic rate constants were 0.123, 0.059, and 0.0095 day^{-1} for each phase. The constants for the anaerobic decomposition were slower (0.118 and 0.024 day^{-1} for the rapid and intermediate phases, respectively). No constant was derived in this study for the slow phase.

These rate constants in the mathematical model are system-dependent, because they are the summation of all environmental parameters that affect the decomposition process in the individual site studied at the time of the experiment. Thus, the specific equation derived from any individual study can be extrapolated only to a system with similar properties, such as temperature, pH, moisture, and soil mineral content. The magnitude of the variation in these constants is not only shown in the preceding study with variation in oxygen tensions but was also revealed in a comparison of the impact of climate on ryegrass decomposition kinetics in England and Nigeria (Jenkinson and Ayanaba, 1977). Decomposition of ryegrass and maize in Rothamsted, England was described by the equation:

$$Y \text{ (percent retention)} = 70.9 \, e^{-2.83t} + 29.1 \, e^{-0.087t} \tag{9.2}$$

where t represents time in years. The decomposition of these substances in Nigeria was described by the same equation, except that each decomposition coefficient was multiplied by 4. Variation in the decomposition constants resulting from localized variation in seasonal soil moisture differences and soil type effects was also detected (Ayanaba and Jenkinson, 1990). Should the experimental objectives include derivation of an ecosystem-independent mathematical representation of biomass decomposition in soil, then more extensive models than the two-term model described herein must be derived (see Tate, 1987).

The data collected in the types of experiments described herein are commonly derived by assessing the evolution of ^{14}C-labeled carbon dioxide from soil samples amended with ^{14}C-labeled plant biomass. Thus, strictly speaking, in such studies the longevity of the carbon atom in soil organic fractions, rather than the survival of the actual plant biomass, is assessed. This distinction is critical in conceptualizing the nature of the organic carbon from which the carbon dioxide is evolved in the latter stages of such a study. Initially, the vast majority of the carbon dioxide collected is a result of the mineralization of added labile plant carbon. During this time of rapid plant biomass decomposition, a portion of the carbon mineralized is incorporated into soil microbial biomass. As time passes, some of the ^{14}C label is incorporated into other soil organic matter fractions, such as humic substances, *in situ* synthesized polysaccharides, and stabilized proteins. Thus, during the slow phase of the decomposition curve (Fig. 9.4), the ^{14}C evolved from the soil sample could originate in residual (more biodegradation-resistant) plant components of the biomass initially added to the soil as well as from the mineralization of soil microbial biomass and humic substances synthesized *in situ* subsequent to amendment. For example, Sørensen (1987) found that after 8 to 15 years incubation in soil, the a portion of the carbon originally added to soil sample as barley straw was distributed in the soil amino acid fraction (21 percent) and microbial biomass (2.7 percent).

9.4. CONCLUSION: MANAGEMENT OF THE SOIL CARBON CYCLE

The function of the soil carbon cycle has regional as well as worldwide implications. For example, ecosystem productivity and stability relies to a large part on the activity of the soil microbial community, whose nutrients and energy are provided by mineralization of organic carbon compounds. The magnitude of soil organic matter reservoirs and their mineralization rate are modulators of atmospheric carbon dioxide loadings. In each of these situations, the rate and direction of change of the carbon forms is of greater importance than its mere presence. In soils at steady state, total soil organic carbon contents are relatively constant (chapter 4), yet large quantities of carbon may be cycling between mineral and organic forms. The primary environmental effect of this nutrient cycling is in enhancement of soil properties conducive to plant community development, thereby resulting in increased system productivity. On a global scale, the primary soil impact on world carbon balances originates from ecosystems in transition, primarily disturbed or managed sites. Such perturbation of soil properties induce attainment of an altered equilibrium concentration of organic matter. A new steady-state carbon content (and, therefore, carbon dioxide flux rate) is achieved. In other words, each soil ecosystem attains an optimal level of organic matter.

Soil organic matter contents may range from nearly zero (e.g., desert soils) to approaching 100 percent. Furthermore, mismanagement of a soil site (i.e., excessive cultivation to destroy soil structure) may result in a nearly total depletion of colloidal organic matter pools. In contrast, under swampy situations, soils (histosols) may be formed consisting of essentially 100 percent organic carbon. With most soils, a level of a few percent organic matter is usually attained and maintained, both in properly managed and native soil sites. The quantity of organic matter retained in a particular soil is the result of achieving a balance between inputs and mineralization as impacted by the physical and chemical properties of the site.

Because organic matter synthesis and decomposition are primarily biologically controlled, they can be managed. Modulators of soil biological activity (see chapter 4), which may be anthropogenically controlled, include pH, temperature, nutrient concentration, and moisture. Furthermore, as the capacity to develop genetically engineered microorganisms improves, inoculation of a particular soil site with microbes specifically designed for optimal function therein will become a management tool. The soil carbon cycle will be maintained to maximize societal benefits and stewardship of a renewable resource, our soil ecosystems.

REFERENCES

Aguilar, R., E. F. Kelly, and R. D. Heil. 1988. Effects of cultivation on soils in Northern Great Plains rangeland. Soil Sci. Soc. Am. J. 52:1081–1085.

Alegre, J. C., and D. K. Cassel. 1986. Effect of land-clearing methods and post clearing management on aggregate stability and organic carbon content of a soil in the humid tropics. Soil Sci. 142:289–295.

Alexander, M. 1977. Introduction to Soil Microbiology. John Wiley & Sons. New York. 467 pp.

Amundson, R., and H. Jenny. 1991. The place of humans in the state factor theory of ecosystems and their soils. Soil Sci. 151:99–109.

Andreyuk, E. I., and S. A. Gordienko. 1978. Transformation of humic acids by soil actinomycetes. Mikrobiol. Zh. (Kiev) 40:690–697 (Russian).

Andriiuk, K. I., S. O. Hordienko, I. N. Havrysh, H. I. Konotop, and V. A. Martynenko. 1973. Decomposition of peat humic acids by associative cultures of microorganisms. Mikrobiol. Zh. 35:554–559 (Russian).

Armentano, R. V. 1980. Drainage of organic soils as a factor in the world carbon cycle. Bioscience 30:825–830.

Ayanaba, A., and D. S. Jenkinson. 1990. Decomposition of carbon-14 labeled ryegrass and maize under tropical conditions. Soil Sci. Soc. Am. J. 54:112–115.

Barber, S. A. 1979. Corn residue management and soil organic matter. Agron. J. 71:625–627.

Béquin. 1990. Molecular biology of cellulose degradation. Annu. Rev. Microbiol. 44:219–248.

Blondeau, R. 1989. Biodegradation of natural and synthetic humic acids by the white rot fungus *Phanerochaete chrysosporium.* Appl. Environ. Microbiol. 55:1282–1285.

Bossert, I., W. M. Kachel, and R. Bartha. 1984. Fate of hydrocarbons during oily sludge disposal in soil. Appl. Environ. Microbiol. 47:763–767.

Boyle, M. 1990. Biodegradation of land-applied sludge. J. Environ. Qual. 19:640–644.

Crawford, R. L. 1981. Lignin Biodegradation and Transformation. John Wiley and Sons, New York. 154 pp.

Dibble, J. T., and R. Bartha. 1979a. Rehabilitation of oil-inundated agricultural land: A case history. Soil Sci. 128:56–60.

Dibble, J. T., and R. Bartha. 1979b. Effect of environmental parameters on the biodegradation of oil sludge. Appl. Environ. Microbiol. 37:729–739.

Douglas, B. F., and F. R. Magdoff. 1991. An evaluation of nitrogen mineralization indices for organic residues. J. Environ. Qual. 20:368–372.

Epstein, E., J. M. Taylor, and R. L. Chaney. 1976., Effects of sewage sludge compost applied to soil on some soil physical and chemical properties. J. Environ. Qual. 5:422–426.

Eswaran, H., E. van den Berg, and P. Reich. 1993. Organic carbon in soils of the world. Soil Sci. Soc. Am. J. 57:192–194.

Fedorov, M. V., and T. K. Il'ina. 1963. Utilization of humic acid by soil actinomycetes as a sole source of carbon and nitrogen. Microbiology 32:234–237.

Gale, P. M., and J. T. Gilmour. 1988. Net mineralization of carbon and nitrogen under aerobic and anaerobic conditions. Soil Sci. Soc. Am. J. 52:1006–1010.

Gupta, U. C. 1967. Carbohydrates *in* A. D. McLaren and G. H. Peterson (eds.) Soil Biochemistry. pp. 91–118. Dekker, New York.

Harden, J. W., E. T. Sundquist, R. F. Stallard, and R. K. Mark. 1992. Dynamics of soil carbon during deglaciation of the Laurentide ice sheet. Science 258:1921–1924.

Hatcher, P. G., M. Schnitzer, L. W. Dennis, and G. E. Marciel. 1981. Aromaticity of humic substances in soil. Soil Sci. Soc. Am. J. 45:1089–1094.

Havlin, J. L., D. E. Kissel, L. D. Maddux, M. M. Claassen, and J. H. Long. 1990. Crop rotation and tillage effects on soil organic carbon and nitrogen. Soil Sci. Soc. Am. J. 54:448–452.

He, X.-T., S. J. Traina, and T. J. Logan. 1992. Chemical properties of municipal solid waste composts. J. Environ. Qual. 21:318–329.

Jenkinson, D. S., and A. Ayanaba. 1977. Decomposition of carbon-14 labeled plant material under tropical conditions. Soil Sci. Soc. Am. J. 41:912–915.

Jenkinson, D. S., and J. H. Rayner. 1977. The turnover of soil organic matter in soil of the Rothamsted classical experiments. Soil Sci. 123:298–305.

Jenny, H. 1980. The Soil Resource, Origin and Behavior. McGraw-Hill Book Co. New York. 377 pp.

Kassim, G., J. P. Martin, and K. Haider. 1982. Incorporation of a wide variety of organic substrate carbons into soil biomass as estimated by fumigation procedure. Soil Sci. Soc. Am. J. 45:1106–1112.

Kern, J. S., and M. G. Johnson. 1993. Conservation tillage impacts on national soil and atmospheric carbon levels. Soil Sci. Soc. Am. J. 57:200–210.

Kirk, T. K., and R. L. Farrell. 1987. Enzymatic "combustion": The microbial degradation of lignin. Annu. Rev. Microbiol. 41:465–505.

Kögel-Knaber, I. 1993. Biodegradation humification and processes in forest soils. In J.-M. Bollag and G. Stotzky, (eds.), Soil Biochemistry, Vol. 6, pp. 101–135. Marcel Dekker. New York

Malcolm, R. L. 1990. The uniqueness of humic substances in each of soil, stream and marine environments. Anal. Chim. Acta 232:19–30.

Mann, L. K. 1986. Changes in soil carbon storage after cultivation. Soil Sci. 142:279–288.

Martin, J. P., and K. Haider. 1979. Biodegradation of ^{14}C-labeled model and cornstalk lignins, phenols, model phenolase humic polymers, and fungal melanins as influenced by a readily available carbon source and soil. Appl. Environ. Microbiol. 38:283–289.

Mays, D. A., G. L. Terman, and J. C. Duggan. 1973. Municipal compost: Effects on crop yields and soil properties. J. Environ. Qual. 2:89–92.

McMurtrie, R. E., H. N. Comins, M. U. F. Kirschbaum, and Y.-P. Wang. 1992. Modifying existing forest growth models to take account of effects of elevated CO_2. Aust. J. Bot. 40:657–677.

Mitchell, M. J., R. Hartenstein, B. L. Swift, E. F. Neuhauser, B. I. Abrams, R. M. Mulligan, B. A. Brown, D. Craig, and D. Kaplan. 1978. Effects of sewage sludges on some chemical and biological characteristics of soil. J. Environ. Microbiol. 7:551–559.

Moorhead, K. K., D. A. Graetz, and K. R. Reddy. 1986. Decomposition of fresh and anaer-obically digested plant biomass in soil. J. Environ. Qual. 16:25–28.

Murayama, S. 1984. Decomposition kinetics of straw saccharides and synthesis of microbial saccharides under field conditions. J. Soil Sci. 35:231–242.

Norris, D. J. 1980. Landspreading of oily and biological sludges in Canada. Proc. Ind. Waste Conf. 35:10–16.

Oades, J. M. 1988. The retention of organic matter in soils. Biogeochem. 5:33–70.

Ouyang, Y., and L. Boersma. 1992a. Dynamic oxygen and carbon dioxide exchange between soil and atmosphere: I. Model development. Soil Sci. Soc. Am. J. 56:1695–1702.

Ouyang, Y., and L. Boersma. 1992b. Dynamic oxygen and carbon dioxide exchange between soil atmosphere: II. Model simulations. Soil Sci. Soc. Am. J. 56:1702–1710.

Piccolo, A., and J. S. C. Mbagwu. 1990. Effects of different organic waste amendments on soil microaggregates stability and molecular sizes of humic substances. Plant Soil 123: 27–37.

Polglase, P. J., and Y. P. Wang. 1992. Potential CO_2-enhancing carbon storage by the terrestrial biosphere. Aust. J. Bot. 40:641–656.

Rasmussen, P. E., and C. R. Rohde. 1988. Long-term tillage and nitrogen fertilization effects on organic nitrogen and carbon in a semiarid soil. Soil Sci. Soc. Am. J. 52:1114–1117.

Richter, D. D., L. I. Barbar, M. A. Huston, and M. Jaeger. 1990. Effects of annual tillage on organic carbon in a fine-textured udalf: The importance of root dynamics to soil carbon storage. Soil Sci. 149:78–83.

Rogers, H. H., J. F. Thomas, and G. E. Bingham. 1983. Response of agronomic and forest species to elevated atmospheric carbon dioxide. Science 220:428–429.

Scott, H. D., and L. S. Wood. 1989. Impact of crop production on the physical status of a Typic Albaqualf. Soil Sci. Soc. Am. J. 53:1819–1825.

Shanks, R. E., and J. S. Olson. 1961. First year breakdown of leaf litter in southern Appalachian forests. Science (Washington, DC) 134:194–195.

Sørensen, L. H. 1987. Organic matter and microbial biomass in a soil incubated in the field for 20 years with ^{14}C-labelled barley straw. Soil Biol. Biochem. 19:39–42.

Sørensen, L. H., and E. A. Paul. 1971. Transformation of acetate carbon into carbohydrate and amino acid metabolites during decomposition in soil. Soil Biol. Biochem. 3:173–180.

Stevenson, F. 1982. Humus Chemistry: Genesis, Composition, Reactions. John Wiley & Sons. New York. 443 pp.

Stott, D. E., J. P. Martin, D. D. Focht., and K. Haider. 1983. Biodegradation, stabilization in humus, and incorporation into soil biomass of 2,4-D and catechol carbons. Soil Sci. Soc. Am. J. 47:66–70.

Stout, J. D., K. M. Goh, and T. A. Rafter. 1981. Chemistry and turnover of naturally occurring resistant organic compounds in soil. In E. A. Paul and J. N. Ladd (eds.), Soil Biochemistry, Vol. 5, pp. 1–73. Dekker. New York.

Sundquist, E. T. 1993. The global carbon dioxide budget. Science 259:936–961.

Tate, R. L., III. 1980. Microbial oxidation of organic matter of histosols. In M. Alexander (ed.), Adv. in Microbial Ecology, Vol. 4, pp. 169–201. Dekker. New York.

Tate, R. L., III. 1987. Soil Organic Matter: Biological and Ecological Effects. John Wiley & Sons. New York. 291 pp.

Theng, K. G., K. R. Tate, and P. Sollins. 1989. Constituents of organic matter in temperate and tropical soils In D. C. Coleman, J. M. Oades, and G. Uehara (eds.), Dynamics of Soil Organic Matter in Tropical Ecosystems, pp. 5–32. University of Hawaii Press. Honolulu.

Vitousek, P. M. 1991. Can planted forests counteract increasing atmospheric carbon dioxide? J. Environ. Qual. 20:348–354.

Winogradsky, S., 1924. Sur la microflora autochtone de la terre arable. Comptes rendus ebdomadaire des séances de l'Academie des Sciences (Paris) D, 178:1236–1239.

Wood, C. W., R. J. Mitchell, B. R. Zutter, and C. L. Linn. 1992. Loblolly pine plant community effects on soil carbon and nitrogen. Soil Sci. 154:410–419.

Woodwell, G. M., J. E. Hobbie, R. A. Houghton, J. M. Melillo, B. Moore, B. J. Peterson, and G. R. Shaver. 1983. Global deforestation: Contribution to atmospheric carbon dioxide. Science 222:1081–1086.

10 The Nitrogen Cycle: Soil-Based Processes

The soil nitrogen cycle can easily be divided into two "minicycles": a) a predominantly soil-resident portion consisting of those processes (mineralization, ammonification, nitrification, and immobilization) associated primarily with the conversion of organic nitrogenous compounds to plant-available mineral forms and their return to biomass (plant, microbial, and, indirectly, animal); and b) the more inclusive (soil as well as the greater biosphere resident processes) cycle entailing the conversion of atmospheric dinitrogen into forms available to the living components of the ecosystem (i.e., nitrogen fixation), the transformations within the soil described previously, and the return of the fixed nitrogen to atmospheric dinitrogen (i.e., denitrification). The interrelationships of these two aspects of the nitrogen cycle are depicted in Fig. 10.1.

The significance of this positional differentiation of the nitrogen cycle in native environments is readily exemplified and contrasted by comparing the situation in a climax forest ecosystem and a newly developing soil system (e.g., a highly impacted mineland site in which all naturally occurring organic matter has been destroyed). In the forest system, external fixed nitrogen sources (atmospheric deposition and nitrogen fixation) is minimal. The majority of the fixed nitrogen used for plant biomass synthesis is derived from incorporating mineral nitrogen compounds produced by *in situ* mineralization of native organic matter. In the pioneer system, indigenous nitrogen resources are minimal. Thus, plant biomass productivity is totally reliant on external resources, such as would be provided by nitrogen-fixing populations or through amendment with organic residues.

In those situations where the bulk of the nitrogen used for plant biomass synthesis is the product of recycling soil resident nitrogen compounds between organic and mineral forms, homeostasis of the ecosystem can be maintained only if nutrient losses are minimized. Thus, it is advantageous for the majority of the fixed nitrogen contained in soil reservoirs to exist in nonmobile forms (e.g., organic or biomass nitrogenous compounds) and, to some degree, ammonium ion. Fortunately, most of the fixed nitrogen in soil systems is retained in organic nitrogen pools. Typical data demonstrating the distribution of soil nitrogen in organic and mineral forms is presented in Table 10.1. Although soil nitrate and ammonium concentrations are highly variable with time and position in the soil profile, note that these compounds at most comprise a few percent of the total soil nitrogen. In the examples cited, quantities of nitrogen in soil organic matter ranged from about 10- to 500-fold the mass of nitrogen in mineral forms. (For further discussion of distribution of nitrogen in these various soil reservoirs, see Tate, 1987, or Stevenson, 1986, for a review of

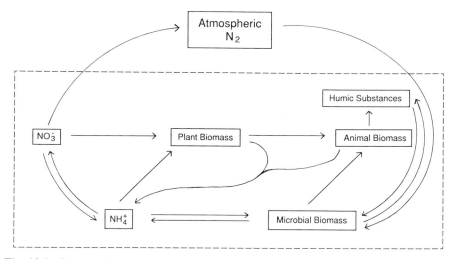

Fig. 10.1. Conceptual model of the nitrogen cycle. Nitrogen reservoirs enclosed in the dashed lines reside in soil, whereas dinitrogen is replinished from atmospheric sources.

the topic. See Bolton et al., 1990; Cabrera and Kissel, 1988a; Cassman and Munns, 1980; Clarholm et al., 1981; or Serna and Pomares, 1992, for examples of primary research involving nitrogen transformations involving these nutrient pools.)

Documenting the predominance of organic nitrogen in soil ecosystems reveals a potential indicator of plant biomass production potential. Since plant biomass synthesis is frequently controlled by the availability of soil mineral nitrogen resources (other potential delimiters of plant productivity include soil moisture, phosphate or trace mineral levels, and a variety of soil physical and chemical properties), the

TABLE 10.1. Examples of Nitrogen Distribution in Mineral and Organic Forms: Agricultural Soils from Valencia, Spain,[1] and Soils Beneath Shrub-Steppe Vegetation (South Eastern Washington, USA)[2]

Soil	Organic N (mg N kg^{-1} soil)	Mineral N (mg N kg^{-1} soil)	% Mineral N
Sandy Soils (Spain) (n = 9)	300–2470	0–86	0–7.1
Loamy Soils (Spain) (n = 9)	710–2400	3–153	0.4–6.7
Clay Soils (Spain) (n = 9)	720–3680	3–85	0.2–1.8
Shrub-Steppe Soils (n = 15)	1500–3300[3]	3.71–7.00	0.2–0.3

[1]From Serna and Pomares (1992).
[2]From Bolton et al. (1990).
[3]Total Nitrogen.

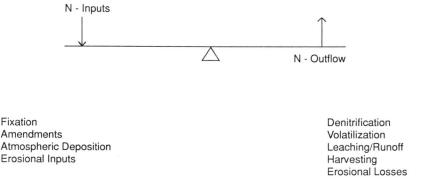

Fixation   Denitrification
Amendments   Volatilization
Atmospheric Deposition   Leaching/Runoff
Erosional Inputs   Harvesting
                                  Erosional Losses

Fig. 10.2. Processes contributing to soil nitrogen balance.

assessment of net nitrogen mineralization rates of the internal soil organic nitrogen pools, and/or the amendment rates of external fixed-nitrogen resources can provide a valuable estimation of potential productivity within a particular ecosystem. Nitrogen inputs from sources external to the soil biological community (fixation or soil amendment) are the primary determinants of ecosystem productivity in soils containing low levels of biologically decomposable organic matter or in cropped soils receiving inputs of industrially fixed nitrogen or organic residues.

A conceptual model of the dependence of plant community productivity and structure on nitrogen mineralization is depicted in Fig. 10.2. Nitrogen mineralization provides a pivotal base for total biomass production. In an ecosystem with large reservoirs of organic matter (microbial, plant and animal biomass, plus colloidal soil organic matter), plant biomass synthesis is controlled by the balance between nitrogen mineralization and immobilization (i.e., the net mineral nitrogen available for uptake by plant roots). Some fixed nitrogen is lost through denitrification, volatilization, leaching, plant biomass removal (harvesting by animals in native sites and cropping in agricultural systems), and soil erosion. In a sustained ecosystem, these nitrogen outputs are necessarily balanced by fixation, plant biomass inputs, atmospheric deposition, and erosional deposition. In a steady-state system, a balance exists between the rates at which mineral nitrogen is generated and removed. Modification of these rates results in the establishment of a new biomass productivity steady state wherein these factors are again balanced. Continued losses of fixed nitrogen eventually results in the creation of a biomass desert.

An appreciation of the pivotal role of nitrogen mineralization on the nature and productivity of the plant community of an ecosystem is essential in managed systems. Both excessive nutrient inputs and losses have negative consequences in regard to system sustainability. Fixed-nitrogen decrements in agricultural systems commonly arise from crop removal and enhanced leaching of mineral nitrogen. System overloads may be encountered through excessive organic matter amendments as may be encountered through disposal of sludges, composts, or animal waste materials or overuse of nitrogenous fertilizers.

Based on the essential nature of mineralization, immobilization, and nitrification on the stability of soil ecosystems, this chapter evaluates the primary nitrogen-cycling processes in soil and the properties of soil that control these reactions. Biological nitrogen fixation is evaluated in chapter 11 and denitrification is examined in chapter 12.

10.1. NITROGEN MINERALIZATION

Simply stated, **nitrogen mineralization** is the conversion of organic nitrogen to inorganic nitrogen. Because, in soil, ammonium is usually rapidly oxidized to nitrate by nitrifiers, net nitrogen mineralization in an ecosystem is estimated by quantifying ammonium and nitrate concentrations. A more limited term, **ammonification,** is used to designate the conversion of organic nitrogen to ammonium. Mineralization is the sole, soil-resident, biological means of generating the nitrogen forms usable by green plants.

10.1.1. Soil Organic Nitrogen Resources

Soil organic nitrogen is distributed between plant, animal, and microbial biomass, litter, and humic substance pools. Thus, determining which biochemicals are mineralized by the soil microbial community entails a complete understanding of the nitrogenous substances contained in biomass plus the more exotic humified substances. All nitrogenous compounds commonly contained in living cells are found in the soil organic fraction. Humified nitrogen, i.e., nitrogen linked to humic and fulvic acids, may be in recognizable biochemical forms (e.g., humified proteins and amino-sugar-containing polysaccharides) or in heterocyclic ring compounds incorporated into the basic structure of humic or fulvic acids. (For more information on the chemistry of soil nitrogen associated with humic substances, see Stevenson, 1982, for a review of the topic or Bondietti et al., 1972; Malcolm, 1990; Piccolo et al., 1990; or Schnitzer and Kernodorff, 1980 for examples of primary research.) It is interesting that Schnitzer and Kerndorff (1980) found that humic and fulvic acids from industrially and environmentally polluted soils contained more nitrogen and sulfur and less oxygen than did these substances isolated from unpolluted soils.

The complex heterocyclic aromatic compounds contained in humic substances are of interest for developing a complete picture of the soil organic nitrogen component, but it must be emphasized that they contribute minimally to soil nitrogen cycling. Nitrogen mineralization rates are determined essentially in totality by the quantity and availability of the more readily metabolized organic nitrogen compounds found in soil. These substances are composed primarily of amino acids and amino sugars and their oligomers and polymers. Other contributors include nucleic acid components plus, in managed systems, such societal products as various heterocyclic-ring-containing pesticides. Again, except in the more limited situations where these latter compounds enter the soil system in the high concentrations associated with accidental spills or mismanagement, they usually contribute minimally to overall soil nitrogen cycling.

This short list of classes of nitrogen-containing biochemicals leaves an implication of simplicity of nitrogen mineralization processes that does not occur in native soils. Indeed, biochemically the process is the result of catalysis of a limited number of compounds by a few relatively commonly occurring enzymes (e.g., proteases and amidases). The predominant nitrogenous compounds (amino acids and amino sugars—monomers, dimers, oligomers, and polymers) are all relatively rapidly decomposed by a vast variety of soil microbes. Yet, it is rather easily demonstrated that nitrogen mineralization is not proportional to the quantities of these compounds present in the soil site as could naively be predicted. For example, Fox and Piekielek (1978a, 1978b) reported that after a 2-year period, a variety of chemical soil nitrogen parameters correlated with nitrogen mineralization in soils from corn fields; yet, with a longer testing period (4 years) and greater variability in soil and weather, none of the tests were found to be accurate predictors of soil nitrogen mineralization (Fox and Piekielek, 1983). Serna and Pomares (1992) found that prediction of soil nitrogen availability to plants in growth chamber studies required analysis of initial mineral nitrogen compounds of the soils plus several chemical and biological assays. Similarly, Juma and Paul (1984) evaluated nitrogen mineralization dynamics using an ^{15}N-labeled Weirdale, loam soil. Their study of ^{15}N enrichment in mineral nitrogen products supported the conclusion that extraction of a highly labeled organic N pools from the soil only partly explained the substrates used for nitrogen mineralization in the soil. Their data supported the conclusion that a variety of nitrogen pools interact in nitrogen mineralization processes.

Furthermore, the diversity in mineralization rates (or availability of nitrogenous compounds to attack by the microbial community) is derived more from location within various soil components than from chemical structural variation of the compounds mineralized. Nitrogen-containing biochemicals may be bound to soil organic matter or particles, contained in plant, animal, or microbial biomass or debris or be dissolved in soil interstitial water. Each of these situations provides an unique impediment to access to the biochemical for liberation of the nitrogen contained therein. For example, humified compounds are protected by the inability of the enzyme to approach the substrate, whereas microbial cells present a reasonably resistant cell wall barrier to protect internal biodegradable compounds.

Another portion of the difficulty in using concentrations of decomposable organic nitrogen compounds as predictors of nitrogen mineralization potential is the fact that nitrogen mineralization may not be the primary reason for converting organic nitrogen to ammonium or nitrate. Organic nitrogenous compounds found in soil may also serve as sources of carbon and energy for the microbial community. Thus, any mineral nitrogen produced by these processes is a by-product of carbon and energy metabolism. For example, Learch et al. (1992) evaluated the mineralization of organic sewage sludge constituents in soils. The quantity of proteins extracted from the sludge were highly correlated with carbon mineralization but had a limited relationship to the nitrogen mineralization rate, supporting the hypothesis that the soil microbial community was using the protein as a carbon rather than nitrogen source. In these situations, the carbon/nitrogen ratio tends to serve as a better indicator of nitrogen mineralization potential in that with a narrow ratio significant production of mineral nitrogen can be expected to occur whenever the microbial community uses the substrate for a carbon/energy source (e.g., DeLuca and Keeney, 1993; Learch et al., 1992). (See the following discussion regarding the interaction of carbon/nitrogen ratio of the substrates metabolized by soil microorganisms and the resultant impact of mineral nitrogen production.)

10.1.2. Assessment of Nitrogen Mineralization

Nitrogen mineralization in soil traditionally has been studied based on its contribution to plant biomass production—generally in agricultural systems. It must be stressed that not all the nitrogen mineralized in soil is available to the plant community. Mineral nitrogen is removed from soil through assimilation by plants and microbes, it may be denitrified, and it may be leached from the root zone. Hence, change in soil mineral nitrogen could be represented as follows:

$$\Delta N_i = N_m - (N_a + N_p + N_v + N_l + N_d) \tag{10.1}$$

where N_i represents inorganic nitrogen, N_a is nitrogen assimilated by microbes, N_p is nitrogen assimilated by plants, N_l is nitrogen leached from the soil, N_v is nitrogen volatilized, and N_d is nitrogen denitrified. The number of terms in this relationship indicates the complexity of tracing the flux of nitrogen atoms between various soil nitrogen reservoirs in a field situation, but the movement of this atom is simplified in laboratory-incubated soil samples that are commonly used to estimate soil nitrogen mineralization potential.

This parameter, usually measured in laboratory-incubated soil samples, is most commonly estimated by assessing net mineral nitrogen production over a given time period in a closed system (batch soil samples or soil columns) in the absence of growing plants. Because such a system would be physically isolated, leaching, volatilization, and denitrification losses are also minimized. Denitrification is generally reduced by maintaining aerobic soil conditions. Therefore, in these closed soil systems

$$\text{Net } N_m = \Delta N_i = N_m - N_a \tag{10.2}$$

Thus, it is apparent that, to understand the dynamics of mineral nitrogen production for plant biomass synthesis, the nuances of nitrogen immobilization must be evaluated.

10.2. NITROGEN IMMOBILIZATION

10.2.1. Process Definition and Organisms Involved

Nitrogen immobilization is the microbial assimilation of mineral nitrogen compounds. Any actively growing microbe in soil contributes to this incorporation of soil ammonium and nitrate into microbial biomass. Thus, nitrogen immobilization occurs under such diverse environmental conditions as those associated with aerobic or anaerobic soils, acid-impacted sites, soils receiving hot spring or even nuclear power plant waste waters, as well as situations where the primary metabolic activity results from autotrophs. In other words, among the microbes commonly encountered that catalyze nitrogen mineralization are aerobes, anaerobes, thermophiles, mesophiles, psychrophiles, acidophiles, neutrophiles, and alkalinophiles.

Historically, the primary justification for studying immobilization has related to its role in the loss of plant nutrient nitrogen from agricultural soils. In unmanaged ecosystems, studies of nitrogen immobilization have been mainly associated with evaluation of basic nitrogen-cycling processes (e.g., the quantification total nitrogen mineralization and the proportion of soil mineral nitrogen reservoirs cycled through microbial biomass). In contrast, in agricultural soils or other managed sites (e.g., reclamation projects), interest in nitrogen immobilization kinetics acquires an economic aspect, because quite frequently the major justification of soil amendment with anthropogenically fixed nitrogen is to stimulate aboveground plant biomass synthesis. Excessive incorporation of the amended nitrogen into microbial biomass necessarily impacts plant community development.

10.2.2. Impact of Nitrogen Immobilization Processes on Plant Communities

The quantities of mineral nitrogen incorporated into the soil microbial population is dependent upon the community activity, which in turn in most aerobic soil systems is controlled by the availability of carbon and energy resources. In anaerobic or oxygen-limited soils, moisture-limited sites, or soils of extreme acidity, microbial activity is usually sufficiently controlled by these factors that nitrogen immobilization is not an impediment to plant biomass productivity. In carbon-controlled communities, the impact of an influx of carbon-rich material on mineral nitrogen resources (i.e., ammonium and nitrate) depends upon the carbon/nitrogen ratio of the soil amendment. If the carbonaceous compounds providing carbon and energy to the active portion of the soil microbial community contain sufficient nitrogen to meet the needs of the soil microbes decomposing them, then a nitrogen deficiency does not occur. Unfortunately, with most commonly utilized carbon and energy source in soil, plant biomass, this is not the situation. Plant biomass has carbon/nitrogen ratios as high as 60/1 to 80/1. Thus, because soil bacteria have carbon to nitrogen ratios of 4/1 to 6/1 and fungal ratios generally range from 10/1 to 12/1, synthesis of microbial biomass from these substrates necessarily results in a soil nitrogen deficit. In other words, some soil nitrogen must be incorporated into microbial biomass to balance the carbon inputs. In soils where insufficient mineral nitrogen exists to compensate for the deficiencies of the carbon and energy source, the microbial growth rate is reduced by the nitrogen limitation. In either situation, a reduction in total soil mineral nitrogen resources by immobilization results in a loss of plant biomass productivity.

The differential fate of carbon and nitrogen contained in the organic matter mineralized by the soil microbial community is reflected by the change in the carbon/nitrogen ratio of the organic matter as it is transformed from plant biomass into soil organic matter. With the decomposition of biomass, the carbon contained therein is oxidized primarily to carbon dioxide with a small portion being incorporated into microbial biomass components. The nitrogen in the substance being decomposed is conserved in the soil site. Thus, the effective carbon/nitrogen ratio of soil humus decreases with time. Field data indicate that after the carbon/nitrogen ratio of the biomass residues amended to soil reaches about 30/1, net nitrogen immobilization is no longer a problem. At this point, instead of consuming mineral nitrogen, the catabolism of the plant biomass becomes a mineral nitrogen source; that is, there is net nitrogen mineralization within the system.

The obvious question in this situation relates to the apparent contradiction associated with the accumulation of mineral nitrogen in the field when the carbon/nitrogen ratio of the decomposed substrate approaches the 30/1 carbon/nitrogen ratio of the growth substrate when the 10/1 mean carbon/nitrogen ratio of the microbial biomass predicts that a mineral nitrogen deficit should occur until the ratio approaches that characteristic of the microbial biomass. That is, if the microbes decomposing the plant residue have an average carbon/nitrogen ratio of 10/1, why is nitrogen not limiting until the metabolized substrate carbon/nitrogen ratio approaches 10/1 rather than 30/1? At least a partial answer to this question is derived from analyzing the basic biochemical makeup of the plant biomass. Remember that a portion of the plant biomass entering the soil system is of limited metabolic availability to the soil microbes. These carbonaceous substances include lignin and lignified substances, easily metabolized substances protected by the more slowly decomposed plant cell walls, plus substances that become associated with and thereby become protected by soil mineral matter such as soil clays. Thus, the effective C/N ratio of the substance actively metabolized by the soil microbial community may be much lower than the overall ratio detected by gross analysis of the decomposing organic matter.

The rate of decomposition of even the more readily decomposed organic substance providing carbon and energy to the soil microbial community affects the impact of the amendment on nitrogen immobilization. This fact is exemplified by two studies, one using relatively simple carbon compounds, the other with complex plant components. Kelley and Stevenson (1987) added glucose, a phenolic glycoside, or catechol with ^{15}N-labeled ammonium sulfate to soil and evaluated nitrogen immobilization. Most rapid immobilization occurred in glucose-amended soils (maximum incorporation at 6 days) compared to maximum incorporation times of 10 days with the phenolic glycoside and 20 days with catechol. In their study, Kelly and Stevenson found that, at maximum immobilization, the nitrogen incorporated into microbial biomass was 3.5- to 6-fold more susceptible to mineralization than was native soil nitrogen. Thus, more recently synthesized, most likely less humified, organic nitrogen compounds may be more metabolically available to the soil microbial community. Similarly, Mengel and Schmeer (1985) evaluated the effect of straw, cellulose, or lignin soil amendments on nitrogen immobilization with the overall objective of the project being to increase the retention of nitrogen in soil as organic nitrogen thereby reducing denitrification losses. These latter workers found that straw and cellulose stimulated greater nitrogen immobilization than did lignin in the 36-day incubation period. This observation relates to the biodegradation susceptibility of the substrates in that little microbial biomass synthesis resulted from amendment of the soil with lignin due to its inherent resistance to microbial catabolism. The complexity of the soil nitrogen cycle interactions was shown in this study in that they found that straw and cellulose amendment also stimulated the denitrification rate in amended soils compared to the uncropped soil. Thus, incorporation of added nitrogen into the crop was higher in the unamended soils than occurred in those receiving exogenous carbon sources.

These observations suggest that the percent nitrogen of organic material entering soil declines or increases until it approaches the percent nitrogen content of the soil microbial biomass. In reality, this change can be predicted and is observed in soil systems receiving large influxes of organic material, but the final carbon/nitrogen ratio achieved in a soil system at equilibrium is considerably greater than that detected in axenic cultures of bacteria and fungi. This limitation results from the fact that, in a biologically active mineral or organic soil, there is a significant concentration of soil humic substances (humic and fulvic acids plus humin) that have very wide (60/1 to 80/1) carbon/nitrogen ratios. Recall that these soil organic components are only minimally susceptible to catabolism by the biological community. The presence of these substances obscures changes induced by the amendment of soil with plant biomass.

The carbon/nitrogen ratio of native soil organic matter is a mean of the average 10/1 ratio of the microbial community and the elevated value of the soil humic substances. For example, Post et al. (1985) in a study of a variety of soil ecosystems varying in climate and aboveground vegetation communities found that all but a sample of subtropical wet forest (carbon/nitrogen ratio = 3.3/1) and two soil samples from tropical thorn wood land (carbon/nitrogen ratio = 9.2/1) had carbon/nitrogen ratios ranging from 10.2/1 to 30.2/1 with a mean of 16.0/1 for all soil types studied and a mean of 17.8/1 when the values were weighted for the variable number of samples per ecosystem type studied. Wet and dry ecosystems from a variety of climates were evaluated. Although neither of these mean values for the carbon/nitrogen ratios of soils can be considered to be representative of all soils, they do provide an indication of the situation in steady-state native soil ecosystems.

10.2.3. Measurement of Soil Nitrogen Immobilization Rates

Assessment of total nitrogen mineralization requires determination of the proportion of soil nitrogen immobilized. This generally requires labeling of nitrogen atoms so that their movement between various soil nitrogen reservoirs can be traced. This task is accomplished through the use of ^{15}N. Mineral nitrogen compounds enriched in ^{15}N may be added to soil samples and its incorporation into organic nitrogen pools measured or the dilution of this labeled inorganic nitrogen pool by mineralization of unlabeled organic nitrogen can be assessed. Limitations of this procedure aside from availability of the mass spectrometer necessary to detect the ^{15}N are involved with mixing of the labeled substances into soil. Calculations of nitrogen mineralized or immobilized by this technique are based on the assumption that a distinct nitrogen pool is not created by the soil amendment. In other words, the added labeled nitrogenous substrate is of the same susceptibility to the soil microbial community responsible for its decomposition as the indigenous nitrogen reservoirs. Due to the heterogeneity of soil, this assumption is likely rarely applicable, but this procedure yields data that are a reasonable estimate of the rate of *in situ* processes.

There has been some interest in developing soil extraction procedures that are specific for recently immobilized nitrogen compounds. For example, He et al. (1988) and Kelley and Stevenson (1985) demonstrated that milder extraction procedures (e.g., hot water, hot 10 mM $CaCl_2$) removed more of the recently immobilized nitrogen than did the more intensive procedures (e.g., acidified permanganate). Although some enrichment of immobilized nitrogen was achieved with these procedures, the nitrogen contained in the extractants was still derived from a variety of soil pools. Serna and Pomares (1992) evaluated a combination of biological methods (plant uptake and aerobic soil incubations) and chemical extraction procedures (autoclave, 0.5 M $KMnO_4$, 6M HCl, or 0.01 M $NaHCO_3$) as predictors of nitrogen availability to maize in calcareous soils. They found no extraction method was an effective indicator of the concentrations of organic nitrogen compounds that can be metabolized by the soil microbial community, but predictive capabilities were improved by combining results from the assessment of initial mineral nitrogen contents of the soils plus several chemical and biological assays in regression analyses.

These observations document that nitrogen immobilization is an essential function in soil microbial communities in that it represents the proportion of nutrients incorporated into the microbial community. From the view of the total ecosystem, nitrogen immobilization pathway represents an alternate fate (e.g., plant biomass vs. soil microbial biomass) of mineral nitrogen. This concept contrasts to the viewpoint of the agriculturist and, to some the degree, the reclamation specialist where nitrogen immobilization is viewed as a process detrimental to aboveground plant productivity. With each nitrogen atom incorporated into the microbial biomass, additional external nitrogen inputs are necessitated to maximize the achievement of the desired ends of an attractive and/or productive aboveground community development.

10.3. QUANTITATIVE DESCRIPTION OF NITROGEN MINERALIZATION KINETICS

Net nitrogen mineralization in native soil systems is generally observed to follow first-order kinetics (see Section 9.5.4). That is, it can be described by the equation

$$\frac{dN}{dt} = -kN \tag{10.3}$$

where N is the concentration of mineralizable nitrogen, k is a rate constant, and t is time. Thus, a commonly used integrated form of this equation used to calculate net nitrogen mineralization follows:

$$N_m = N_0(1 - e^{-kt}) \tag{10.4}$$

where N_m represents the net nitrogen mineralized, N_0 is the organic nitrogen at zero time, and k and t are a rate constant and time, respectively. It must be noted that situations may also exist where zero-order net nitrogen mineralization kinetics are detected; that is, nitrogen mineralization is independent of the organic nitrogen concentration. In this situation, nitrogen mineralization kinetics are described by the equation

$$\frac{dN}{dt} = k \tag{10.5}$$

where k is a rate constant. Occurrence of zero-order kinetics can be predicted to occur in soils amended with high concentrations of nitrogen rich organic substances (e.g., bacterial or fungal biomass waste substances from fermentative, industrial processes).

The most commonly occurring alternative mathematical model used to describe nitrogen mineralization in soil samples is the double exponential model. As indicated previously, organic nitrogen occurs in a variety of forms and physical situations in native soil samples. Some organic nitrogen compounds are easily decomposed by the microbial community and are easily encountered by the organisms or their enzymes, whereas others are inherently biochemically resistant or physically occluded. In reality, each class or organic nitrogenous compound has a characteristic decay rate, but field data indicate that these substances can rather easily be grouped into two categories, those relatively rapidly metabolized and a more resistant reservoir. Thus, the preceding model can be expanded to include both organic nitrogen reservoirs as follows:

$$N_t = N_0 S \left(1 - e^{-ht} \right) + N_0 \left(1 - S \right) \left(1 - e^{-kt} \right) \tag{10.6}$$

where N_t is the nitrogen mineralized in time t, N_0 is the potentially mineralizable nitrogen, S and $(1 - S)$ represent the labile and less labile organic nitrogen fractions of the total organic nitrogen pool, and h and k are decomposition rate constants. Deans et al. (1986) applied this double exponential model to a variety of historical studies and found that it provided a good description of the net nitrogen mineralization kinetics. Other mathematical representations of nitrogen mineralization include empirical models (e.g., see Broadbent, 1986) and a variety of double exponential and mixed-order representations (e.g., see Bonde and Rosswall, 1987, and Bonde and Lindberg, 1988). A caveat in deriving nitrogen mineralization constants from application of these mathematical expressions is that the values derived may vary depending upon the methods used to fit the data to the model (e.g., see Ellert and Bettany, 1988; Juma et al., 1984; Talpaz et al., 1981).

Extrapolation of Laboratory-Derived Net Nitrogen Mineralization Rates to Field Situations: Independent of which mathematical model is used to analyze nitrogen mineralization data, applying laboratory-derived results to field situations is dependent upon a number of assumptions (see Section 8.5.4). The primary assumptions that must be considered are a) soil organic nitrogen mineralization occurs in the laboratory-incubated samples at rates comparable to those existing in the field, b) preparation of soil samples for laboratory analysis does not alter the availability of various soil-organic-nitrogen reservoirs to the microbial community, and c) (many workers assume that) nitrogen mineralization is a first-order process; that is, alternative models are not evaluated in the data analysis process. In reality, all these assumptions are of questionable validity, at best.

The difficulties with the assumption that laboratory nitrogen mineralization kinetics reflect field rates are readily exposed when we consider that nitrogen mineralization kinetics vary widely with temperature and moisture (e.g., see Addiscott, 1983; Cabrera and Kissel, 1988b; Cassman and Munns, 1980; Kladivko and Keeney, 1987; Myers et al., 1982; Smith et al., 1977; Stanford et al., 1975), whereas in the most commonly implemented laboratory incubation methods, these soil ecosystem properties are controlled within a limited range. Cassman and Munns (1980) derived mathematical models to compensate for field variation in temperature (15, 20, 25, and 30°C) and moisture (0.1, 0.3, 0.7, 2, 4, and 10 bars). Marion and Black (1987) used empirically derived models to include the effect of time and temperature on nitrogen mineralization kinetics in Arctic tundra soils. Temperature variability may be compensated by using the buried bag procedure (Poovarodom et al. 1988; Westermann and Crothers, 1980), but field moisture variation is limited within the bag. Using empirically derived relationships or even widely applied models, such as Q_{10} or the Arhenius relationship, are useful in adapting laboratory parameters to the more variable field situation.

The second assumption (collection and preparation of soils for laboratory incubation has minimal effect on nitrogen mineralization kinetics) is invalidated when we consider the soil properties controlling the availability of the various soil organic nitrogen pools to the microbial community. It has already been stated that a primary controller of nitrogen mineralization is the accessibility of the substrates to the microbes and enzymes. Physical handling of soils, which occurs during collection and sieving of the samples, clearly alters the occlusion of organic compounds within soil aggregates and thereby their susceptibility to biological decomposition. For examples of studies of the impact of this liberation of soil organic matter in sample preparation on kinetics and the development of mathematical models to account for this laboratory-generated artifact, see Beauchamp et al. (1986), Craswell and Waring (1972a, 1972b), Cabrera (1993), and Cabrera and Kissel (1988a, 1988b). Because we understand that even the minimal handling of soil samples associated with the buried bag procedure results in the alteration of nitrogen availability and thus the nitrogen mineralization parameters derived from the study, we can conclude that currently no method exists to negate the impact of sample handling on our estimation of soil nitrogen mineralization potential.

A final consideration in estimating field nitrogen mineralization parameters is their spatial and temporal variation. Among the many factors limiting laboratory analysis of this soil community property is the reality that only a limited number of soil samples can be analyzed and that the incubations require several weeks for completion. Insufficient numbers of soil samples are generally assayed to overcome the impact of variability of nitrogen mineralization rates in the field in space and time. Nitrogen mineralization is highly variable across the landscape and changes significantly on a microscale with variation of limiting physical and chemical parameters. Goovaerts and Chiang (1993) found that microscale nitrogen mineralization variability (< 1 m) was explained to a large degree by the distribution of soil oxidizable carbon, which correlated with gravimetric water content. This microscale variation impacts the nature of a representative soil sample. Goovaerts and Chiang estimated that 24 to 37 samples were needed to estimated nitrogen mineralization potential (with a precision of 0.1 at $p = 0.05$) for a 1600-m^2 plot. To see the impact of variation with time, we must incorporate data describing variability of primary soil properties controlling nitrogen mineralization into the analysis procedure. Similar results were presented by Starr et al. (1992) in determining soil sample size and number requirements for assessing field variation of soil nitrate concentrations.

The basis for this spatial variation in nitrogen mineralization potential relates to properties of both the macrosoil ecosystem as well as the microstructure of the system. The former is exemplified by studies of the impact of plant distribution on soil organic nitrogenous compounds and nitrogen mineralization potential (e.g., see Bolton et al., 1990). Typically in such studies, soil organic nitrogen as well as associated biological activities are concentrated within the portions of the soil profile that are under the direct influence of the growing plant (i.e., the rhizosphere and areas impacted by plant litter). The effect may be direct through the addition of plant-derived organic matter to the soil system or indirect as a result of the improvement of soil properties conducive to microbial population development. The latter may include moisture preservation due to elevated soil organic matter levels, temperature modulation on hot summer days under plant canopies, as well as improved soil structure due to encouragement of aggregate formation by plant roots.

Variation in the distribution of soil organic nitrogen reservoirs is also associated with soil structural development. Separation of soil fractions by particle density or size results in fraction of soil organic matter into more readily decomposed and more biodegradation-resistant portions. Mineralizable carbon and nitrogen tend to be associated with smaller particle sizes (Anderson et al., 1981; Cameron and Posner, 1979; Schnitzer and Kodama, 1992). These fractions tend to have a lower carbon/nitrogen ratio, with most of the mineralizable soil nitrogen found in the fractions with smaller diameter particles. Similarly, separation of soil based on density fractionation results in enrichment of readily decomposable organic matter (hence, an enrichment for higher nitrogen mineralization potential) in light fractions (i.e., specific gravity greater than 1.2 g cm^{-3}) (e.g., see Sollins et al., 1984).

10.4. MICROBIOLOGY OF MINERALIZATION

Nitrogen mineralization is essential to the development of a sustainable ecosystem. Hence, it is reasonable to conclude that organic nitrogenous compounds are mineralized to mineral nitrogen in any situation where life is possible. Were this not a basic truth, the biotic community would require a continued input of mineral nitrogen and would have to adapt continuously to increasing levels of accumulated organic matter. This situation is rare, if it ever occurs, in natural soil systems. Plant biomass accumulates in swampy ecosystems, but aboveground plant production is balanced with the slow nitrogen mineralization rate occurring in waterlogged soils associated with the swampy conditions. In our society, soils that receive constant inputs of mineral nitrogen are either fertilized agricultural or landscaped situations (wherein the fertilization regime presumably balances plant production needs) or pollution-impacted situations (e.g., soils receiving nitrogen-laden industrial effluents).

Thus, based on the widespread occurrence of nitrogen mineralization, we can predict that an equally diverse microbial community can catalyze the process. Nitrogen is mineralized by a diverse group of bacteria, fungi, and actinomycetes. These organisms are aerobes, anaerobes, thermophiles, mesophiles, psychrophiles, as well as metabolic diverse in regards to their energy needs. Specific microbial species active in this process represent all microbial genera growing and reproducing in soil. Indeed, quantification of nitrogen-mineralizing populations in soil varies or is limited not by the number present in the soil sample but rather by the ingenuity of the individual devising the growth medium for the microbial populations. Common nitrogen substrates used to isolate organisms involved in nitrogen mineralization include protein, chitin, and amino sugars.

10.5. ENVIRONMENTAL INFLUENCES ON NITROGEN MINERALIZATION

Because the processes associated with nitrogen mineralization and immobilization are basic to microbial life in general, it is reasonable to assume a priori that these biologically catalyzed portions of the nitrogen cycle occur under essentially any environmental condition wherein microbial growth is possible. Simply stated, nitrogen immobilization is the incorporation of mineral nitrogen into cell biochemicals. Thus, nitrogen immobilization is easily concluded to be a major by-product of microbial growth. In contrast, the direct association of nitrogen mineralization with carbon mineralization provides an indirect link between the former process and biological energy generation. Thus, the questions involving environmental influences on nitrogen mineralization and immobilization relate not to whether the processes occur but rather to the impact of modification of soil physical and chemical properties on the rate of these nitrogen conversions.

The primary ecosystem properties that are usually evaluated for their impact on nitrogen immobilization and mineralization rates are nitrogen concentration, soil moisture, pH, and temperature. Nitrogen concentration effects were evaluated previously with the discussion of net nitrogen mineralization rates. The impact of moisture, temperature, and pH variation on these processes follows a pattern similar to that which has been observed for most soil biologically catalyzed processes (e.g., Fig. 4.3). There is a minimum and maximum level of the stress factor wherein no activity is possible. Between these limiting values, the biological activity tends to increase to a maximum level at the optimal adaptation point of the stress factor after which the biological activity again declines. This generalized reaction of the micro-bial community is exemplified best by analyzing the impact of soil moisture on nitrogen mineralization. Again, a generalized description of the process is presented, because individual reaction rates at specific soil moisture levels vary with the nature of the microbial community existent in a particular soil ecosystem. With no other limitations, microbial growth, and thereby nitrogen mineralization and immobiliza-tion, tends to increase as water availability increases. An optimal level of activity is generally reached at approximately water saturation; that is, the point where ade-quate soil water is available for the microbial to interact with their growth substrates and for by-products of their metabolism to be removed from the cells' environs. As the soils become flooded, oxygen diffusion becomes a limiting factor to microbial respiration. Once the oxygen tension is reduced by an imbalance between consump-tion and influx of free oxygen, cellular processes become limited. Nitrogen mineral-ization and immobilization decline. Both of these processes continue under anaer-obic conditions, albeit at a generally reduced rate due to the lower metabolic rates of the anaerobic organisms compared to those occurring under oxygenated condi-tions. Optimal soil moisture conditions for aerobic transformations of nitrogen generally range between 0.3 bar and saturation (60 to 90 percent water-filled pore space) (e.g., see Walters et al., 1992), whereas the optima for pH and temperature are near neutrality and within the thermophillic range (40–60°C), respectively. Clearly, although most soils are not thermophillic and major active ecosystems have soils with pH values substantially below neutrality (e.g., tropical and temperate forest soils), these processes occur at rates sufficient to sustain highly productive ecosystems.

10.6. NITRIFICATION

Nitrification is the biological formation of nitrate and nitrite from compounds containing reduced nitrogen. These oxidative reactions are catalyzed by two, mutu-ally exclusive, groups of organisms, ammonium and nitrite oxidizers. The primary reactions involved follow:

$$2\ NH_4^+ + 3O_2 \rightarrow 2NO_2^- + 4H^+ + 2H_2O$$

$$2NO_2^- + O_2 \rightarrow 2NO_3^-$$

These processes involve the direct incorporation of molecular oxygen into the nitrogen oxide products, and hydrogen ions are produced by ammonium oxidation. Conversion of ammonium to nitrate results in a change in the oxidation state of the nitrogen from its most reduced state (-3) to the most oxidized form ($+5$) (Table 10.2). As will be explained later, the primary implications of these observations are a) nitrification is an obligatorily aerobic process and b) oxidation of ammonium results in acidification of the soil environment.

TABLE 10.2. Oxidation State of Nitrogen Compounds Associated with Soil Nitrogen Cycling.

Compound	N Oxidation State
NH_4^+	-3
NH_2OH	-1
NO	$+2$
N_2O	$+1$
NO_2^-	$+3$
NO_3^-	$+5$

10.6.1. Identity of Bacterial Species that Nitrify

All nitrifiers are closely related by their specialized biochemical reactions. All oxidize reduced nitrogenous compounds for energy and fix carbon dioxide for their carbon source (see Section 3.2.2). The primary taxonomic difference at the genus level is based on cellular morphology. Common ammonium oxidizers, nitrite oxidizers, their source, and morphology are listed in Table 10.3. Note that the ammonium oxidizers are distinguished by the prefix *Nitroso-* for the genus name, whereas the nitrite oxidizers have the comparable prefix *Nitro-*.

Some confusion may develop when studying the classical literature because past dogma has a decided impact on our concept of the role of these various species in soil. Prior to the early 1980s, it was commonly held that, of the species of nitrifiers, only *Nitrosomonas* spp. and *Nitrobacter* spp. occurred commonly in soil and that these two genera were *undoubtedly* the major nitrifying chemoautotrophs. Hence, when population densities of ammonium oxidizers were estimated, the numbers were presented as populations densities of *Nitrosomonas*. Similarly, the epithet *Nitrobacter* was used globally for nitrite oxidizers. With the application of fluorescent antibodies to quantify nitrifier populations in soil at the species level, it has become evident that a variety of different species were found in soil systems (e.g., see Belser and Schmidt, 1978a, 1978b; Fliermans et al., 1974; Stanley and Schmidt, 1981). As more specific enumeration procedures such as polymerase chain reaction application (e.g., Navarro et al., 1992) and fluorescent antibodies specific for various genera of nitrifiers are applied to the study of the diversity of nitrifier populations in soil, an improved understanding of the variety of these organisms *in situ* will emerge. At this point, it is evident that sufficient variation in the identity of active nitrifiers in soils occurs to justify the use of the epithets ammonium and nitrite oxidizers in place of the more specific genera designations.

10.6.2. Benefits to the Microorganism from Nitrification

The primary organisms responsible for ammonium oxidation to nitrate in soil are chemolithotrophs. (See Section 10.6.6 for a discussion of the possible role of heterotrophic nitrifiers in soil.) The organisms drive their energy from the oxidation of the

TABLE 10.3. Common Species of Autotrophic Nitrifiers, Their Sources, and Morphologies.

Bacterial Species	Source	Morphology
Ammonium Oxidizers		
Nitrosomonas europea	Soil, water, sewage	Ellipsoidal, short rods
Nitrosospira briensis	Soil	Spiral
Nitrosococcus oceanus	Marine	Spherical
Nitrosococcus mobilis	Marine	Spherical
Nitrosococcus nitrosus	Marine/soil	Spherical
Nitrosolobus multiformis	Soil	Lobate, pleomorphic
Nitrosovibrio tenuis	Soil	Comma shaped
Nitrite Oxidizers		
Nitrococcus mobilis	Marine	Spherical
Nitrobacter gracilis	Marine	Short rods
Nitrobacter winogradski	Soil	Short rods
Nitrobacter agilis	Soil, water	Short rods

nitrogen atom. Conversion of ammonium to nitrite yields 66 kcal per mole, whereas nitrite oxidation to nitrate produces 20 kcal. The recovery of this energy by the microbes ranges from about 5 to 15 percent. Thus, large quantities of ammonium or nitrite must be oxidized for each carbon fixed by the microbes. Ammonium oxidizers typically oxidize between 14 and 70 ammonium-nitrogens per carbon incorporated into cellular biomass, whereas between 76 to 135 nitrite-nitrogens must be oxidized for the comparable task.

10.6.3. Quantification of Nitrifiers in Soil Samples

Difficulties associated with culture of the bacterial species involved in nitrification have resulted in the development of indirect culture methods and procedures for direct observation of cells within the soil sample or in extracts. The two most commonly employed techniques are the most-probable-number method using nitrifier-specific growth media and direct enumeration of specific organisms using fluorescent antibodies.

Two requirements of the most-probable-number method are a) media and growth conditions allowing any and all nitrifying cells present in the soil sample to grow to a population density that can be detected must be used and b) the cells must be sufficiently separated from any particulate matter in soil so that each cell is individually and independently dispersed in the diluent (i.e., nitrifier cells and not soil particles must be enumerated). No most-probable-number technique for quantification of nitrifiers meets either of these requirement. All media are selective in that only a portion of the nitrifier population can grow. This situation is the most likely reason for the development of the central dogma cited previously in that the dominant species growing in the commonly used most-probable-number media are *Nitrosomonas* sp. Further complications associated with essentially all procedures based on selective media is that the more rapidly growing organisms limit or preclude development of populations of more fastidious organisms. Support for the role of this phenomenon with nitrifiers is seen in the work of Matulewich et al. (1975), where it was found that the much shorter incubation times commonly employed for nitrifier assays (e.g., Alexander and Clark, 1965) did not allow for maximal population development. Incubation times of up to 55 and 100 days were required for ammonium and nitrite oxidizers, respectively. Difficulties associated with using the longer incubation periods include delays in data acquisition as well as the generally encountered problems involving evaporation of the growth media. With longer incubation times, the increased salt concentration in the growth media due to the loss of water may itself induce artifacts in population selection. Therefore, analysis of data procured through this procedure must always be predicated on the limitations imposed by the selectivity of the media, the artifacts resulting from the long incubation times necessary for population development as well as the failure of some investigators to use sufficient incubation periods for their studies, and the low statistical precision inherent in such procedures. The latter limitation precludes use of the procedure for studies where we must be able to detect slight changes in population densities.

In light of these methodological problems, the question may be asked, Why is the most-probable-number procedure still the most commonly used procedure to estimate field populations of nitrifiers? Perhaps, even more crucial is the question, Why is its continued common usage likely? The answer to both questions relates to the practicality of the procedure. Most-probable-number assays of nitrifiers are still the only methods available to most investigators in the field.

All the problems inherent in the most-probable-number procedures are overcome with the use of fluorescent antibody techniques. Data are produced rapidly, specific populations are quantified, and small changes in populations of the nitrifiers are easily measured. Yet, as with any scientific procedure, methodological limitations of the fluorescent antibody technique governs the wide implementation of the procedure. A major barrier to implementation is the availability of equipment necessary for the assays. Whereas most-probable-number incubations can be conducted with equipment available in essentially any field microbiology laboratory, fluorescent antibody procedures require at least the availability of a fluorescent microscope. Should in-house production of the species-specific antibodies be necessary, the investigator also must have the necessary animal care facilities. Specific data interpretation difficulties associated with the use of fluorescent antibodies are a) many serotypes of autotrophic nitrifiers occur (thus, large numbers of fluorescent antibody types may need to be used to quantify the total nitrifier population) and b) it is difficult to isolate and purify nitrifiers.

10.6.4. Discrepancies Between Population Enumeration Data and Field Nitrification Rates

As indicated previously, the energy yield from oxidation of ammonium and nitrite and the free-energy efficiency for the process are known. Thus, we can reasonably assume that some relationship between the amount of nitrate produced in a soil system and the population of nitrifiers detected therein should occur. Unfortunately, the most common observation is that the population density of autotrophic nitrifiers in soil samples is insufficient to yield the nitrate concentrations detected. A number of reasons have been proposed to explain the difficulties:

- *Inherent inaccuracies of the most-probable-number procedure.* These inaccuracies include the fact that total nitrifier populations are underestimated in many situations because of their inability to grow in the selective media used as well as the imprecision of the procedure.
- *Noncoincidence of nitrifiers and nitrate production.* The nitrate present in the soil sample could have been produced in the past and the population of nitrifiers responsible for its production subsequently declined.
- *Errors in estimating growth yields.* The growth yields in culture used to estimate population production per unit of ammonium or nitrite oxidized probably differ from those yielded in soil.
- *Imported nitrate.* The nitrate detected in a soil sample may not have been produced *in situ*. Nitrate moves freely in the soil profile. The nitrate detected in a particular sample of soil could have been produced elsewhere and transported into the soil sample with water infiltration.

Each of these explanations for the fact that the nitrifier population detected in soil samples is frequently less than 10 percent of that necessary to produce the nitrate present in the samples is plausible. It is highly probable that simple descriptions of the population dynamics in any specific soil site present an idealistic picture of the true situation. That is, nitrifier populations cannot be estimated by *in situ* nitrate or nitrite concentrations because of the ephemeral nature of these substances in soil, population fluctuations controlled by substrate availability, and limitations in our understanding of the complexity of the biological interactions occurring in the system.

10.6.5. Source of Ammonium and Nitrite for Nitrifiers

Ammonification is the primary source of ammonium in most undisturbed and in many disturbed lands (i.e., nonfertilized systems). Exogenous sources include point sources (e.g., sewage or comparable waste substances amended to soil) as well as nonpoint sources (e.g., nitrogenous fertilizers leaching from or applied to cropped lands). Continuous enrichment of nitrifiers may occur downstream from a sewage plant, or intermittent population stimulation may occur from irregular flow from cropped soils due to variation in rainfall patterns.

Any nitrite found in soil is presumed to arise from ammonium oxidation. Under normal situations, nitrite does not accumulate in soils. Thus, the limiting factor in nitrification in most soils is the conversion of ammonium to nitrite (low levels of ammonium are detected in most soils). The nitrite is usually converted to nitrate as rapidly as it is formed. Under alkaline conditions, nitrite may accumulate, because

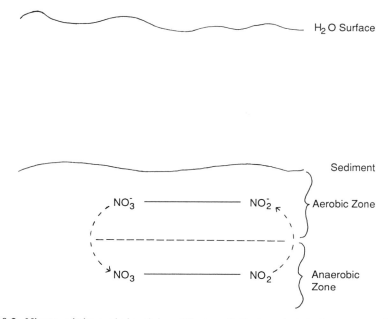

Fig. 10.3. Nitrate–nitrite cycle involving diffusion of nitrate and nitrite between aerobic and anaerobic zones in sediments.

ammonia is toxic to *Nitrobacter* sp. An equilibrium between ammonium and ammonia normally exists in soils with ammonium predominating in neutral and acidic conditions and ammonia predominating under alkaline conditions. Thus, under alkaline conditions in the presence of ammonia, *Nitrobacter* spp. are inhibited but the ammonium oxidizers remain active. Hence, nitrite accumulates. This phenomenon also occurs when anhydrous ammonium is added to soil.

Nitrite oxidizers may also derive a portion of their growth substrate from the reduction of nitrate (e.g., see Belser, 1977). In a flooded system, nitrate may be reduced to nitrite in anaerobic microenvironments (Fig. 10.3). The nitrite may then diffuse to aerobic sites where the nitrite oxidizers can convert it to nitrate. In these situations, populations of nitrite oxidizers are elevated in comparison to the ammonium oxidizers.

10.6.6. Environmental Properties Limiting Nitrification

In contrast to the situation with mineralizers—both carbon and nitrogen—where a wide diversity of soil microbes can catalyze the process, the kinetics and occurrence of nitrification in soil reflects the fact that a limited number of bacterial species are involved in the process. Primary concerns in predicting the extent and rate of nitrification in an ecosystem are a) substrate availability, b) temperature, c) pH, d) oxygen tension, and e) soil moisture.

Substrate Availability: As indicated previously, most soils contain low concentrations of nitrate and ammonium, with detectable nitrite levels occurring occasionally (e.g., see Jones and Schwab, 1993). Thus, it can be concluded that the nitrite oxidizers are generally controlled by the presence of their growth substrate, which indirectly is determined by ammonium availability.

The kinetics of nitrification processes necessarily depend upon population density of nitrifying bacteria present and efficiency of nitrifying enzymes. Nitrifiers are highly efficient organisms at recovering ammonium and nitrite from their environment. That is, the enzymes responsible have low Michaelis constants. Properties of the nitrifier community that can be deduced from the reaction kinetics follow:

- Nitrate production may increase proportionally to ammonium concentration in the soil sample. In this situation, the ammonium-oxidizing enzymes in the existing nitrifier population are substrate-limited. Thus, nitrate production in these systems is described by Michaelis–Menten kinetics (e.g., See Malhi and McGill, 1982).

- Nitrate production in some cases is described by zero-order kinetics (i.e., nitrification rate is not proportional to the quantity of substrate present). Indigenous ammonium production is adequate to provide all the substrate necessary to support growth of the indigenous nitrifier population. Nitrification kinetics in this situation may or may not deviate from zero order with time. If the nitrifiers are growth-substrate-limited, then continued elevated levels of the energy-providing substrate can induce a population increase. The nitrifier population increases until another environmental factor (e.g., space, oxygen, pH, species interactions) becomes limiting. Amendment of an ecosystem wherein the nitrifier population is controlled by one of these latter site properties rather than growth substrate availability will not induce population growth; therefore, deviation from zero-order kinetics will not occur.

- Nitrification may not occur in an ecosystem until several days following amendment. In this case, for some reason a functional nitrifier population has not developed. With amendment of the soil site with ammonium, a nitrifier population slowly develops. An example of this phenomenon was provided in a study of a pine forest soil by Vitousek and Matson (1985).

In conclusion, the *in situ* nitrification rate varies with the degree of saturation of the existent enzymes in the soil system and the environmental limitations controlling nitrifier population development. Changes in the nitrification rate are dependent upon the capacity of the nitrifier population density to increase.

Toxic concentrations of ammonium and nitrite can be reached in axenic cultures of nitrifiers. Generally, normally encountered environmental levels of the nitrification substrates are far below these inhibitory concentrations, but elevated ammonium concentrations have been shown to inhibit the process. For example, Malhi and McGill (1982) in a study of three Canadian soils found that concentrations of 200 $\mu g \ g^{-1}$ NH_4^+-N supported rapid nitrification but that elevation of the ammonium–nitrogen concentration to 300 $\mu g \ g^{-1}$ inhibited the process. Malhi and

McGill suggested that the inhibition could result from a combined effect of low pH and salt content.

Temperature: In the laboratory, ammonium oxidation to nitrite occurs at temperatures from near freezing to 65°C, whereas nitrite oxidation occurs from near freezing and is completely inhibited at about 40°C. The latter process has an optimum generally reported to be between 30 and 35°C. Since nitrification occurs in tropical and subtropical soils at temperatures far above the limits for nitrite oxidation, some compensation must occur in the microbe's native environment that is not provided under axenic culture.

pH: In culture, both the rate of nitrification and the number of nitrifiers declines below a pH of about 6.0. Negligible nitrification is detected below a pH of 5.0 to 5.5. The generally detected optimum for this process is between 6.6 and 8.0. This observation has led to the dogma that autotrophic nitrification cannot occur in acidic environments; yet, it can be shown readily that nitrification does occur in acid soils, such as is commonly encountered in forest soils. Explanations of this discrepancy between the laboratory and real-world observations include a) microsite variation in soil pH, b) invoking the possibility of the occurrence of nonculturable, acidophilic, autotrophic nitrifiers in these soils, c) selection of heterotrophic nitrifier populations under these acidic conditions, and d) adaptation of the growth properties of commonly encountered nitrifiers to accommodate the acidic conditions. Each of these hypothesized mechanisms for nitrification in the acid forest soils may operate under some situations, but none is totally satisfying. (See Dommergues et al., 1978, for a review of the pros and cons of these mechanisms.) For example, the possibility of nonculturable, acid-loving organisms is plausible, and some acid-loving autotrophic nitrifiers have been cultured (e.g., see Hankinson and Schmidt, 1988). The presence of humic acid has also been shown to provide some relief of pH inhibition of nitrification in soil columns (Bazin et al., 1991). The postulation of more neutral microsites in otherwise acidic soils is plausible in soils whose predominant pH is near the limit for autotrophic nitrifiers, but prediction of the occurrence of such islands of neutrality in extremely acidic soils is implausible. The potential for the formation of microaggregates of nitrifying bacteria in acidic soils is interesting and a potentially useful resolution of this laboratory–field anomaly (De Beer et al., 1993; De Boer et al., 1991). The size of the aggregates is limited by the penetration depth of the oxygen necessary for this obligatorily aerobic process.

The most commonly encountered (and perhaps debated) interpretation of the field data relating to nitrification in acidic soils involves the role of heterotrophic nitrifiers. Heterotrophic nitrification is the oxidation of reduced nitrogen forms to nitrite or nitrate by heterotrophic bacteria and fungi. The reduced nitrogen can be either organic or inorganic. Heterotrophic nitrification can be easily demonstrated in culture. Large numbers of soil bacteria, actinomycetes, and fungi that are capable of oxidizing ammonium to nitrite or nitrate (e.g., Doxtander and Alexander, 1966; Eylar and Schmidt, 1959; Schimel et al., 1984; Tate, 1977; Verstraete and Alexander, 1973) have long been known to exist in soil, but their function in native soil ecosystems at rates to affect significantly *in situ* nitrate concentrations is questioned. Ammonium or organic nitrogen oxidation by these organisms generally occurs after the active growth phase of the organism in media with a low carbon/nitrogen ratio. More nitrogen is usually present than the organisms need for growth. The quantities of nitrate or nitrite produced are small compared to those produced by the auto-trophic population (less than 1 vs. 2000 or greater μg ml^{-1} solution for the heterotrophs and autotrophs, respectively). Because the oxidation of the reduced nitrogen occurs subsequent to replication of the heterotrophic cells, it is unlikely that the organisms are recovering the energy from the oxidation. Although there are reports of existence of large populations of heterotrophic microbes in some ecosystems, their environmental significance is unknown.

Oxygen Tension: Nitrification is an obligatorily aerobic process. Molecular oxygen is directly incorporated into the final product. Therefore, both soil moisture and soil structure impact the nitrification rate indirectly through control of oxygen diffusion. Nitrification can occur at oxygen concentrations as low as 0.3 μg ml^{-1}, thus the key in occurrence of this process in waterlogged or otherwise diffusion-limited systems is the availability of a means for oxygen import. Influx of oxygen into soils where it would normally not be found can be through such processes as mass movement of oxygenated water into flooded soils or even transport of air through the plant into the rhizosphere as occurs with rice plants.

Moisture: As indicated previously, limitations of nitrification in elevated mois-ture situations are derived primarily from the indirect control of availability of molecular oxygen by its diffusion rate in water. At the other extreme, desiccated soils, preclusion of nitrification in extremely dry soils, and the direct proportionality of nitrification rate with moisture level between this totally inhibitory level and moisture saturation results from the fact that the soil microbial community is essen-tially aquatic. That is, microbes require a thin film of water on the soil particles within which to respire and grow. In the absence of this layer of water, no nitrifica-tion can occur. As the thickness of the coating of water on soil particles increases, the "aquatic" soil nitrifier bacterial population becomes active.

Other Limitations to Nitrification: Two site characteristics—with the potential to alter soil nitrification potential—that should be considered in land reclamation or management plans are metal contamination and salinity. Both have been demon-strated to have a significant impact on nitrification rates.

Liang and Tabatabai (1978) assessed the inhibition of nitrification by a variety of elements including silver, mercury, cadmium, nickel, arsenic, chromium, boron, aluminum, selenium molybdenum, manganese, lead, cobalt, copper, tin, iron (II and III), and zinc. Soil pHs ranged from 5.8 to 7.8, clay contents ranged from 39 to 50 percent, and organic matter ranged from 2.58 to 5.45 percent. When added at a concentration of 5 $\mu moles$ g^{-1} soil, all 19 metals inhibited nitrification, but the amount of inhibition varied with metal added and the soil. The potential exists for the microbial population to adapt to the toxicant in situations where the active mi-crobial population is not killed. This possibility was suggested by Morrissey et al. (1974) when they recorded an initial inhibition of nitrification in a Cheshire fine sandy loam soil by cadmium, chromium, copper, lead, manganese, nickel, and zinc and a subsequent adaptation to the presence of the metal over a several week period.

Thus, these workers concluded that, although with elevated metal concentrations a transient inhibition was observed, in the long run there was little significant inhibition of nitrification by the metals in their system.

McClung and Frankenberger (1985) noted that increased salinity generally decreased nitrification rate in three diverse soils. Inhibition of the nitrification rate varied from 8 to 83 percent, dependent upon the soil type and the nature of the salt added (Na_2SO_4, NaCl, or $CaCl_2$). Generally, the sulfate salt was less inhibitory than were the chloride salts. As with the metal-contaminated soils cited previously, the potential exists for microbial adaptation to increased salt content. Somville (1984) found that freshwater nitrifiers can adapt to increased salinity in an estuarine environment.

10.7. CONCLUDING OBSERVATIONS: CONTROL OF THE INTERNAL SOIL NITROGEN CYCLE

In nonmanaged or minimally managed soil ecosystems (e.g., native forests, "wild lands"), nitrogen cycling could be described as being in balance. In other words, organic nitrogen is mineralized to ammonium, a portion is nitrified to nitrate, and most of the mineral nitrogen is returned to biomass. This balance does not imply that no mineral nitrogen is lost from the site. Aside from denitrification processes (as discussed in chapter 12), some nitrate can be leached below the root zone through soil-saturating rainfalls. In balanced systems, this small nitrogen loss is balanced by atmospheric inputs and nitrogen fixation (see chapter 11).

Greater potential for disruption of the balance between nitrogen mineralization and incorporation into nascent biomass occurs in managed sites as exemplified by systems amended with external supplies of organic matter or receiving significant mineral fertilizer inputs. In these situations, the internal nitrogen cycle can be said to have been disrupted or skewed by the anthropogenic intervention. Rates of soil biological processes generally adjust in these situations (usually through enhanced biomass production, perhaps through increased denitrification), but the potential also exists for enhanced negative ecosystem impact. Because ammonium is relatively nonmobile in the soil profile (see Stevenson, 1986), most environmental contamination difficulties are derived from the synthesis of nitrate (i.e., nitrification).

The process of nitrification in soil can be likened to a double-edged sword. To maximize crop production and plant community development in reclaimed sites, fertilization with nitrogenous fertilizers is desirable, if not essential. Nitrification of this added nitrogen is necessary to maximize plant incorporation, but this oxidation of ammonium to nitrate greatly increases the mobility of this essential nutrient. Hence, nitrate may be leach into groundwaters or be transported through overland flow into surface waters. A desirable soil substituent is transformed into a water contaminant.

The best means of reducing the environmental impact of nitrification in a managed system is to attempt to maintain nitrate availability to that level required by the growing plant. This control is generally achieved by using slow-release fertilizers or nitrification inhibitors.

Ammonium concentration is the primary nutrient limitation of nitrification in soil. Once the ammonium is formed, it is rapidly transformed to nitrate—a trait desirable in balanced systems where plant biomass synthesis is tightly linked to organic matter mineralization. Such is not the case in fertilized sites. To slow ammonium release from fertilizers, the fertilizer particles (e.g., urea) can be coated with a substances that slowly dissolves in soil, thus gradually releasing the ammonium or ammonium precursor. The objective is commonly achieved by using slow-release fertilizers or through the timed mixing of plant growth nutrients with irrigation water (e.g., fertigation of turf or golf grasses or in drip irrigation systems for row crops). Dissolution of fertilizer nitrogen can be achieved by varying fertilizer particle size, by coating the particles with slowly dissolved or metabolized substances such as sulfur-coated urea (e.g., Landschoot and Waddington, 1987; Severson and Mahler, 1988) or by incorporating the fertilizer in polymers (e.g., see Gandeza et al., 1991), which are commonly used in urban systems for yard and turf fertilization or in agricultural systems, as well as by using urease inhibitors (e.g., see Bremner and Douglas, 1971, Gould et al., 1978; Krogmeier et al., 1989). Urease is an enzyme that produces ammonium through the hydrolysis of urea. Similarly, a variety of compounds have been shown to be efficient and effective inhibitors of the conversion of ammonium, once it is produced, to nitrate (e.g., see Baldwin et al., 1983; Belser and Mays, 1980; Belser and Schmidt, 1981; Bremner and McCarty, 1993; Goring, 1962; McCarty and Bremner, 1986). Note that the role of plant-derived toxicants in the inhibition of nitrification in soil is questionable, in that although these compounds may inhibit nitrifier activity in culture, there is little evidence supporting their function at concentrations naturally detected in soil systems (Bremner and McCarty, 1993).

REFERENCES

Addiscott, T. M. 1983. Kinetics and temperature relationships of mineralization and nitrification in Rothamsted soils with differing histories. J. Soil Sci. 34:343–353.

Alexander, M., and F. E. Clark. 1965. Nitrifying bacteria. In C. A. Black (ed.), Methods of Soil Analysis, pt. 2, pp. 1477–1483. American Society of Agronomists. Madison, WI.

Anderson, D. W., S. Saggar, J. R. Bettany, and J. W. B. Stewart. 1981. Particle size fractions and their use in studies of soil organic matter: I. The nature and distribution of forms of carbon, nitrogen, and sulfur. Soil Sci. Soc. Am. J. 45:767–772.

Baldwin, I. T., R. K. Olson, and W. A. Reiners. 1983. Protein binding phenolics and the inhibition of nitrification in subalpine balsam fir soils. Soil Biol. Biochem. 15:419–423.

Bazin, M. J., A. Rutili, A. Gaines, and J. M. Lynch. 1991. Humic acid relieves pH-inhibition of nitrification in continuous-flow columns. FEMS Microbiol. Ecol. 85:9–14.

Beauchamp, E. G., W. D. Reynolds, D. Brasche-Villeneuve, and K. Kirby. 1986. Nitrogen mineralization kinetics with different soil pretreatments and cropping histories. Soil Sci. Soc. Am. J. 50:1478–1483.

Belser, L. W. 1977. Nitrate reduction to nitrite, a possible source of nitrite for growth of nitrite-oxidizing bacteria. Appl. Environ. Microbiol. 34:403–410.

Belser, L. W., and E. L. Mays. 1980. Specific inhibition of nitrite oxidation by chlorate and its use in assessing nitrification in soils and sediments. Appl. Environ. Microbiol. 39:505–510.

Belser, L. W., and E. L. Schmidt. 1978a. Diversity of ammonia oxidizing nitrifier population of a soil. Appl. Environ. Microbiol. 36:584–588.

Belser, L. W., and E. L. Schmidt. 1978b. Serological diversity within a terrestrial ammonia-oxidizing population. Appl. Environ. Microbiol. 36:589–593.

Belser, L. W., and E. L. Schmidt. 1981. Inhibitory effect of nitrpyrin on three genera of ammonia-oxidizing nitrifiers. Appl. Environ. Microbiol. 41:819–821.

Bolton, H., Jr., J. L. Smith, and R. E. Wildung. 1990. Nitrogen mineralization potentials of shrub-steppe soils with different disturbance histories. Soil Sci. Soc. Am. J. 54:887–891.

Bonde, T. A., and T. Lindberg. 1988. Nitrogen mineralization kinetics in soil during long-term aerobic laboratory incubations: A case study. J. Environ. Qual. 17:414–417.

Bonde, T. A., and T. Rosswall. 1987. Seasonal variation of potentially mineralizable nitrogen in four cropping systems. Soil Sci. Soc. Am. J. 51:1508–1514.

Bondietti, E., J. P. Martin, and K. Haider. 1972. Stabilization of amino sugar units in humic-type polymers. Soil Sci. Soc. Am. Proc. 36:597–602.

Bremner, J. M., and L. A. Douglas. 1971. Inhibition of urease activity in soils. Soil Biol. Biochem. 3:297–307.

Bremner, J. M., and G. W. McCarty. 1993. Inhibition of nitrification in soil by allelochemicals derived from plants and plant residues. Soil Biochemistry 8:181–218. Dekker. New York.

Broadbent, F. E. 1986. Empirical modeling of soil nitrogen mineralization. Soil Sci. 141:208–213.

Cabrera, M. L. 1993. Modeling the flush of nitrogen mineralization caused by drying and rewetting soils. Soil Sci. Soc. Am. J. 57:63–66.

Cabrera, M. L., and D. E. Kissel. 1988a. Potentially mineralizable nitrogen in disturbed and undisturbed soil samples. Soil Sci. Soc. Am. J. 52:1010–1015.

Cabrera, M. L., and D. E. Kissel. 1988b. Evaluation of a method to predict nitrogen mineralization from soil organic matter under field conditions. Soil Sci. Soc. Am. J. 52:1027–1031.

Cameron, R. S., and A. M. Posner. 1979. Mineralisable organic nitrogen in soil fractionated according to particle size. J. Soil Sci. 30:565–577.

Cassman, K. G., and D. N. Munns. 1980. Nitrogen mineralization as affected by soil moisture, temperature, and depth. Soil Sci. Soc. Am. J. 44:1233–1237.

Clarhom, M., B. Popovic, T. Rosswall, B. Soderstrom, B. Sohlenius, H. Staaf, and A. Wiren. 1981. Biological aspects of nitrogen mineralization in humus from a pine forest podzol incubated under different moisture and temperature conditions. Oikos 37:137–145.

Crasswell, E. T., and S. A. Waring. 1972a. Effect of grinding on the decomposition of soil organic matter. I. The mineralization of organic nitrogen in relation to soil type. Soil Biol. Biochem. 4:427–433.

Crasswell, E. T., and S. A. Waring. 1972b. Effect of grinding on the decomposition of soil organic matter. II. Oxygen uptake and nitrogen mineralization in virgin and cultivated cracking clay soils. Soil Biol. Biochem. 4:435–442.

Deans, J. R., J. A. E. Molina, and C. E. Clapp. 1986. Models for predicting potentially mineralizable nitrogen and decomposition rate constants. Soil Sci. Soc. Am. J. 50:323–326.

De Beer, D., J. C. Van Den Heuvel, and S. P. P. Ottengraf. 1993. Microelectrode measurements of the activity distribution in nitrifying bacterial aggregates. Appl. Environ. Microbiol. 59:573–579.

De Boer, W., P. J. A. Klein Gunnewiek, M. Veenhuis, E. Bock, and H. J. Laanbroek. 1991. Nitrification at low pH by aggregated chemolithotrophic bacteria. Appl. Environ. Microbiol. 57:3600–3604.

DeLuca, T. H., and D. R. Keeney. 1993. Soluble organics and extractable nitrogen in paired prairie and cultivated soils in central Iowa. Soil Sci. 155:219–228.

Dommergues, Y. R., L. W. Belser, and E. L. Schmidt. 1978. Limiting factors for microbial growth and activity in soil. *In* M. Alexander (ed.) Advances in Microbial Ecology, Vol. 2, pp. 49–104. Plenum Press. New York.

Doxtander, K. G., and M. Alexander. 1966. Nitrification by heterotrophic soil microorganisms. Soil Sci. Soc. Am. Proc. 30:351–355.

Ellert, B. H., and J. R. Bettany. 1988. Comparison of kinetic models for describing net sulfur and nitrogen mineralization. Soil Sci. Soc. Am. J. 52:1692–1702.

Eylar, O. R., and E. L. Schmidt. 1959. A survey of heterotrophic micro-organisms from soil for ability to form nitrite and nitrate. J. Gen. Microbiol. 20:473–481.

Fliermans, C. B., B. B. Bohlool, and E. L. Schmidt. 1974. Autecological study of the chemoautotroph *Nitrobacter* by immunofluorescence. Appl. Microbiol. 27:124–129.

Fox, R. H., and W. P. Piekielek. 1978a. Field testing of several nitrogen availability indexes. Soil Sci. Soc. Am. J. 42:747–750.

Fox, R. H., and W. P. Piekielek. 1978b. A rapid method for estimating the nitrogen-supplying capability of a soil. Soil Sci. Soc. Am. J. 42:751–753.

Fox, R. H., and W. P. Piekielek. 1983. Response of corn to nitrogen fertilizer and the prediction of soil nitrogen availability with chemical tests in Pennsylvania. Pa. Agric. Exp. Stn. Bull. 843.

Gandeza, A. T., S. Shoji, and I. Yamada. 1991. Simulation of crop response to polyolefin-coated urea: I. Field dissolution. Soil Sci. Soc. Am. J. 55:1462–1467.

Goovaerts, P., and C. N. Chiang. 1993. Temporal persistence of spatial patterns for mineralizable nitrogen and selected soil properties. Soil Sci. Soc. Am. J. 57:372–381.

Goring, C. A. I. 1962. Control of nitrification by 2-chloro-6-(trichloro-methyl) pyridine. Soil Sci. 93:211–218.

Gould, W. D., F. D. Cook, and J. A. Bulat. 1978. Inhibition of urease activity by heterocyclic sulfur compounds. Soil Sci. Soc. Am. J. 42:66–72.

Hankinson, T. R., and E. L. Schmidt. 1988. An acidophilic and a neutrophilic *Nitrobacter* strain isolated from the numerically predominant nitrite-oxidizing population of an acid forest soil. Appl. Environ. Microbiol. 54:1536–1540.

He, X.-T., F. J. Stevenson, R. L. Mulvaney, and K. R. Kelley. 1988. Extraction of newly immobilized ^{15}N from an Illinois Mollisol using aqueous phenol. Soil Biol. Biochem. 20:857–862.

Jones, R. D., and A. P. Schwab. 1993. Nitrate leaching and nitrite occurrence in a fine-textured soil. Soil Sci. 155:272–282.

Juma, N. G., and E. A. Paul. 1984. Mineralizable soil nitrogen: Amounts and extractability ratios. Soil Sci. Soc. Am. J. 48:76–80.

Juma, N. G., E. A. Paul, and B. Mary. 1984. Kinetic analysis of net nitrogen mineralization in soil. Soil Sci. Soc. Am. J. 48:753–757.

Kelly, K. R., and F. J. Stevenson. 1985. Characterization and extractability of immobilized ^{15}N from the soil microbial biomass. Soil Biol. Biochem. 17:517–523.

Kelly, K. R., and F. J. Stevenson. 1987. Effects of carbon source on immobilization and chemical distribution of fertilizer nitrogen in soil. Soil Sci. Soc. Am. J. 51:946–951.

Kladivko, E. J., and D. R. Keeney. 1987. Soil nitrogen mineralization as affected by water and temperature interactions. Biol. Fert. Soil. 5:248–252.

Krogmeier, M. J., G. W. McCarty, and J. M. Bremner. 1989. Potential phytotoxicity associated with the use of soil urease inhibitors. Proc. Nat. Acad. Sci. 86:1110–1112.

Landschoot, P. J., and D. V. Waddington. 1987. Response of turfgrass to various nitrogen sources. Soil Sci. Soc. Am. J. 51:225–230.

Learch, R. N., K. A. Barbarick, L. E. Sommers, and D. G. Westfall. 1992. Sewage sludge proteins as labile carbon and nitrogen sources. Soil Sci. Soc. Am. J. 56:1470–1476.

Liang, C. N., and M. A. Tabatabai. 1978. Effects of trace elements on nitrification in soils. J. Environ. Qual. 7:291–293.

Malcolm, R. L. 1990. The uniqueness of humic substances in each of soil, stream, and marine environments. Anal. Chem. Acta 232:19–30.

Malhi, S. S., and W. B. McGill. 1982. Nitrification in three Alberta soils: Effect of temperature, moisture and substrate concentration. Soil Biol. Biochem. 14:393–399.

Marion, G. M., and C. H. Black. 1987. The effect of time and temperature on nitrogen mineralization in Arctic tundra soils. Soil Sci. Soc. Am. J. 51:1501–1508.

Matulewich, V. A., P. F. Strom, and M. S. Finstein. 1975. Length of incubation for enumerating nitrifying bacteria present in various environments. Appl. Environ. Microbiol. 29:265–268.

McCarty, G. W., and J. M. Bremner. 1986. Effect of phenolic compounds on nitrification in soil. Soil Sci. Soc. Am. J. 50:920–923.

McClung, G., and W. T. Frankenberger, Jr. 1985. Soil nitrogen transformations as affected by salinity. Soil Sci. 139:405–411.

Mengel, K., and H. Schmeer. 1985. Effect of straw, cellulose, and lignin on the turnover and availability of labelled ammonium nitrate. Biol. Fert. Soils 1:175–181.

Morrissey, R. F., E. P. Dugan, and J. S. Koths. 1974. Inhibition of nitrification by incorporation of selected heavy metals in soil. Abst. Ann. Mtg. Am. Soc. Microbiol. 1974:2.

Myers, R. J. K., C. A. Campbell, and K. L. Weier. 1982. Quantitative relationship between net nitrogen mineralization and moisture content of soils. Can. J. Soil Sci. 62:111–124.

Navarro, E., P. Simonet, P. Normand, and R. Bardin. 1992. Characterization of natural populations of *Nitrobacter* spp. using PCR/RFLP analysis of ribosomal intergenic spacer. Arch. Microbiol. 157:107–115.

Piccolo, A., L. Campanella, and B. M. Petronio. 1990. Carbon-13 nuclear magnetic resonance spectra of soil humic substances extracted by different mechanisms. Soil Sci. Soc. Am. J. 54:750–756.

Poovarodom, S., R. L. Tate III, and R. A. Bloom. 1988. Nitrogen mineralization rates of the acidic, xeric soils of the New Jersey Pinelands: Field rates. Soil Sci. 145:257–263.

Post, W. M., J. Pastor, P. J. Ainke, and A. G. Stangenberger. 1985. Global patterns of soil nitrogen storage. Nature (London) 317:613–616.

Schimel, J. P., M. K. Firestone, and K. S. Killham. 1984. Identification of heterotrophic nitrification in a Sierran forest soil. Appl. Environ. Microbiol. 48:802–806.

Schnitzer, M., and H. Kerndorff. 1980. Effects of pollution on humic substances. J. Environ. Sci. Health. 15B:431–456.

Schnitzer, M., and H. Kodama. 1992. Interactions between organic and inorganic components in particle-size fractions separated from four soils. Soil Sci. Soc. Am. J. 56:1099–1105.

Serna, M. D., and F. Pomares. 1992. Evaluation of chemical indices of soil organic nitrogen availability in calcareous soils. Soil Sci. Soc. Am. J. 56:1486–1491.

Severson, G. R., and R. L. Mahler. 1988. Influence of soil water potential and seed-banded sulfur-coated urea on Spring barley emergence. Soil Sci. Soc. Am. J. 52:529–534.

Smith, S. J., L. B. Young, and G. E. Miller. 1977. Evaluation of soil nitrogen mineralization potentials under modified field conditions. Soil Sci. Soc. Am. J. 41:74–76.

Sollins, P., G. Spycher, and C. A. Glassman. 1984. Net nitrogen mineralization from light- and heavy-fraction forest soil organic matter. Soil Biol. Biochem. 16:31–37.

Somville, M. 1984. Use of nitrifying activity measurements for describing the effect of salinity on nitrification in the Scheldt Estuary. Appl. Environ. Microbiol. 47:424–426.

Stanford, G., M. H. Frere, and R. A. V. Pol. 1975. Effect of fluctuating temperatures on soil nitrogen mineralization. Soil Sci. 119:222–226.

Stanley, P. M., and E. L. Schmidt. 1981. Serological diversity of *Nitrobacter* spp. from soil and aquatic habitats. Appl. Environ. Microbiol. 41:1069–1071.

Starr, J. L., T. B. Parkin, and J. J. Meisinger. 1992. Sample size consideration in the determination of soil nitrate. Soil Sci. Soc. Am. J. 56:1824–1830.

Stevenson, F. J. 1982. Humus Chemistry: Genesis, Composition, Reactions. John Wiley & Sons. New York. 443 pp.

Stevenson, F. J. 1986. Cycles in Soil: Carbon, Nitrogen, Phosphorus, Sulfur, Micronutrients. John Wiley & Sons. New York. 380 pp.

Talpaz, H., P. Fine, and B. Bar-Yosef. 1981. On the estimation of N-mineralization parameters from incubation experiments. Soil Sci. Soc. Am. J. 45:993–996.

Tate, R. L., III. 1977. Nitrification in histosols: A potential role for the heterotrophic nitrifier. Appl. Environ. Microbiol. 33:911–914.

Tate, R. L., III. 1987. Soil Organic Matter: Biological and Ecological Effects. John Wiley & Sons. New York. 291 pp.

Verstraete, W., and M. Alexander. 1973. Heterotrophic nitrification in samples of natural ecosystems. Environ. Sci. Tech. 7:39–42.

Vitousek, P. M., and P. A. Matson. 1985. Causes of delayed nitrate production in 2 Indiana USA forests. For. Sci. 31:122–131.

Walters, D. T., M. S. Aulakh, and J. W. Doran. 1992. Effects of soil aeration, legume residue, and soil texture on transformations of macro- and micronutrients in soil. Soil Sci. 153:100–107.

Westermann, D. T., and S. E. Crothers. 1980. Measuring soil nitrogen mineralization under field conditions. Agron. J. 72:1009–1012.

11 Nitrogen Fixation: The Gateway to Soil Nitrogen Cycling

A reliable source of fixed nitrogen is an obligatory requirement for sustained bio-mass productivity in all terrestrial ecosystems. In sites with significant accumulations of organic matter, the majority of this major biomass nutrient may be provided by cycling nitrogen between organic and plant-available inorganic forms as described in chapter 10. This situation is best exemplified by climax ecosystems (e.g., a mature forest site). In contrast, the biological communities of more nutrient-limited soil ecosystem types may be totally reliant upon external sources to support living community development. Such conditions are characteristic of pioneer sites (e.g., volcanic soils), which contain essentially no reservoirs of fixed nitrogen. Although there is a clear cut difference between climax and pioneer systems, care must be taken in not undervaluing the importance of fixed nitrogen inputs in either system. Because of the inevitable losses of fixed nitrogen from *all* soils (biological and abiological processes), some external supply of fixed nitrogen is always required in terrestrial sites to support long-term productivity. Otherwise, soil fixed-nitrogen reserves slowly decline until biomass productivity is reduced or even precluded. Therefore, the basis for considering soil nitrogen reservoirs as a renewable nutrient resource in soil is the existence of a functional nitrogen fixing microbial population *in situ*.

Fortunately for the development of highly productive, sustainable terrestrial systems, fixed-nitrogen inputs balance losses. Total losses of fixed nitrogen from land ecosystems of approximately 160–225 Tg year^{-1} are replaced by inputs of 214–262 Tg year^{-1} (Knowles, 1981). The magnitude of various pathways of fixed nitrogen into and out of land ecosystems are demonstrated in Fig. 11.1 The primary routes for nitrogen flow from soils are volatilization, runoff, and denitrification, whereas the largest source is biological nitrogen fixation. Fertilizer application (anthropogenic input) is a distant second. Understandably, most of the anthro-pogenically fixed nitrogen entering the soil ecosystem (estimated to be 30 Tg year^{-1} by Burns and Hardy, 1975) is applied to agricultural and urban soils (i.e., yards and recreational areas). A less obvious fact is that the majority of the biolog-ically fixed nitrogen enters the land-based nitrogen reserves in cropped lands be-cause of the intensive management of nitrogen-fixing crops (Burns and Hardy, 1975).

Symbiotic nitrogen fixation may be exploited to reduce use of industrially fixed nitrogen (see chapter 12 for discussion of symbiotic nitrogen fixation). Although

Nitrogen Fixation (149-180)
Anthropogenic Inputs (40-57)
Atmospheric Deposition (5)
Total Inputs (214-262)

Volatilization [26-53]
Runoff (10-40)
Denitrification (150-185)
Total Losses (160-225)

Soil Fixed Nitrogen Balance

Fig. 11.1. Major nitrogen inputs and losses contributing to the soil fixed-nitrogen balance. Values in parentheses are Tg N yr^{-1} as estimated by Knowles (1981). Similarly, values in brackets were derived by Söderlund and Svensson (1976).

intensive agricultural systems are usually sustained through liberal use of industrially fixed nitrogen, economic and environmental pressures are dictating reduction of fertilizer use and maximization of *in situ* biological nitrogen fixation. Most commonly, this entails cultivation of legumes, crops commonly involved in symbiotic associations with nitrogen-fixing bacteria. Similar plant–microbe partnerships involved in nitrogen fixation are actinorhizal associations. In contrast to the legume interaction that tend to be identified with temperate agricultural systems (although their worldwide contributions to nitrogen cycling extend far beyond this limited food and fiber category), actinorhizal symbioses are more generally linked to native or "wild land" situations. For these latter ecosystems, the ability to balance the use of industrial and biological nitrogen sources is of limited importance because the plants involved in the symbioses are rarely harvested.

With legume crops, the greatest gain in fixed nitrogen results from the return of all new plant biomass to the soil. Removal of the aboveground portion of the plant leads to little increase in soil nitrogen. Some benefit results from retention of the legume root system within soil (Alexander, 1977) but the bulk of the fixed nitrogen is contained within the harvested crop. In cropped systems, the best situation that can be anticipated would be retention of sufficient fixed nitrogen through incorporation of the root systems alone to maintain the existing levels of soil nitrogen or, at least, to minimize fixed-nitrogen losses.

These introductory comments reveal the critical nature of nitrogen fixation in maintaining native as well as managed ecosystems. Understanding the balance between fixed-nitrogen supplies, looses, and inputs in a soil ecosystem is especially important for the development of management or reclamation plans for damaged or mismanaged soil sites. Thus, this chapter evaluates nitrogen fixation as a process and explains the nuances of varying ecosystem properties on nitrogen-fixation rates in managed and native soil systems.

11.1. BIOCHEMISTRY OF NITROGEN FIXATION

An in-depth examination of the biological, biochemical, and genetic aspects of biological nitrogen fixation exceeds the purview of primary interest of most soil microbiologists. Yet, a foundational understanding of some essential properties of the enzymes and the organisms involved is necessary to comprehend the properties of the soil ecosystem that control the dynamics of nitrogen fixation in native ecosystems. The primary traits of biological nitrogen fixation that control the yields of fixed nitrogen in native soils and the kinetics of this production to be considered herein are the complexity of the enzyme (nitrogenase) catalyzing the reduction of dinitrogen N_2 to ammonium, the quantities and sources of the energy required to drive the process, and the diversity of microbes responsible for the process. For more detailed analyses of the biochemical and genetic aspects of nitrogen fixation processes, see basic biochemistry and general microbiology textbooks and recent reviews of the topic (e.g., Brewin, 1991; Burris, 1991; Caetano-Anollés and Gresshoff, 1991).

11.1.1. The Process

Dinitrogen, because of the triple bond between the two nitrogen atoms, is a nearly inert molecule with respect to the metabolic potentialities of the biological community. The only enzymologically catalyzed reaction involving dinitrogen is accomplished by a limited number of bacterial strains. As will become evident when individual genera are discussed, these bacteria constitute a highly diverse physiological group with the capacity to reduce dinitrogen to ammonium ion as their primary common trait. It should also be stated that the ability to transform dinitrogen to ammonium is limited to a small number of bacterial genera; that is, no fungi, plants, or animals posses this capability.

In nitrogen fixation, the nitrogen atom is reduced from its most oxidized state (N_2) to its most reduced form (NH_4^+) by the following reaction:

$$N_2 + 8H^+ + energy \rightarrow 2NH_3 + H_2$$

Considering the variety of intermediate oxidation levels that can exist for nitrogen between dinitrogen and ammonium (see chapter 10) and the environmental importance of these compounds, an important question regarding biological nitrogen fixation relates to the potential for producing partially reduced nitrogen intermediates during the conversion of dinitrogen to ammonium by nitrogenase. No free intermediates between dinitrogen and ammonia are produced. Thus, the only direct product of nitrogen fixation encountered by soil microbes is ammonium.

Energy Consumption During Nitrogen Fixation: Considering the bonding energy associated with the triple bond of dinitrogen, it is reasonable to conclude that nitrogen fixation is an energy-intensive process. Under ideal conditions, a minimum of 16 molecules of MgATP are required for each atom of dinitrogen cleaved and reduced. Because ideal conditions are rarely achieved in native ecosystems, the energy source requirement is generally between 20 and 30 molecules of MgATP (See Burris, 1991). To put this quantity of energy into perspective, recall that the total oxidation of one molecule of glucose yields a maximum of 38 ATP molecules (see chapter 3). Thus, a microbe-fixing nitrogen must devote nearly all the energy provided by the oxidation of a glucose molecule to the reduction of one dinitrogen molecule to two ammonium ions. The ATP molecules supporting nitrogen fixation are derived primarily either directly from photosynthetic processes (photoautotrophs that fix nitrogen) or from decomposition of organic compounds (nitrogen-fixing heterotrophs supported indirectly from photosynthetic processes).

Knowledge of this energy requirement allows the soil microbiologist to deduce the sites of activity and the relative importance of the various metabolic groups of nitrogen-fixing bacteria. For example, as is discussed later, this energy input limits the quantities of dinitrogen that can be reduced by microbes catalyzing soil organic matter as their energy source. For maximal conversion of dinitrogen to ammonium, a stable, abundant energy source is necessary. Such reliable supplies occur naturally in the rhizosphere (i.e., photosynthetically fixed carbon contained in root exudates or provided by symbiotic associations) or through direct conversion of light energy for the process (cyanobacteria and photosynthetic bacteria).

This energy requirement essentially dictates that the primary sites for nitrogen fixation in soil are limited to rhizosphere. Potential exceptions to this observation could be soil sites receiving large quantities of fixed carbon; that is, spill sites or systems receiving large quantities of industrial effluents. For example, Neilson and Sparell (1976) isolated nitrogen-fixing enterobacteria from paper mill process waters. In conclusion, in native ecosystems, highest nitrogen fixation contributions to soil ecosystems are associated with situations where energy inputs are provided by photosynthesis (nitrogen fixation by blue-green bacteria) or through transfer of plant photosynthate directly to nitrogen-fixing bacteria or actinomycetes (symbiotic nitrogen fixation).

Hydrogen Production and Hydrogenase Activity: A related concern involving the stoichiometry of reducing dinitrogen to ammonium involves the inefficiency associated with loss of reducing power (H^+) through the formation of hydrogen gas. It appears to be logical that the most efficient process would involve incorporating all reducing power into the nitrogenous product; but, that does not occur. Formation of hydrogen gas appears to be illogical also from the fact that hydrogen is a specific competitive inhibitor of nitrogenase. Although there has been considerable research concerning the optimized use of reducing power in dinitrogen reduction, the synthesis of hydrogen gas appears to be an obligatory part of the nitrogen fixation reaction (Burris, 1991).

The quantities of electron flow lost to the microbe through hydrogen ion reduction can be highly significant. For example, nodules formed by *Rhizobium meliloti* or *Rhizobium trifolii* on their respective hosts have been shown to loose at least 17 percent of their electrons through this mechanism (Ruiz-Argüeso et al., 1979). The average quantity of reducing power lost to hydrogen gas for each of these organisms was considerably greater than this minimum value. For the alfalfa nodulator, *R. meliloti,* the average proportion of electron lost was 25 percent. The comparable value for the clover-nodulating organism (*R. trifolii*) was 35 percent.

Although the formation of this hydrogen gas during nitrogen fixation can not be prevented, the efficiency of the nitrogen-reducing process can be improved by recovering the energy contained in hydrogen gas. Hydrogenase is an enzyme that catalyzes the oxidation of hydrogen gas with water as the final product. This oxidative reaction is coupled with ATP formation or with reduction of ferredoxin or flavodoxin. These electron acceptors are of particular interest because their reduction can lead to increased ammonium ion yields. The electrons transferred to ferredoxin and flavodoxin are returned to the dinitrogen reductase portion of the nitrogenase enzyme. Examples of the occurrence of hydrogenase activity in nitrogen-fixing bacteria include *Rhodopseudomonas capsulata* (Colbeau et al., 1980), where high hydrogenase activities were found in cells containing active nitrogenase, *Bradyrhizobium japonicum* (Keyser et al. 1984; Merberg and Maier, 1983), and *R. leguminosarum* (Nelson and Salminen, 1982). The proportion of each of these species producing hydrogenase is extremely variable. For example, in the study conducted by Keyser et al. (1984), a small portion of the *Bradyrhizobium japonicum* strains produced hydrogenase (i.e., were Hup$^+$). Of the 972 *Bradyrhizobium japonicum* isolates from 65 soybean fields located in 12 states (USA) studied, only 20 percent produced hydrogenase. Frequency of occurrence of these hydrogenase producers varied considerably between strains. None of the isolates in serogroup 135 were Hup$^+$, whereas 93 percent were positive in serogroup 122.

11.1.2. The Enzyme, Nitrogenase

Reduction of dinitrogen to ammonium in nitrogen-fixing bacteria is catalyzed by an enzyme system consisting of two distinct proteins, dinitrogenase (also known as MoFe protein or protein I) and dinitrogenase reductase (alias, Fe protein or protein II). Dinitrogenase is a large protein (220 to 240 kdaltons), which binds and reduces dinitrogen. Dinitrogenase reductase transfers the electrons to dinitrogenase.

Dinitrogenase and dinitrogenase reductase are highly conserved proteins. This fact is supported by the observation that purified dinitrogenase and dinitrogenase reductase produced from a variety of nonrelated bacterial species can be combined to produce active nitrogen fixation. Thus, a limited set of basic properties can be attributed to all nitrogenase enzymes.

All nitrogenases are extremely oxygen-sensitive. Because nitrogen fixation is catalyzed by strictly anaerobic bacteria, facultative aerobes, microaerophilic organisms, as well as strictly aerobic bacteria, a major proportion of the organisms involved must posses a mechanism to protect the oxygen-labile protein. Two overall adaptive mechanisms to protect nitrogenase can be postulated. Either molecular oxygen can be excluded from sites of active nitrogenase or nitrogenase molecules, which are inherently resistant to oxygen can be produced. Throughout the world of nitrogen-fixing bacteria, the organisms uniformly protect nitrogenase enzymes from encounters with toxic levels of molecular oxygen.

Seven individual means for excluding molecular oxygen from the environment of nitrogenase enzymes have been observed. These include:

- **Respiratory Protection:** The rate at which individual cells consume oxygen may be enhanced so that the oxygen levels in its microenvironment are reduced to acceptable levels. For example, *Azotobacter* sp. cells have been observed to oxidize more fixed carbon in the presence of oxygen than is needed to satisfy microbial energy requirements. This protective mechanism appears to operate in a variety of free-living organisms including such metabolically diverse organisms as *Azotobacter vinelandii* (e.g., Shah et al., 1973) and cyanobacteria (e.g., Murry et al., 1984; Peschek et al., 1991).

- **Conformational Protection:** An alternative to developing an enzyme that is inherently resistant to oxygen is to develop a nitrogenase that changes to a protective conformation in the presence of oxygen. This has been observed with *Azotobacter.* In the protected state, nitrogen fixation is totally precluded, but the activity rapidly returns when the oxygen tension is reduced to acceptable levels. This alteration of nitrogenase structure results from association of a small protein (approx. 24,000 dalton molecular weight) with the nitrogenase protein.

- **Oxygen Regulation of Nitrogenase Synthesis:** Nitrogenase is a complex of large protein molecules. Synthesis of these molecules thus requires expenditure of significant amounts of cell energy. One means of protecting the nitrogen-fixing apparatus is to prevent its production under unfavorable conditions. Molecular oxygen represses nitrogenase synthesis in *Klebsiella pneumoniae, Azotobacter chroococcum,* and some strains of rhizobia.

- **Gum Production:** For oxygen to interfere with dinitrogen reduction, the inhibitor and susceptible enzyme must interact. Nitrogenase can be protected by production of a physical barrier to molecular oxygen. Gums or polysaccharides accumulated external to the cell wall may provide such protective wall. Microorganisms may synthesize extracellular polysaccharides that reduce the diffusion rate of oxygen in the vicinity of the cell wall. This synthesis has been proposed to occur for *Azotobacter* spp. and *Derxia gummosa.*

- **Heterocyst Production:** Nitrogen fixation can be an especially incompatible process in photosynthetic organisms in that molecular oxygen is a product of carbon fixation in aerobic, nitrogen-fixing cyanobacteria. With these organisms, to provide protection for oxygen-labile enzymes, they may be isolated in structures with thick cell walls (heterocysts) to reduce contact of nitrogenase with molecular oxygen. For further discussion of the protection of nitrogenase in cyanobacteria, see Fay (1992).

- **Leghemoglobin Production:** In *Rhizobium*-legume nodules, molecular oxygen is complexed by substances with high oxygen affinities in order to reduce intracellular oxygen concentrations. A hemoglobin-like molecule (leghemoglobin) is synthesized in the cytosol around the packets of bacteria in the nodule (see chapter 12 for a description of nodule structure). This complexing of molecular oxygen results in reduction of oxygen at the bacteroid surface to about 10 nM. Further protection of nitrogenase in the bacteroids is provided by a terminal oxidase system that has an unusually high affinity for molecular oxygen.

- **Migration to Suitable Environment:** Aerotactic organisms may migrate to positions in their microenvironment suitable for nitrogen fixation. This has been observed with *Azospirillum* spp. (Barak et al., 1982).

- **Other Mechanisms:** In aquatic ecosystems, as exemplified by cyanobacteria associated with rice cultivation, nitrogen-fixing cells may be protected by clumping of the cells or oxygen-generating and nitrogen-fixing processes may be separated temporally. For example, cyanobacterial cells have been shown to form clusters. Molecular oxygen consumption on the surface of the cluster is sufficient to reduce oxygen concentrations inside the grouping to noninhibitory levels.

11.1.3. Measurement of Biological Nitrogen Fixation in Culture and in the Field

Accurate quantification of fixed-nitrogen inputs into terrestrial systems due to the biological nitrogen fixation is difficult to achieve. Even though large quantities of dinitrogen are reduced to ammonium by the soil biological community annually (e.g., Knowles, 1981), the actual change in quantities of fixed nitrogen in individual soil microsites is relatively small compared to the background concentrations of both ammonium and dinitrogen, which are common chemical components of soil air and water as well as the general atmosphere. This situation necessitates sensitive and accurate measurement of small changes in reactants and products. Sensitive procedures useful for study of complex soil samples are not always available.

Although there are a variety of techniques that may be adapted to quantify dinitrogen or ammonium ions in terrestrial ecosystems, their use to evaluate changes in fixed-nitrogen concentrations in native soil samples or in growth media of axenically grown microbial cultures is problematic. To a large degree, using direct measurements of changes in ammonium or dinitrogen concentrations for quantification of fixed-nitrogen production yields equivocal data. It is not unusual for the gains or losses of these nitrogenous substances in soil samples, where nitrogen fixation is actively occurring, to be less than or approximately equal to the deviation of the experimental procedure used to quantify them. Thus, the sought after value is obscured within the background "noise" associated with the data (i.e., the standard deviation or standard error of the mean of the results).

To overcome the methodological limitations imposed by the ubiquity of dinitrogen and ammonium, a variety of indirect as well as more direct procedures have historically been employed to estimate quantities of nitrogen fixed in culture and in soil samples. The methods used have included growth in microbiological media lacking a fixed-nitrogen source to determine the capability of microbial isolates to fix nitrogen, to assess changes in the distribution of fixed nitrogen in known soil nitrogen pools (nitrogen balance studies), to trace movement of nitrogen atoms with ^{15}N-labeled substrates, and to use acetylene reduction as an indicator of nitrogen fixation capacity.

Nitrogen-Free Cultivation and Nitrogen Balance Procedures: Evaluating the capacity of a microorganism to grow in media lacking a source of fixed nitrogen and assessing the changes in nitrogen pools in a soil site were the most commonly used early methods for assessing nitrogen fixation potentials. These techniques, although applied to very different systems, are examined together because they are both supported by the same faulty assumption. This assumption is that all nitrogen inputs and losses in a soil site or sample or a culture medium are understood and can be accurately quantified. Thus, increases in quantities of inorganic fixed nitrogen in the field or the microbial biomass in laboratory cultures were assumed to have resulted from biological nitrogen fixation. The tenets of this assumption are rarely achieved in either the field or the laboratory.

From the viewpoint of assessing nitrogen-fixing activity of axenically grown microbial cells, it has long been possible to prepare laboratory media with nitrogen-free chemicals and reasonably pure water. Because we know that a fixed nitrogen source is essential for cell growth and reproduction, it is apparently reasonable to conclude that if a microbe grows in a growth medium composed of nitrogen-free water and chemicals, it must be fixing nitrogen. It is not difficult to find a variety of papers published prior to the 1950s that document nitrogen-fixing capabilities of a number of bacterial species by using this method. Unfortunately, conclusions reached using nitrogen-free growth media for microbial culture are equivocal at best and are commonly erroneous.

Assuming that the fixed nitrogen has been adequately removed from substituents of the growth media or that the quantities of microbial biomass produced exceeds those allowed by the trace nitrogenous contaminants of the substituents of growth medium, it seems apparent that the only other source of fixed nitrogen for microbial replication is biological fixation. Unfortunately, fixed nitrogen is a common component of the atmosphere of laboratories and incubators. The tendency to disregard this air pollution resulted in a misplaced confidence in the laboratory results. Volatile nitrogenous compounds are easily transferred via the atmospheric route into the nitrogen-free media in laboratory situations. It is easy to envision conditions where a culture of a microbe producing a volatile nitrogenous compound is growing adjacent to a culture containing a nitrogen-free medium. Furthermore, biological products of neighboring cultures were not the only source of contamination for studies using nitrogen-free growth media. Although scrupulously clean chemicals and water were generally employed in preparing for culture of the test microbes, incorporation of ammonium into the microbial culture media easily occurred at any time ammonia-based cleaning solvents were used in the laboratory or when a bottle of ammonium hydroxide was opened in the laboratory.

In the more complex field situations, the extent of biological nitrogen fixation has been historically estimated by assessing changes in quantities of nitrogen contained within the various soil nitrogen pools; that is, nitrogen balance sheets were prepared. With this method, fixed-nitrogen concentrations are measured directly in soil samples. Ammonium is usually extracted from the sample, steam distilled, and titrated. Nitrate and nitrite ions are then reduced to ammonium and similarly assayed. Finally, organic nitrogen is converted to ammonium using the Kjeldahl digestion procedure. Two requirements basic to any nitrogen balance-sheet analysis are a) a diligence in analyzing all nitrogen sources and b) a capacity to assess accurately small changes in generally sizable nitrogen pools. Neither condition is met in these analyses. Error results from difficulties in quantifying fixed-nitrogen losses through denitrification and erosion as well as inputs from atmospheric sources (e.g., inputs of volatized ammonium). The problems associated with quantification of slight changes in preexisting fixed-nitrogen pools were discussed previously. Applicability of nitrogen balance data to total ecosystems is further reduced by the inaccuracies introduced by extrapolation of data with high standard deviations collected by analyzing a single or a limited number of field samples to account for all nitrogen inputs into the study site. Using nitrogen balance-sheet type studies are most commonly limited to well-defined systems experiencing major changes in the size of the nitrogen reservoirs.

Use of ^{15}N Tracers: A nearly ideal means of compensating for the tendency of background nitrogen concentrations to obscure small changes in the distribution of atoms among the various reservoirs of soil nitrogen is to trace the movement of labeled atoms within the system of interest. Sufficient labeled dinitrogen can be added to the soil sample to overcome any difficulties associated with the natural background levels of the nitrogen isotope contained therein. No radioactive nitrogen isotopes have a half-life long enough to be practical for use in studies of nitrogen fixation in soil environments, but the heavy isotope ^{15}N is useful to trace nitrogen transformations in natural soil samples. Assays for ^{15}N in soil nitrogen pools are sensitive and accurate.

The primary assumption underlying use of ^{15}N for quantifying nitrogen fixation is that the microbes use the labeled atom indiscriminately; that is, the microbes must not be capable of differentiating between the ^{15}N label and the more abundant ^{14}N isotope. A further consideration is that an apparent microbial discrimination in nitrogen atom sources must not be imposed on the system by incomplete homogenization of the amended labeled nitrogenous substrate with indigenous soil nitrogen pools. That is, the ^{15}N-labeled dinitrogen molecules must be reduced to ammonium in proportion to their contribution to the total dinitrogen pool, not by their different locations within soil microsites. Unfortunately, when using ^{15}N to study nitrogen fixation, we must remember that some discrimination favoring ^{15}N atoms by nitrogenase in leguminous root nodules has been shown (see Burris, 1991) and that it is not practically possible to distribute evenly an externally supplied ^{15}N-labeled mixture within the heterogeneous, complex soil structure. Even with these difficulties, ^{15}N has commonly been used as a tracer for nitrogen in nitrogen fixation studies.

One means of counteracting the isotope discrimination problem is to use ^{15}N-depleted dinitrogen. For this method, the reduction in the atom percent ^{15}N in the ammonium present in the sample is measured in samples incubated in an atmosphere containing dinitrogen from which the ^{15}N has been removed or highly reduced.

Other difficulties associated with large-scale use of ^{15}N to quantify nitrogen fixation in environmental samples are the requirement of a mass spectrometer to quantify enrichment or depletion of the ^{15}N in the ammonium pool extracted from the test system, the cost involved with the purchase of the isotopically labeled substances, and the rather long incubation times in gas-tight systems required to ensure that the labeled dinitrogen sources are not diluted by atmospheric sources.

Acetylene Reduction: To overcome the methodological limitations involved with using ^{15}N-labeled dinitrogen, a surrogate compound, reduced by nitrogenase at

$$2H^+ + 2e^- \longrightarrow H_2$$

$$N_2 + 8H^+ + 8e^- \longrightarrow 2NH_3 + H_2$$

$$N_2O + 2H^+ + 2e^- \longrightarrow N_2 + H_2O$$

$$CN^- + 7H^+ + 6e^- \longrightarrow CH_4 + NH_3$$

$$N_3^- + 3H^+ + 2e^- \longrightarrow N_2 + NH_3$$

$$C_2H_2 + 2H^+ + 2e^- \longrightarrow C_2H_4$$

Fig. 11.2. Some alternate substrates reduced by the nitrogenase system.

rates comparable to those associated with dinitrogen reduction, could be useful. Ideally, this substance should be quantified easily by methods that are both readily available in most laboratories and reasonably inexpensive. Fortunately, nitrogenase is not totally specific for dinitrogen as a substrate. This enzyme can also reduce nitrous oxide, cyanide, methyl isocyanide, azide, acetylene, and cyclopropene (Fig. 11.2). The reduction of these compounds can serve as a measure of nitrogenase. The most useful alternate substrate for nitrogenase is acetylene in that it is not normally present in atmospheric samples, it is relatively inexpensive, and it can be quantified using gas chromatographs, which are more common in laboratories than are mass spectrometers.

Because most studies of nitrogen fixation have an objective of assessing or predicting the quantities of dinitrogen that could be reduced in the samples, it would be useful if the quantities of ethylene produced from acetylene could be extrapolated to nitrogen fixation potential. The stoichiometry of the reduction of dinitrogen or acetylene by nitrogenase follows:

$$N_2 + 8H^+ \rightarrow 2NH_3 + H_2$$

$$3C_2H_2 + 6H^+ \rightarrow 3C_2H_4$$

Considering the hydrogen ions incorporated into the acetylene and dinitrogen by nitrogenase, reducing three acetylene atoms to three ethylene atoms should be equivalent to reducing one dinitrogen to two ammonia atoms; that is, production of three ethylene molecules should represent the reduction of one dinitrogen.

Unfortunately, extrapolating data collected from analyzing acetylene to estimate field nitrogen fixation rates is more complicated than this stoichiometric comparison predicts. Comparing data derived from assessing nitrogen fixation in parallel samples using ^{15}N-labeled dinitrogen and acetylene suggests that this ratio of 3/1 (ethylene to dinitrogen) can be attained at times in culture and soil, but more commonly the ratio ranges from 0.75/1 to 4.5/1 or greater. Therefore, caution is necessary in interpreting the acetylene-derived data. Because of this data compromise, nitrogen fixation potential measured by the acetylene procedure is generally designated nitrogen fixation (acetylene) so that readers may realize that the results are derived from extrapolating the results from an acetylene-based assay rather than from an assessment of dinitrogen reduction directly.

This discrepancy between predicted equivalence of acetylene and dinitrogen reduction from basic chemical principles and actual field results can be explained, to a large degree, by the different solubilities of dinitrogen and acetylene in water. Dinitrogen is poorly soluble in water. Thus, nitrogenase is rarely saturated by its primary substrate, whereas acetylene is extremely water-soluble. An acetylene concentration of 0.2 atmospheres is usually sufficient to saturate the enzyme. Therefore, when acetylene is used as a substrate for nitrogenase, all nitrogenase enzyme molecules in the sample are measured. (Because the substrate is at saturating concentrations in the reaction mixture, the reaction rate is proportional to the quantities of enzyme present.) When the nonsaturating concentrations of dinitrogen are used to assess nitrogen-fixing capacity, some of the enzyme is inactive and therefore undetected. (In the latter situation, the reaction rate is substrate-, not enzyme-concentration limited.)

A further complication in using the acetylene reduction procedure to estimate biological nitrogen fixation rates in field samples can result from reduction of the ethylene produced by nitrogenase. To accurately quantify nitrogenase activity, all the ethylene produced from reducing acetylene must be detected. In some soil ecosystems, the ethylene produced by nitrogenase is oxidized by soil microorganisms that use it as a growth substrate. Thus, to ensure that the ethylene in the assay vessel represents the total produced by nitrogenase, a control sample for the disappearance of ethylene must be included in the study.

11.2. GENERAL PROPERTIES OF TERRESTRIAL NITROGEN-FIXING ORGANISMS

The Microbes: Nitrogen fixation is catalyzed solely by bacteria either living among the general soil microbial population or in symbiotic associations (i.e., *Rhizobium*-legume or actinorhizal symbioses [see chapter 12] or with fungi [lichens]). Rhizosphere-associated nitrogen-fixing bacteria (e.g., *Azospirillum* sp. growing in and around the root) can also be defined as existing in a loosely symbiotic relationship with the plant. Nitrogen-fixing microorganisms, which are called **diazotrophs,** constitute a large group of marginally related bacteria. In contrast to the situation with nitrifying bacteria where the capacity to nitrify is sufficiently definitive that it can be used as a taxonomic trait, the extreme physiological and structural diversity of diazotrophs precludes using a single metabolic property, synthesis of nitrogenase, as a generic defining trait. Grouping nitrogen-fixing organisms into one or a few bacterial genera based on their capacity to reduce dinitrogen to ammonium is counterproductive. These organisms are found in such diverse groups as eubacteria, the photosynthetic cyanobacteria (formerly known as the blue-green algae), and actinomyctes.

The only clear commonalties between the diazotrophs is their general classification as bacteria and their ability to fix nitrogen. Diazotrophs are aerobes, facultative

TABLE 11.1. Examples of Diazotrophs Commonly Found in Soil, Rhizosphere, and Soil-Linked Aquatic Systems.

Habitat and Metabolic Grouping	Microbial Groups
Nonassociated, aerobes, heterotrophs	*Azotobacter*
	Bacillus
	Beijerinckia
	Derxia
Nonassociated, aerobes, lithotrophs	*Thiobacillus*
Nonassociated, anaerobes, heterotrophs	*Clostridium*
	Desulfovibrio
Aquatic (swamps, bogs, rice fields), aerobes, phototrophs	Cyanobacteria [blue-green algae]
Aquatic (swamps, bogs, rice fields), anaerobes, phototrophs	*Chaetomium*
	Rhodopseudomonas
	Thiospirillum
Rhizosphere (loosely associated), facultative aerobes, heterotrophs	*Enterobacter*
	Erwinia
	Klebsiella
Rhizosphere (loosely associated), aerobes, heterotrophs	*Azospirillum*
	Pseudomonas
Root nodules (legumes or angiosperms)	*Frankia*
	Rhizobium

anaerobes, or anaerobes. They may derive their energy from organic substances, inorganic compounds, or solar energy directly. Diazotrophs grow as single cells or chains of individual cells (e.g., rhizobia, clostridia), as loosely associated colonies (e.g., some blue-green algae), and as myclial structures (e.g., the actinomycetes). This metabolic and physical variability enables these organisms to contribute to fixed-nitrogen pools in essentially any ecosystem where microbes are capable of growth and reproduction (e.g., see Table 11.1).

Population Densities of Diazotrophs in Soils: Although they are highly adaptable to soil physical and chemical conditions, diazotrophic bacteria are rarely the dominant populations in terrestrial ecosystems. Exceptions are found in situations where the development of a heterotrophic soil microbial community is limited by availability of fixed nitrogen. This dominance of nitrogen-fixing bacteria would prevail until sufficient fixed-nitrogen accumulates in the site to support an active heterotrophic microbial population. Thus, high population densities of biological nitrogen fixers are generally localized and transient, but we can reasonably assume that at least minimal populations of nitrogen-fixing procaryotes are found in all soil sites where life processes are possible. This conclusion is supported by the diversity of metabolic groupings of organisms capable of fixing nitrogen and the essential role of fixed nitrogen in biological community development.

Physical and Chemical Limitations to Diazotrophes: The type of biological-nitrogen-fixing community involved and the quantities of ammonium ion produced in an ecosystem is controlled by temperature, pH, moisture saturation of the ecosystem, and the extent of the energy supply available to support diazotroph metabolism. Each of these soil properties must be considered when designing experiments to study *in situ* diazotrophs as well as when developing site reclamation plans where biological nitrogen fixation is to be encouraged. For example, aeration is an important consideration in predicting nitrogen fixation potential at the site in that free-living anaerobic diazotropic activity exceeds that of aerobes. Yet, because plants are sensitive to oxygen deprivation of their roots, optimization of nitrogen fixation in anaerobic microsites of soil aggregates would be the method of choice for augmenting ammonium production by free-living diazotrophs.

Similarly, for heterotrophic diazotrophs living among the general soil microbial populations, high concentrations of carbonaceous substrates are critical. Therefore, in organic-matter-deficient soils, a low population of nitrogen-fixing heterotrophic bacteria is supported. The relative contributions of each of these groups of organisms to total nitrogen fixed is proportional to the quantities of energy available to them.

An example of the control of free-living diazotroph population density with available energy source is provided by Azam et al. (1988), where a high correlation between amounts of glucose added to soil and nitrogen fixation rates was detected. The relationship of nitrogen contributions by the various diazotrophs based on the relationship to their energy source is graphically demonstrated in comparing relative fixed-nitrogen contributions in agricultural soils. According to Burns and Hardy (1975) 100–300 kg N ha^{-1} year^{-1} is produced in these soils from the *Rhizobium*-legume association compared to 0.4–0.8 kg N ha^{-1} year^{-1} from free-living soil bacteria. Furthermore, the efficiency of converting energy available in the ecosystem to fixed nitrogen is necessarily less than would be measured in axenic culture because some of the energy in the growth substrates is lost to competing microbes in the soil and to the mechanisms, which are necessary to deal with the stresses of the soil environnment.

Availability of Newly Fixed Nitrogen to Plants: A further consideration regarding the impact of nitrogen fixation activity by free-living diazotrophs on total ecosystem biomass productivity relates to the availability of the limited amounts of ammonium ion produced to the higher plant community. Nitrogen fixed by diazotrophs is only indirectly available to the general community. The nitrogen fixed by the diazotrophs is initially incorporated into their biomass. For this fixed nitrogen to become generally available to the soil community, the diazotroph cell must die and be mineralized. Because newly incorporated biologically fixed nitrogen appears to be more available to the soil microbial community than is native soil nitrogen (e.g., Azam et al., 1988), free-living diazotrophs may contribute more to the general soil fertility than directly to the stimulation of plant growth.

11.3. FREE-LIVING DIAZOTROPHS

From a consideration of the wide variety of ecosystems in which free-living bacteria that fix nitrogen are found, we can conclude that a vast diversity of diazotrophs must exist. Indeed, bacteria capable of fixing nitrogen while growing independently in soil are distributed among at least 26 genera, 11 families, and 3 orders of bacteria. Note that within each individual genus, the number of species capable of fixing nitrogen is small and that not every strain of a diazotrophic species necessarily can fix nitrogen; that is, as indicated previously, nitrogen fixation is not a distinctive taxonomic trait.

To indicate the limitations to growth and their contributions to total fixed nitrogen in soil ecosystems, representative examples of aerobic, anaerobic, and autotrophic nitrogen-fixing bacteria will be presented. The relationship of some of these organism to the properties of their natural habitats are analyzed in the next section. The individual species examined are not necessarily the most important in any given ecosystem or in soil ecosystems in general. In many cases, they simply represent those more commonly studied. Quite commonly, the frequency of isolation and the evaluation of individual diazotrophs is determined by the ease of isolation and their propensity for reproducible growth in axenic culture rather than the proportion of the soil nitrogen-fixing population that they comprise.

Aerobic, Free-Living Diazotrophs: The three most studied free-living, strict aerobes studied are *Azotobacter chroococcum, Azotobacter beijerinckia (Beijerinkia* sp.), and *Derxia gumosa.* From the number of reports in the literature, the conclusion could easily be reached that *Azotobacter chroococcum* is the most widespread species in neutral or alkaline soils. This conclusion is tentative at best and may easily be in error. The frequency with which this organism is the topic of study may relate to the fact that it has been commonly studied in general bacteriology classes. Thus, soil microbiologists were well versed in the methods for isolating and growing this bacterial species.

Azotobacter chroococcum is a strict aerobe with a temperature optimum of about 30°C. The numbers of these organisms in soil range from undetectable to several thousand per gram of soil. It is unusual to find large populations of this organism in soil, thus *Azotobacter chroococcum* is unlikely to be an important contributor to total ecosystem fixed-nitrogen inputs, except in some specialized situations. Exceptions to this conclusion would be in sites in which this species normally exists receiving inputs of fixed-carbon substances that can provide energy for growth of the *Azotobacter. Azotobacter beijerinckia* is dominant in acidic soils—pH values as low as 3.0. This organism is common in tropical soils and is rarely found in temperate soils. It has also been reported to occur in high Arctic soils (Jordan and McNicol, 1978), whereas both *Azotobacter chroococcum* and *Azotobacter beijerinckia* populations have been detected in Antarctic soils (see Vishniac, 1993). Similarly, *Derxia gummosa* is common in tropical soils of South America with pH values ranging from about 4.5 to 6.5.

Azospirillum lipoferum, which was originally described by Beijerink in 1925 (Breed et al. 1957) and named *Spirillum lipoferum,* forms loose symbiotic associations on grasses. These spiral-shaped, gram-negative organisms typically grow around roots as well as penetrate the roots to grow intercellularly. These organisms use root exudates for their carbon and energy source while fixing nitrogen. As is discussed in section 11.4.3, small increases in plant biomass productivity have resulted from inoculation of roots with various stains of *Azospirillum lipoferum.*

Anaerobes and Facultative Bacteria: The most commonly encountered anaerobic diazotrophs are *Clostridium* spp. These organisms are nearly universally present in poorly drained soils as well as in anaerobic microsites of arable soils. *Klebsiella* spp. are the most commonly observed facultative nitrogen-fixing bacteria. These organisms are widely distributed in aquatic and terrestrial environments, especially in rhizosphere soils. In arable lands, clostridial populations commonly range from 10^2 to 10^5 per gram of soil although under optimal conditions, populations as high as 10^6 g^{-1} have been detected. In contrast, *Klebsiella* sp. have been found in rhizosphere samples at population densities of 10^6-10^8 g^{-1} root tissue (Evans et al., 1972). The nitrogen fixation efficiency of these organisms is comparable to that reported for aerobic organisms (2–20 mg N fixed g^{-1} carbohydrate consumed).

Cyanobacteria: Soil microbiologists tend to disregard the input of algae to soil biogeochemical cycles, but blue-green algae can provide significant inputs of fixed nitrogen into land-based ecosystems. These photosynthetic organisms are found primarily in surface soil crusts. Significant algal populations may develop on soil surfaces, especially in the presence of a thin layer of free-standing water. Approximately one-tenth of the established genera of cyanobacteria contain diazotrophs. The absolute contribution of these organisms to soil nitrogen economy is difficult to assess because their distribution is frequently localized, but significant contributions have been measured in rice field soils (see section 11.4).

Due to the photosynthetic nature of these organisms, the quantities of nitrogen fixed are independent of that available in an external chemical energy supply. Yields of fixed nitrogen due to cyanobacteria can be highly significant. In rice paddy soils, contributions up to 30 kg N ha^{-1} $year^{-1}$ are commonly reported. Rychert and Skujins (1974) found that algae and lichens in soils crusts of the Great Basin Desert could produce fixed nitrogen at rates up to 84 g N ha^{-1} $year^{-1}$ (acetylene) in the laboratory. The rates they observed would extrapolate to 10–100 kg N ha^{-1} $year^{-1}$ if comparable efficiency were achieved in field samples.

11.4. FUNCTION OF DIAZOTROPHS IN NATIVE ECOSYSTEMS

The relative contributions of general free-living, nitrogen-fixing bacteria to overall soil ecosystem productivity can be gleaned from the foregoing discussion, whereas the importance of symbiotic associations is detailed in chapter 12. Properties of rhizosphere-associated nitrogen fixation by free-living bacteria and interactions in wetland-type ecosystems are emphasized here. The latter analyses include native swamps, bogs, and marshes, as well as agriculturally managed rice soils. These examples are particularly instructional in that the conditions in each of these systems types exemplify extremes in soil properties conducive to the function of nitrogenase. Both wetland soils and rhizosphere sites can contain elevated concentrations of readily metabolizable organic matter in comparison to generally arable soils, and oxygen tensions are reduced in both sites with increased probability of development of anaerobic microsites.

11.4.1. Nitrogen-Fixing Bacteria Among Rhizosphere Populations

The rhizosphere is home for a variety of heterotrophic diazotrophs exemplified by members of the family *Enterobacteriaceae* including *Enterobacter agglomerans, Klebsiella pneumoniae, Enterobacter cloacae* and *Erwinia herbicola* (e.g., Haahtela et al., 1981; Lindberg and Granhall, 1984; Pedersen et al., 1978), *Bacillus* species (e.g, Lindberg and Granhall, 1984), and *Azospirillum lipoferum* (e.g., Haahtela et al., 1981). The presence of the enterobacteria and the *Azospirillum* species conform to a priori notions regarding this ecosystem; that is, the organisms are heterotrophs and capable of functioning under reduced oxygen tensions or in the total absence of this gas. Indeed, because both aerobic and anaerobic microsites occur in the vicinity of the growing root, even the isolation of strict aerobes (e.g., the example of a *Bacillus* species cited previously) from this environment is not problematic.

Perhaps a more significant concern for students of the environment relates to the significance of the contribution by diazotrophs residing in the rhizosphere to the fixed-nitrogen resources of the growing plant. The role of free-living diazotrophs in the rhizosphere in plant nitrogen metabolism is most commonly described as minimal, but significant. Acetylene and ^{15}N-labeled dinitrogen are reduced in the rhizosphere albeit at minimal levels. For example, Giller et al. (1988) found little to no nitrogen fixation in the rhizosphere of sorghum (*Sorghum bicolor*) and millet (*Pennisetum americanum*). With the sorghum, measurable nitrogen fixation was detected only on one occasion, whereas with the millet, nitrogen fixation accounted for less than 1 percent of the plant nitrogen accumulated. Berestetskii and Vasyuk (1983) found weak nitrogen fixation activity with spring wheat (*Triticum aestivum*), *Poa pratensis,* and *Phleum pratense* but higher activity with millet. Pedersen et al. (1978) estimate that the maximum rate of nitrogen fixation (extrapolated from acetylene reduction data) in spring wheat and sorghum was 2.5-g nitrogen ha^{-1} day^{-1}. Interestingly, although diazotrophs functioning in rhizosphere or rhizoplane ecosystems rely on root exudates to meet fixed carbon and energy requirements, these nitrogen fixation rates are within the range generally observed for free-living diazotrophs in less nutritionally luxuriant soil sites. This observation suggests that, although a priori it could be anticipated that the rhizosphere-residing, nitrogen-fixing bacteria should not be energy-limited, the actual nitrogen fixation rates detected indicate that competition with the general rhizosphere microbial community for carbon and energy supplies limits the capacity of the diazotrophs to exploit nutrients in root exudates.

The contribution of rhizosphere nitrogen fixation to the nitrogen budget of higher plants is minimal. (See Giller and Day, 1985, for a further discussion of this topic.) From the view of managing soil biologically fixed nitrogen to decrease anthropogenic intervention and increase system sustainability, the question of whether the nitrogen-producing capacity of the rhizosphere ecosystem can be increased emerges. Although optimal conditions for function of nitrogenase can be anticipated to occur in the rhizosphere, either microsite conditions vary from these anticipated situations or the efficiency of nitrogen fixation by the diazotrophs in the fields is not sufficient to lead to meaningful fixed-nitrogen contributions to the plant. (Selection and use of highly efficient diazotrophs strains are discussed later.)

Difficulties with the heterogeneity of the microenvironment of the rhizosphere encountered by nitrogen-fixing bacteria is exemplified by evaluating the impact of variation of oxygen tension of the root environment on *in situ* nitrogen fixation. For example, Haahtela et al. (1983) found that anaerobic conditions were required for maximal expression of nitrogenase activity of *Klebsiella* sp. and *Enterobacter* sp. isolated from plants. Similarly, in roots of corn (*Zea mays*), maximal acetylene reduction was noted at partial pressures of oxygen of 1–2 kPa. The activity was strongly inhibited with oxygen tensions above or equal to 6 kPa (Alexander et al., 1987). Alexander and Zuberer (1989) found that with ^{15}N-labeled dinitrogen ammonium production was 200-fold greater at 2-kPa molecular oxygen than at 10 kPa. Thus, we can predict that a controlling factor for nitrogen fixation in the rhizosphere is the development of anaerobic microsites. Total anaerobiosis in this habitat is clearly precluded by the requirement for oxygen for respiration of the root tissue.

A potential means of reducing fertilizer expense for cropped systems and anthropogenic intervention in reclamation projects is to optimize *in situ* inputs of biologically fixed nitrogen through manipulation of rhizosphere populations. Because habitats favorable to growth and development of diazotrophs exist around the plant root, the objective of such studies can be reduced to selecting efficient nitrogen-fixing strains of rhizosphere bacteria that can compete with indigenous microbes when returned to the root ecosystem. An apparently optimistic expectation for achieving such goals is possible in that selecting natural genetic variants and producing geneticly engineered strains with improved nitrogen fixation efficiency can readily be accomplished in the laboratory. Furthermore, the potential exists for the manipulating rhizosphere populations for the successful inoculation of these laboratory-cultivated strains into field situations. Application of these procedures is exemplified by studies of *Azospirillum* spp. These organisms are normal components of the rhizosphere microbial community, fix nitrogen, and are amenable to laboratory culture.

As with any inoculation procedure, albeit soil or rhizosphere inoculation, the initial concerns with amending axenically cultivated strains are a) Can the desired effect be observed in the ecosystem following addition of the test strain (e.g., stimulation of plant biomass synthesis)? and b) Does the alteration of the ecosystem result from the anticipated activity of the amended microbe or is an alternative mechanism operative? Once we acquire responses to these questions, then considerations regarding reproducibility, stability, and magnitude of the amendment response must be addressed.

Inoculation of a variety of grass varieties with *Azospirillum* strains increases plant biomass production (e.g., Gaskins et al., 1977; Smith et al., 1976; Subba-Rao et al., 1978). Several mechanisms may be proposed to explain this increase in plant growth by inoculation with cultures of *Azospirillum*. Explanations for the stimulation of biomass production include a) augmented ammonium supplies produced by the *Azospirillum* sp. (i.e., nitrogen fixation), b) plant hormone synthesis by the bacterium (e.g., Tien et al., 1979), c) alteration of root structure (e.g., Fallik et al., 1988; Hadas and Okon, 1987; Jain and Patriquin, 1984), and d) response to increased plant nutrients provided by mineralization of dead *Azospirillum* cells.

It is probable that all the mechanisms function in part in regards to *Azospirillum* inoculation of grasses because nitrogenase synthesis and activity appear not to be the primary explanation for increases in plant productivity. Estimates of fixed nitrogen produced by *Azospirillum* sp. in the rhizosphere range from 2 kg N ha^{-1} (e.g., Von Bülow and Döbereiner, 1975) to as much as 90 kg N ha^{-1} (Döbereiner et al., 1972, 1973) with inoculated grasses in South America. Other studies have reported low levels of nitrogen fixation in temperate grasses (e.g., Barber et al., 1976; Harris et al., 1989; Okon et al., 1983). Alternative explanations of the effect of *Azospirillum* inoculation include synthesis of plant hormones by the bacterium that would stimulate root biomass production (e.g., Tien et al., 1979) and alteration of plant cation metabolism. Cation accumulation by the plant may (Lin et al., 1983) or may not (Bashan et al., 1990) be affected by root *Azospirillum* populations.

Although these data suggest limitations in yields of fixed nitrogen in inoculated rhizospheres, as understanding of the genetics mechanisms associated with this process in *Azospirillum* and of the properties of the bacterium controlling its capacity to compete with indigenous rhizosphere populations improves, the probability of developing an economic and effective rhizosphere inoculation procedure will increase. Also, along with using genetic modification to improve the competitive ability of *Azospirillum* cells introduced into the rhizosphere, fungicide or bacteriocides may be added with the diazotroph to increase the probability of establishing the culture (Bashan, 1986a). With this procedure, the inhibitors are used to limit the activity of indigenous populations until the foreign bacterial strain can become established, thereby giving the amended organism a competitive advantage.

Associated with the selection of appropriate strains for inoculation of root or seed tissue is the development of an effective inoculation method. Viable cells must be delivered to the appropriate habitat on the root at a time of root development when maximal benefits to the plant are accrued and the conditions in the microsite wherein the microbes are anticipated to function are optimal for nitrogen fixation. *Azotobacter* populations can become established on grass roots through inoculation of either seeds or root tissue, although seed inoculation appears to be the most effective means of delivery (Bashan, 1986b). One advantage of seed inoculation is that the amended bacterial strain is present as the nascent root tissue is produced. Thus, instead of competing with established bacterial populations as would be the situation when roots are inoculated directly, the competition is reduced to a race between seed and soil flora to colonize the newly emergent root tissue.

Should the objective be to inoculate root tissue directly, the capability of the added cells or their progeny to spread from inoculated to noninoculated roots becomes important. Roots of infected plants can serve as a vector for transmission of the exogenous bacterial strain to noninfected populations (Bashan and Levanony, 1987; Bashan and Levanony, 1989). In fact, weeds growing between rows of crop plants can transport the added bacterial strain (Bashan and Levanony, 1989). Thus, although a time delay in spreading the inoculant must occur with direct inoculation of plant roots in the field, diffusion of inoculated bacterial strains throughout a field can be anticipated to result eventually. For this process, active bacterial cultures can be added to the soil surface adjacent to the plant stem and washed into the soil (rhizosphere), plant roots can be dipped into the inoculum prior to planting, or the inoculum can be banded adjacent to the plant in a region where the roots of the growing plant would be anticipated to penetrate. As has been shown with rhizobial inoculation of legumes (see chapter 12), a variety of carriers can be used to introduce *Azospirillum* inoculum into soil (e.g., Bashan, 1986c). Each of these techniques would be less efficient than seed inoculation (i.e., from the viewpoint of time as well as quantity of inoculum that must be used), but with highly competitive and nitrogen fixation-efficient diazotroph strains, such procedures may become more feasible.

11.4.2. Nitrogen Fixation in Flooded Ecosystems

Wetland soils are sites of limited but active biological nitrogen fixation. The quantities of fixed nitrogen gained by this source are within the range commonly reported to be produced by free-living microbes in soils in general. For example, Waugmen and Bellamy (1980) estimate nitrogen fixation inputs due to heterotrophic bacteria in German mires (using the acetylene reduction method) to range from 0.07 to 2.1 g N m^{-2} year^{-1}. Slightly higher fixed-nitrogen contributions have reported for Gulf coast salt marshes (USA) (4.5 and 15 g N m^{-1} year^{-1}, Casselman et al., 1981). Because biomass productivity is low in swampy environments as a result of the prevailing flooded conditions, biologically fixed nitrogen can be a major source of this essential plant nutrient.

Free-living bacteria are responsible for nitrogen fixation in swampy soils. As might be anticipated, considering the oxygen-limiting conditions of marshes, anaerobic bacteria are major participants in this process. For example, Dicker and Smith (1980) found that *Clostridium* sp., *Azotobacter* sp., as well as *Desulfovibrio* sp. accounted for a major portion of the nitrogen-fixing activity in a Delaware (USA) salt marsh, although large populations of *Azotobacter* sp. were present in the samples. Co-occurrence of obligate aerobes and anaerobes in this ecosystem appears to be contradictory, but consideration of the heterogeneity of the soil profile (i.e., water-saturated soil overlain by flowing water) reveals that it is reasonable to anticipate occurrence of both aerobic and anoxic microsites. Oxygen-bearing water flowing into the system provides for the sustenance of aerobic microsites, whereas activity of the aerobic bacteria in the system creates anaerobic sites. Furthermore, sequestering of water into nonflowing situations within the soil profile also results in the development of anaerobic habitats within the soil profile (stagnation).

The diversity of diazotrophs present in swampy soils indicates the existence of a variety of energy sources that support microbial growth. Metabolic energy supply in these ecosystems are provided by a) root exudates (e.g., from the *Spartina alterniflora* Loisel. plants in the salt marshes), b) accumulations of partially decomposed organic matter under anoxic conditions (i.e., products of anaerobic decomposition of biomass substituents), as well as c) solar energy (e.g., growth of cyanobacteria in the aquatic portion of the site).

Rice paddies provide an excellent example of the function of these various diazotroph groups in a managed, flooded ecosystem. A diverse population of autotrophic bacteria are found in the flood waters (e.g., Habte and Alexander 1980a; Kulasooriya and De Silva, 1981). For example, 73 strains of cyanobacteria distributed taxonomically among 21 genera were isolated from rice soils in central Sri Lanka (Kulasooriya and De Silva, 1981). Similarly, the rhizosphere contains an active heterotrophic diazotroph population (e.g., Habte and Alexander, 1980b; Yoo et al., 1986). Also note that these heterotrophic strains may be active on aboveground plant surfaces (Ito et al., 1980). The activities of the latter populations were shown to be approximately 2.5-fold more active at fixing nitrogen than rhizosphere populations (Ito et al., 1980).

Wetland systems with their low, but significant, biologically fixed-nitrogen inputs provide an even more vivid picture of the complexity of diazotrophic communities within a single ecosystem than was provided by the rhizosphere examples discussed previously. This diversity in physiological ability provides a genotypic foundation for long-term sustenance of the system under ever-changing conditions. Balance between fixed-nitrogen inputs from heterotrophic or autotrophic populations, aerobes or anaerobes, and rhizosphere or leaf sheath populations is dependent upon such site properties as water influx rate, depth of flooding (i.e., drained to submerged), diurnal variation of temperature, and light influx.

11.5. CONCLUSIONS

The capacity to fix nitrogen is widely distributed among the bacterial genera, although only a limited number of species are actually capable of catalyzing the process. Nitrogenase, the enzyme responsible, is highly conserved among this diverse group of bacteria. Two characteristics of this reductive process dictate the yield of fixed nitrogen in various terrestrial environments and the microbial adaptations necessary to protect the enzyme. First, nitrogen fixation is an energy-intensive process. Thus, large quantities of dinitrogen are fixed only in situations of high energy input (e.g., symbiotic associations involving formation of root nodules and through action of photosynthetic bacteria). Diazotrophs growing in competition with the general heterotrophic soil microbial community contribute small quantities of fixed nitrogen to the biological community. Second, nitrogenase is an oxygen-labile enzyme. Hence, each diazotroph must expend resources to maintain anoxic conditions in the environment of active nitrogen fixation. These two overall traits of the nitrogen fixation process and the diverse group of bacteria capable of catalyzing the reaction combined with the extremely heterogeneous soil ecosystem select for a highly diverse indigenous population of diazotrophs in any terrestrial ecosystem. For example, the rhizosphere can contain aerobic, facultative anaerobic, and strictly anaerobic diazotrophs when the oxygen sensitivity of the organisms alone is considered. This diversity of *in situ* populations and variations in phenotypic expression of nitrogen-fixing efficiency allows cautious optimism for genetic manipulation and selection of diazotrophs that can be introduced into soil systems to improve contributions to soil-fixed-nitrogen resources.

REFERENCES

Alexander, D. B., and D. A. Zuberer. 1989. $^{15}N_2$ fixation by bacteria associated with maize roots at a low partial O_2 pressure. Appl. Environ. Microbiol. 55:1748–1753.

Alexander, D. B., D. A. Zuberer, and D. M. Vietor. 1987. Nitrogen fixation (C_2H_2 reduction) associated with roots of intact *Zea mays* in fritted clay at reduced oxygen tensions. Soil Biol. Biochem. 19:1–6.

Alexander, M. 1977. Introduction to Soil Microbiology. John Wiley & Sons. New York. 467 pp.

Azam, G., T. L. Mulvaney, and F. J. Stevenson. 1988. Quantification and potential availability of non-symbiotically fixed ^{15}N in soil. Biol. Fert. Soils 7:32–38.

Barak, R., I. Nur, Y. Okon, and Y. Henis. 1982. Aerotactic response of *Azospirillum brasilense*. J. Bacteriol. 152:643–649.

Barber, L. E., J. D. Tjepkema, S. A. Russell, and H. J. Evans. 1976. Acetylene reduction (nitrogen fixation) associated with corn inoculated with *Spirillum*. Appl. Environ. Microbiol. 32:108–113.

Bashan, Y. 1986a. Enhancement of wheat root colonization and plant development by *Azospirillum brasilense* following temporary depression of rhizosphere microflora. Appl. Environ. Microbiol. 51:1067–1071.

Bashan, Y. 1986b. Significance of timing and level of inoculation with rhizosphere bacteria of wheat plants. Soil Biol. Biochem. 18:297–302.

Bashan, Y. 1986c. Alginate beads as synthetic inoculant carriers for slow release of bacteria that affect plant growth. Appl. Environ. Microbiol. 51:1089–1098.

Bashan, Y., S. K. Harrison, and R. E. Whitmoyer. 1990. Enhanced growth of wheat and soybean plants inoculated with *Azospirillum brasilense* is not necessarily due to general enhancement of mineral uptake. Appl. Environ. Microbiol. 56:769–775.

Bashan, Y., and H. Levanony. 1987. Horizontal and vertical movement of *Azospirillum brasilense* CD in the soil and along the rhizosphere of wheat and weeds in controlled and field environments. J. Gen. Microbiol. 133:3473–3480.

Bashan, Y., and H. Levanony. 1989. Wheat root tips as a vector for passive vertical transfer of *Azospirillum brasilense* CD. J. Gen. Microbiol. 135:2899–2908.

Berestetskii, O. A., and L. F. Vasyuk. 1983. Nitrogen-fixing activity in rhizosphere and on roots of nonleguminous plants. Biol. Bull. 10:32–37.

Breed, R. S., E. G. D. Murray, and N. R. Smith et al. Bergey's Manual of Determinative Bacteriology. Williams and Wilkins, New York. 1094 pp.

Brewin, N. J. 1991. Development of the legume root nodule. Annu. Rev. Cell Biol. 7:191–226.

Burns, R. C., and R. W. F. Hardy. 1975. Nitrogen Fixation in Bacteria and Higher Plants. Springer-Verlag. New York. 189 pp.

Burris, R. H. 1991. Nitrogenases. J. Biol. Chem. 266:9339–9342.

Caetano-Anollés, G., and P. M. Gresshoff. 1991. Plant genetic control of nodulation. Annu. Rev. Microbiol. 45:345–382.

Casselman, M. E., W. H. Patrick, Jr, and R. D. DeLaune. 1981. Nitrogen fixation in a gulf coast salt marsh. Soil Sci. Soc. Am. J. 45:51–56.

Colbeau, A., B. C. Kelley, and P. M. Vignais. 1980. Hydrogenase activity in *Rhodopseudomonas capsulata:* Relationship with nitrogenase activity. J. Bacteriol. 144:141–148.

Dicker, H. J., and D. W. Smith. 1980. Enumeration and relative importance of acetylene-reducing (nitrogen-fixing) bacteria in a Delaware salt marsh. Appl. Environ. Microbiol. 39:1019–1025.

Döbereiner, J., J. M. Day, and P. J. Dart. 1972. Nitrogenase activity in the rhizosphere of sugarcane and some other tropical grasses. Plant Soil 37:191–196.

Döbereiner, J., J. M. Day, and P. J. Dart. 1973. Rhizosphere associations between grasses and nitrogen fixing bacteria: Effect of O_2 on nitrogenase activity in the rhizosphere of *Paspalum notatum.* Soil Biol. Biochem. 5:157–159.

Evans, H. J., N. E. R. Campbell, and S. Hill. 1972. Asymbiotic nitrogen-fixing bacteria from the surfaces of nodules and roots of legumes. Can. J. Microbiol. 18:13–21.

Fallik, E., Y. Okon, and M. Fisher. 1988. Growth response of maize roots to *Azospirillum* inoculation: Effect of soil organic matter content, number of rhizosphere bacteria and timing of inoculation. Soil Biol. Biochem. 10:45–50.

Fay, P. 1992. Oxygen relations of nitrogen fixation in cyanobacteria. Microbiol. Rev. 56:340–373.

Gaskins, M. H., M. U. Garcia, T. M. Tien, and D. H. Hubbell. 1977. Nitrogen fixation and growth substance produced by *Spirillum lipoferum* and their effects on plant growth. Plant Physiol. 59:128.

Giller, K. E., and J. M. Day. 1985. Nitrogen fixation in the rhizosphere: Significance in natural and agricultural systems. *In* A. E. fitter (ed.), Ecological Interactions in Soil: Plants, Microbes, and Animals, pp. 127–147. Blackwell Scientific Publications. Boston.

Giller, K. E., S. P. Wani, J. M. Day, and P. J. Dart. 1988. Short-term measurements of uptake of nitrogen fixed in the rhizospheres of sorghum (*Sorghum bicolor*) and millet (*Pennisetum americum*). Biol. Fert. Soils 7:11–15.

Haahtela, K., K. Kari, and V. Sundman. 1983. Nitrogenase activity (acetylene reduction) of root-associated, clod-climate *Azospirillum, Enterobacter, Klebsiella,* and *Pseudomonas* species during growth on various carbon sources and at various partial pressures of oxygen. Appl. Environ. Microbiol. 45:563–570.

Haahtela, K., T. Wartiovaara, V. Sundman, and J. Skujins. 1981. Root-associated N_2 fixation (acetylene reduction) by *Enterobacteriaceae* and *Azospirillum* strains in cold-climate spodosols. Appl. Environ. Microbiol. 41:203–206.

Habte, M., and M. Alexander. 1980a. Nitrogen fixation by photosynthetic bacteria in lowland rice culture. Appl. Environ. Microbiol. 39:342–347.

Habte, M., and M. Alexander. 1980b. Effect of rice plants on nitrogenase activity in flooded soils. Appl. Environ. Microbiol. 40:507–510.

Hadas, R., and Y. Okon. 1987. Effect of *Azospirillum brasilense* inoculation on root morphology and respiration in tomato seedlings. Biol. Fert. Soils 5:241–247.

Harris, J. M., J. A. Lucas, M. R. Davey, G. Lethbridge, and K. A. Powell. 1989. Establishment of *Azospirillum* inoculant in the rhizosphere of winter wheat. Soil Biol. Biochem. 21: 59–64.

Ito, O., D. Cabrera, and I. Watanabe. 1980. Fixation of dinitrogen-15 associated with rice plants. Appl. Environ. Microbiol. 39:554–558.

Jain, D. K., and D. G. Patriquin. 1984. Root hair deformation, bacterial attachment, and plant growth in wheat-*Azospirillum* association. Appl. Environ. Microbiol. 48:1208–1213.

Jordan, D. C., and P. J. McNicol. 1978. Identification of *Beijerinckia* in the high Arctic (Devon Island, Northwest Territories). Appl. Environ. Microbiol. 35:204–205.

Keyser, H. H., D. F. Weber, and S. L. Uratsu. 1984. *Rhizobium japonicum* serogroup and hydrogenase phenotype distribution in 12 states. Appl. Environ. Microbiol. 47:613–615.

Knowles, R. 1981. Denitrification. *In* E. A. Paul and J. N. Ladd (eds.), Soil Biochemistry, Vol. 5, pp. 323–369. Dekker. New York.

Kulasooriya, S. A., and R. S. Y. De Silva. 1981. Mutivariate interpretation of the distribution of nitrogen fixing blue-green algae in rice soils in central Sri Lanka. Ann. Bot. (Lond.) 47:31–52.

Lin, W., Y. Okon, and R. W. F. Hardy. 1983. Enhanced mineral uptake by *Zea mays* and *Sorghum bicolor* roots inoculated with *Azospirillum brasilense*. Appl. Environ. Microbiol. 45:1775–1779.

Lindberg, T., and U. Granhall. 1984. Isolation and characterization of dinitrogen-fixing bacteria from the rhizosphere of temperate cereals and forage grasses. Appl. Environ. Microbiol. 48:683–689.

Merberg, D., and R. J. Maier. 1983. Mutants of *Rhizobium japonicum* with increased hydrogenase activity. Science 220:1064–1065.

Murry, M. A., A. J. Horne, and J. R. Benemann. 1984. Physiological studies of oxygen protection mechanism in the heterocysts of *Anabaena cylindrica*. Appl. Environ. Microbiol. 47:449–454.

Neilson, A. H., and L. Sparell. 1976. Acetylene reduction (nitrogen fixation) by *Enterobacteriaceae* isolated from paper mill process waters. Appl. Environ. Microbiol. 32:197–205.

Nelson, L. M., and S. O. Salminen. 1982. Uptake hydrogenase activity and ATP formation in *Rhizobium leguminosarum* bacteroids. J. Bacteriol. 151:989–995.

Okon, Y., P. G. Heytler, and R. W. F. Hardy. 1983. N_2 fixation by *Azospirillum* and its incorporation into host *Setaria italica*. Appl. Environ. Microbiol. 46:694–697.

Pedersen, W. L., K. Chakrabarty, R. V. Klucas, and A. K. Vidaver. 1978. Nitrogen fixation (acetylene reduction) associated with roots of winter wheat and sorghum in Nebraska. Appl. Environ. Microbiol. 35:129–135.

Peschek, G. A., K. Villgrater, and M. Wastyn. 1991. 'Respiratory protection' of nitrogenase in dinitrogen-fixing cyanobacteria. Plant Soil. 137:17–24.

Ruiz-Argüeso, T. R. J. Maier, and J. H. Evans. 1979. Hydrogen evolution from alfalfa and clover nodules and hydrogen uptake by free-living *Rhizobium meliloti*. Appl. Environ. Microbiol. 37:582–587.

Rychert, R. C., and J. Skujins. 1974. Nitrogen fixation by blue-green-algae-lichin crusts in the great basin desert. Soil Sci. Soc. Am. Proc. 38:768–771.

Shah, V. K., J. L. Pate, and W. J. Brill. 1973. Protection of nitrogenase in *Azotobacter vinelandii*. J. Bacteriol. 115:15–17.

Smith, R. L., J. H. Bouton, S. C. Shank, K. H. Quesenberry, M. E. Tyler, J. R. Milam, M. H. Gaskins, and R. C. Littell. 1976. Nitrogen fixation in grasses inoculated with *Spirillum lipoferum*. Science 193:1003–1005.

Söderlund, R., and B. H. Svensson. 1976. The global nitrogen cycle. *In* B. H. Svenson and R. Söderlund (eds). Nitrogen, Phosphorus, and Sulphur—Global Cycles. SCOPE Rep. 7. Ecol. Bull. (Stockholm) 22:23–73.

Subba-Rao, N. S., K. V. B. R. Tilak, M. Lakshmi Kumari, and D. S. Singh. 1978. Response of crops to *Spirillum lipoferum* inoculation. *In* W. H. Orme-Johnson and W. E. Newton (eds.), Steinbock–Kettering International Symposium on Nitrogen Fixation, p. 20. University of Wisconsin. Madison, WI.

Tien, T. M., M. H. Gaskins, and D. H. Hubbell. 1979. Plant growth substances produced by *Azospirillum brasilense* and their effect on the growth of pearl millet (*Pennisetum americanum L.*). Appl. Environ. Microbiol. 37:1016–1024.

Vishniac, H. S. 1993. The microbiology of Antarctic soils. *In* E. I. Friedmann (ed.), Antarctic Microbiology, pp. 297–241. Wiley-Liss, Inc. New York.

Von Bülow, J. F. W., and J. Döbereiner. 1975. Potential for nitrogen fixation in maize genotypes in Brazil. Proc. Natl. Acad. Sci. USA 72:2389–2393.

Waugman, G. J., and D. J. Bellamy. 1980. Nitrogen fixation and the nitrogen balance in peatland ecosystems. Ecology 61:1185–1198.

Yoo, I. D., T. Fuji, Y. Sano, K. Komagata, T. Yoneyama, S. Iyama, and Y. Hirota. 1986. Dinitrogen fixation of rice and *Klebsiella* associations. Crop Sci. 26:297–301.

12 Symbiotic Nitrogen Fixation

Root-nodule-forming symbiotic associations between diazotrophic bacteria and plants are the primary biological contributors of fixed nitrogen in soil-based ecosystems. This situation results from a) the predominance of the agricultural exploitation of *Rhizobium*–legume symbioses and b) the advantage conferred on these symbiotic diazotrophs by the ready availability of plant-produced photosynthate. As was presented in chapter 11, the abundant carbon and energy supply in the root nodule results in yields of fixed nitrogen severalfold greater than those from heterotrophic diazotrophs living nonassociated or loosely linked with plants. Root-nodule-based nitrogen-fixing associations include *Rhizobium*–legume and actinorhizal symbioses. The latter root-nodule-forming symbioses involve interactions of nonleguminous angiosperms with actinomycetes of the genus *Frankia*. *Rhizobium*–legume symbioses are the most studied, but due to the necessity of reclaiming a variety of damaged or mismanaged soil systems in which growth of this symbiotic association is limited or precluded, considerable interest in enhancement of fixed-nitrogen contributions from actinorhizal plants has developed.

Both actinorhizal and *Rhizobium*–legume symbioses will be analyzed in this chapter. The primary objectives for this study are to elucidate the properties of the bacteria and the host plant that determine the efficacy and predominance of the associations in ecosystems and to use these interactions as models for evaluating the potential for their manipulation for management of biogeochemical cycles in soil ecosystems. Topics were selected for discussion to stress the environmental aspects of nitrogen fixation in nodulated plant root systems. Biochemical and biological aspects of the topic are analyzed only where needed to provide a foundation for understanding their impact on the occurrence of nitrogen fixation by these symbiotic associations in native ecosystems. For a more detailed discussion of the biochemistry or biology of actinorhizal or *Rhizobium*–legume interactions, refer to essentially any basic general microbiology text or biochemistry text. Several review articles that provide an overview of these topics in greater detail than can be provided herein are cited.

12.1. *RHIZOBIUM*–LEGUME ASSOCIATION

Due to its purposeful use in agriculture, the greatest documented contribution of fixed nitrogen in land-based systems results from the infection of legume roots by species of the bacterial genera *Rhizobium* and *Bradyrhizobium*. A tremendous potential for contribution of fixed nitrogen to soil ecosystems exists among the legumes. There are approximately 700 genera and about 13,000 species of legumes, only a portion of which have been examined for the capability to fix nitrogen. Estimates are that the rhizobial symbiosis with the somewhat greater than 100 agriculturally important legumes contribute nearly half the annual quantity of biological fixed nitrogen entering soil ecosystems.

The bacterial symbionts are gram-negative, non-spore-forming, aerobic rods. Some strains are nonmotile, but typically rhizobia are motile. Rhizobia metabolize several carbohydrates, possibly with the production of acid, but no gas. The symbiotic association with legumes is a distinguishing classification trait of the genus. (Rhizobial species are grouped into the genera *Rhizobium, Bradyrhizobium,* and *Azorhizobium*—stem-nodulating organisms—e.g., see Adebayo et al., 1989.) Speciation of these bacteria is historically based on host specificity as defined by cross-inoculation grouping.

12.1.1. Grouping of Rhizobial Strains

From the view of a microbiologist, rhizobia constitute a somewhat nondescript group of bacteria. Classification schemes based on data derived from using traditional physiological tests are inadequate to separate the genus into meaningful groupings. An alternative trait common to rhizobia that provides a reproducible method for their classification that was testable by methods available to early microbiologists is host specificity. The resultant association of strains into groups based on host specificity is called cross-inoculation groups. **Cross-inoculation groups** are by definition a collection of leguminous species that develop nodules when exposed to bacteria obtained from the nodules of any member of that group. More than 20 cross-inoculation groups have been established. Of the seven most studied groups, six are sufficiently described to designate the responsible bacteria as species. The most studied cross-inoculation groups are:

- **The alfalfa group** (legumes include alfalfa, *Medicago* spp., and sweet clover, (*Melilotus* spp.)), which is nodulated by *R. meliloti*
- **The clover group** (clovers, *Trifolium* spp.), which is nodulated by *R. trifolii*
- **The pea group** (e.g., pea, *Pisum* spp., and vetch, *Vicia*), which is nodulated by *R. leguminosarum*
- **The bean group** (beans, *Phaseolus* spp.), which is nodulated by *R. phaseoli*
- **The soybean group** (soybeans, *Glycine* spp.), which is nodulated by *Bradyrhizobium japonicum*. In earlier literature, *Bradyrhizobium japonicum* and *Bradyrhizobium* spp. (symbionts of the cowpea group) were previously classified as slow-growing variants of *Rhizobium japonicum*.
- **The cowpea group.** This group has not achieved a species designation, but these organisms nodulate a variety of legumes including cowpeas (*Vigna* sp.), Kudzu (*Pueraria* sp.), peanuts (*Arachis* sp.), and lima beans (*Phaseolus* sp.).

The cross-inoculation system has provided a useful classification system for rhizobial strains, but some major problems have historically been associated with its application. Some of these are logistical. For example, considerable greenhouse space, which may be required for an accurate determination of the identity of the *Rhizobium* strain of interest, is necessary to grow the legume seedlings. Of perhaps greater concern in using this method to group newly isolated rhizobial strains is the potential for some organisms to nodulate more than one legume species. For example, soybean and cowpea groups contain strains of rhizobia isolated on legume groups that nodulate other legume species. This variation in nodulation specificity is termed **symbiotic promiscuity.** Thus, the validity of the cross-inoculation system has been challenged. The procedure works reasonably well for the frequently studied rhizobia nodulating common, agricultural crops of temperate regions. More difficulties may be encountered as a greater proportion of the as-yet little-studied rhizobia, which nodulate tropical legumes, are classified. Fortunately for the classification system, many rhizobia do nodulate legumes outside of their particular class, *but* these nodules are generally nonfunctional in nitrogen fixation (i.e., they are not effective). Thus, the designation of species or groups based on the formation of effective nodules is sustained. Note that not all the literature reports of symbiotic promiscuity represent true diversity in legume species specificity. In some cases, apparent host diversity has been the result of contamination of the rhizobial cultural with other rhizobia (Leps et al., 1980).

Cross-inoculation groups have provided a reasonably stable philosophical basis for the taxonomic scheme for grouping rhizobial strains, but these groupings are gradually falling into disrepute. A variety of methods, including evaluation of DNA hybridization, numerical taxonomic procedures, DNA base ratios, serology, antibiotic resistance, genetic markers, and nutritional traits, are applied to a taxonomy in transition. Use of strain-specific antibodies allows identification of axenically grown rhizobia. Cross-inoculation groups are serologically distinct. Serological identification of rhizobial strains is quick, inexpensive, and reliable. Similarly, PCR analysis for portions of the cellular genome specific to rhizobia has provided a sensitive means of detecting and quantifying rhizobia in soil and culture (e.g., see Hartmann and Amarger, 1991). These latter methods are particularly useful in that they can be used to identify individual strains occurring at low population densities directly in soil samples. As the nuances of the implications of the application of these procedures to the study of rhizobial taxonomy become better understood, a more inclusive grouping of temperate and tropical strains will emerge. (See Elkan and Bunn, 1992, for an analysis of the current status of rhizobial taxonomy.)

12.1.2. Role of Rhizobial Cells in Nitrogen Fixation

Although many of the early microbiology textbooks state the contrary, rhizobia can fix nitrogen in axenic culture. Pagan et al. (1975) developed a defined medium and culture procedure that allows nitrogen fixation by free-living rhizobial strains. With this procedure, the bacteria are grown to a high density on yeast extract and mannitol agar, suspended in sterile water, and spread on a defined agar medium. Other studies have shown that nitrogenase activity of *Rhizobium japonicum* growing in defined medium is dependent upon the carbon source metabolized (Kurz and Larue, 1975) and oxygen tension of the growth medium (Tjepkema and Evans, 1975). Microaerophilic conditions are required to induce nitrogen-fixing activity by the rhizobial cells. Kiester and Evans (1976) found that the optimal oxygen level for nitrogen fixation by *Rhizobium japonicum* and a *Rhizobium* sp. was approximately 0.1 percent in the gas phase. Carbon dioxide is also required by the rhizobial cells. Aguilar and Favelukes (1982) found an obligatory need for carbon dioxide for nitrogen fixation under microaerophilic conditions. This effect of carbon dioxide is mediated through ribulose bisphosphate carboxylase activity (Manian et al., 1984). Once the environmental conditions are optimized, the rate of nitrogen fixation by free-living cells is comparable to that observed with bacteroides isolated from soybean nodules.

12.1.3. Nodule Formation in *Rhizobium*–Legume Associations

Rhizobia in soil are capable of infecting roots of susceptible plants to produce root nodules (Fig. 12.1). These nodules have a structure characteristic for the various symbiotic associations. Several processes with inputs from the bacterial genome and from the plant are responsible for developing the nodule structure. This summary of the process of legume root nodule formation is divided into three topic areas: a) properties of the plant and microbe facilitating contact and binding of rhizobial cells to the host root, b) recognition of susceptible host roots and attachment, and c) root invasion processes and development of effective nodules. A variety of excellent reviews describing the biological, genetic, and biochemical aspects of nodule formation have been prepared (e.g., see Brewin, 1991; Caetano-Anollés and Gresshoff, 1991a). Thus, only the salient details of the process are outlined here. Aside for the aforementioned reviews, examples of primary research publications are provided as examples of the type of current research involving the interaction of rhizobial strains and their specific legume hosts.

Encounter Between Susceptible Roots and Rhizobia: Although their population densities may be low, rhizobia are found in most soils, even those sites where legumes have not grown for many decades. Hence, limitations in frequency of nodule formation in native ecosystems is derived more from the rate of encounter of infecting bacteria and host rather than presence of the symbiotic partners. Considering the heterogeneity of soil, the size of the bacterial cell, and the low proportion of the soil A horizon occupied by plant roots, random encounters between stationary bacterial cells and growing roots must be considered to be an inefficient means of inducing nodulation. Fortunately, a variety of basic properties of the soil system enhance this naturally low probability. First, in undisturbed soils, rhizobial cells are concentrated in the vicinity of the decaying root from previous growing seasons. The probability of infecting nascent root tissue in native soil systems is therefore enhanced by the propensity of newly developing root systems to grow into preexisting root channels.

Fig. 12.1. Root nodules formed by *Bradyrhizobium japonicum* and soybeans.

A further increase in nodulation frequency could result from the movement or migration of infective bacteria that does occur within the soil profile. This migration of the rhizobial cells is primarily passive; that is, it would not involve directed motility of the cells (Issa et al., 1993a, 1993b). The bacteria instead are carried by water flowing through soil macropores. **Macropores** are large passages in the soil matrix that allow the ready movement of air and water. These pores include the spaces between soil aggregates as well as the channels created by the growth of roots (root channels) and the movement of animals (e.g., earthworm casts). Passive transport of soil bacteria with leaching water disperses cells over relatively large distances (in the A horizon, millimeter to centimeter distances). Due to the small distances over which it occurs, active motility of individual bacterial cells impacts the infection rate primarily in the microsite in response to exudate production by the growing root (chemotaxis, see the following discussion).

This differentiation of impact between passive and active aspects of bacterial transport in soil does not preclude conference of a competitive advantage to rhizobial cells by the capacity for active motility. Note that a number of reports exist in the literature supporting as well as opposing the conclusion that motile strains of rhizobia have an enhanced potential to occupy legume nodules compared to non-motile bacteria. The seemingly contradictory conclusions result to a large degree from varying the properties of the laboratory- or greenhouse-incubated soils used in the experiments as well as from the various population densities used. It is indisputable that motility of rhizobia is not required for nodule formation; yet a number of studies have shown both enhanced occurrence of nodulation by mobile strains (e.g., Gulash et al., 1984; Hunter and Fahring, 1980; Malek, 1992; Napoli and Albersheim, 1980) and no effect of motility (Ames et al., 1980; Issa et al. 1993a, 1993b; Liu et al., 1989).

Many of the contradictory results are attributable to variation in the importance of chemotaxis of infective rhizobia under the various experimental conditions. In situations where both motile and nonmotile strains have equal access to susceptible root tissue, no benefit is gained from mobility. In contrast, if the soil moisture levels and population densities used in the experiment are such that the microorganisms are required to traverse a short distance from their habitat in the soil to the root, then an actively motile organism would gain a benefit over one that must rely on passive diffusion. Thus, we can conclude that directed motility (i.e., chemotactic attraction) is most important to nodulation once the microbe has been moved passively by mass water flow into the vicinity of the susceptible portion of the root.

This model of the relative importance of passive and active motility of bacterial strains in nodule formation rests upon the assumption that rhizobia can respond chemotactically to root exudates. Rhizobia are attracted to root exudate components (e.g., Bhagwat and Thomas, 1982). Exudate components found to attract various rhizobial strains include sugars and amino acids (e.g., Bergman et al., 1988), proteins (e.g., Currier and Strobel, 1977), dicarboxylic acids (e.g., Barbour et al., 1991), aromatic compounds (e.g., Kape et al., 1991), and flavonoids (e.g., Dharmatilake and Bauer, 1992). Note that attraction to the general components of root exudates (e.g., sugars and amino acids and their polymers) is not specific to the rhizobia, but a variety of rhizosphere bacteria are attracted to the root by these compounds. To a major degree, the response of the rhizobial population to root exudates is part of a more generalized stimulation of bacterial population densities in the rhizosphere.

There is an effect of some root exudate components and expression of nodulating genes by the *Rhizobium* cell. Deletions in the *nif–nod* region of the symbiotic megaplasmid eliminated the chemotactic attraction of *Rhizobium meliloti* to plant roots by the flavonoid luteolin (Caetano-Anollés et al., 1988). The *nif–nod* region is constituted of genes involved in nodule formation and nitrogen fixation. For further discussion of the induction of these genes by components of root exudates and their role in nodule initiation, see Brewin (1991).

Host Recognition and Attachment: Reaction to root exudate components allows the rhizobial strains to reach the specific portion of the root structure most susceptible to *Rhizobium* infection (that portion of the root behind the apical meristem at the site of emergence of the root hair, i.e., the zone of root elongation) where recognition and attachment take place. Recognition of susceptible host tissue involves an interaction between proteins (lectins) on the plant root surface and rhizobial exopolysaccharides. This process could be likened to the specificity and interactions associated with an antibody–antigen reaction. Early suggestions of the role of lectins in the recognition process were presented by Bohlool and Schmidt (1974) when they found that a lectin from soybean combined specifically with the soybean-nodulating bacterium [*Bradyrhizobium japonicum*] and that this lectin did not combine with other representative rhizobia. Sherwood et al. (1984) found that the location of the lectin-binding polysaccharides of *Rhizobium trifolii* on the cell surface varied with age of the culture. Random distribution of the binding agent as well as polar concentration of the substance was detected after different periods of cellular growth. The quantity and location of these receptors on the bacteria directly correlated with their attachment in short-term studies of clover root hair binding. For example, cells from 3- or 21-day-old cultures attached nearly exclusively in a polar fashion, whereas 5-day-old cells with the binding factors randomly distributed on the cell surface attached randomly to the root tissue.

Rhizobial exopolysaccharides appear to serve multiple purposes in nodulation. (See Reuber et al., 1991, for a more detailed discussion of the role of exopolysaccharides in nodule formation.) The varied roles include assistance in entangling the bacterial cell in the mucigel of the plant root, inhibition of host defenses through masking of the bacterial surface, and encapsulation of the bacterial cell for its protection in the infection thread. Dazzo et al. (1984) in their study of the phases of attachment of *R. trifolii* to clover root hairs found that, after attachment and orientation of the bacterial cell with the root surface, extracellular microfibrils associated with lateral and polar surfaces of the bacterial cell accumulated. The formation of these fibrils resulted in an attachment of the bacteria to the root hair sufficiently strong to withstand the hydrodynamic shear forces associated with high-speed vortexing of the mixture. (See Brewin, 1991, for further discussion of the role of these substances in nodule formation.)

Nodule Formation: Once the rhizobial cell has attached to the legume root, an infection thread spreading down the root hair is formed. Development of this structure is preceded by deformation of the root hair, which is induced by exopolysaccharides formed by the rhizobial cell (e.g., see Battisti et al., 1992; Ervin and Hubbell, 1985; Higashi and Abe, 1980; van Brussel et al., 1992). The invagination of root hair wall forming the cellulosic infection thread allows the bacterial cells to invade the central portions of what becomes the nodule. The invading rhizobial cells pass through the infection thread to the root cells adjacent to the root hair. These root cells subsequently become infected with the invading rhizobial cells. Only tetraploid cells are infected and become the primordia of the nodule. Note that only a small portion (usually less than 5 percent) of the invaded root hairs develop nodules. Division of the infected cells results in formation of the nodule. Bacteria multiply within the tetraploid cells forming swollen, variable-shaped and branched cells known as **bacteroids,** which are separated into packets within the nodule by a **peribacteroid membrane.** There are typically four to six bacteria per packet. Nitrogen fixation within the root nodules occurs only after formation of the bacteroids.

The nodulation cycle could be said to be completed by death and deterioration of the nodule and release of the bacteroids and bacteria contained therein into the soil.

The bacteroids apparently cannot divide when released into soil by death and decay of the nodule. Data collected by Paau et al. (1980) indicates that *Rhizobium meliloti* bacteroids in alfalfa nodules are degraded during nodule senescence. The nodule apparently always contains small numbers of dormant rod-shaped cells that can exist free in the soil environment.

12.1.4. Plant Controls of Nodulation

The host plant is not a passive participant in this process involving "rearrangement of its architecture" by the *Rhizobium* strain. Expression of several plant genes are necessary to establish the symbiotic interaction. In their survey of the literature, Caetano-Anollés and Gresshoff (1991a) provide a listing of symbiotic legume mutants for soybean (*Glycine max* (L.) Merr.), alfalfa (*Medicago sativa* L.), red clover (*Trifolium pratense* L.), crimson clover (*Trifolium incarnatum* L.), pea (*Pisum sativum* L.), chickpea (*Cicer arietinum* L.), and peanut (*Archis hypogaea* L.). For some of the legumes, only one or two genes impacting nodulation have thus far been elucidated, but a total of 22 genes interacting with the bacteria symbiont in nodule formation were listed for pea. These observations support the conclusion that the genetic-based contribution of the host plant to root nodule formation is as, or more, complex as that of the invading bacterium.

An interesting point of control by the host plant is its determination and optimization of the number of nodules formed. Most obvious among these properties is the impact of untimely fertilization of the crop on nodulation. Such inappropriate addition of fixed nitrogen to the crop suppresses nodule formation.

A less appreciated interaction is autoregulation. With autoregulation of nodulation, the plant controls the quantity and timing of nodule formation. After the first infection of roots, no further increase in number of root nodules occurs. The benefit to the host plant in preventing the formation of unnecessary nodules is obvious, because nodulation requires the commitment of large quantities of plant resources. This inhibition of nodulation is reversible in that removal of nodules can induce a new round of nodule formation. This process requires expression of specific, as yet undescribed, plant genes.

Furthermore, the rhizobia involved in the first round of infection must have the genetic capability to induce nodule formation and grow within the nodules. George and Robert (1991) assessed the capacity of a number of mutants of *Rhizobium leguminosarum* bv. phaseoli to induce an autoregulatory response of *Phaseolus vulgaris* L. They found that all mutants unable to form nodules, independent of the stage of genetic blockage of the nodule development process, failed to induce a suppressive response of the host. Their data indicated that the host response correlated with the ability of the bacteria to grow inside the nodules but not with the capacity to initiate nodule formation or to fix nitrogen. In other words, the host response required that the infecting bacterium be infective, but not effective.

The interaction of autoregulatory responses and bacterial population density to yield optimal nodulation was examined in soybean (*Glycine max* (L.) Merr. Cv. Bragg) following inoculation with *Bradyrhizobium japonicum* USDA 110 (Caetano-

Anollés and Gresshoff, 1991b). It was shown that the roots were initially susceptible to infection 3–4 days after germination. The number of nodules formed was proportional to the bacterial density up to an optimal concentration. This maximal population density resulting in an increase in the quantities of nodules formed was greater for a super nodulating soybean variant than was observed with the wild type. Thus, the plant genome can be manipulated to increase nodulation of this agronomically important crop.

12.2. MANIPULATION OF *RHIZOBIUM*–LEGUME SYMBIOSES FOR ECOSYSTEM MANAGEMENT

The foregoing evaluation of the *Rhizobium*–legume symbiosis suggests the existence of a relationship that is not only of interest for basic science studies but also a variety of practical implications are implied. The ubiquity of legumes as well as their significance as food crops suggests both economic and environmental benefits from management of this interaction. The prime questions in such considerations follow:

• Do *Rhizobium*–legume symbioses add significant quantities of fixed nitrogen to a soil ecosystem?
• Is there potential to improve the nitrogen inputs in a manner that also augments ecosystem productivity?
• Are these improvements feasible with currently available methods?

Encouragement of *Rhizobium*–legume symbioses development definitely increases the incorporation of biological fixed nitrogen into soil ecosystems. Analysis of the balance of nitrogen inputs and outputs to land systems and the sources of this fixed nitrogen developed by Burns and Hardy (1975) lends support to the conclusion that *Rhizobium*–legume symbioses are the primary source of fixed nitrogen in land-based systems. About 64 percent of the 139×10^9 kg N_2 biologically fixed in terrestrial systems is produced in agricultural systems. Because the bulk of biological nitrogen fixation in agricultural systems is derived from the cultivation of legumes, these symbioses can be concluded to provide well over half of the biological source of fixed nitrogen. Data collated by Alexander (1977) provide a prospective on fixed-nitrogen contributions of individual crops on a more localized bases. Inputs of fixed nitrogen for alfalfa, red clover, pea, soybean, cowpea, and vetch range from 65 to 335 kg N_2 fixed ha^{-1} year^{-1}.

Once it has been concluded that these symbiotic associations are major contributors of fixed nitrogen in soil systems, the point of concern must be redirected to consider the potential to increase fixed-nitrogen inputs by optimizing this process. That is, the possibility of improving plant biomass yields or plant community sustainability through the use of more efficient combinations of rhizobia and legumes must be examined. A diverse population of rhizobial strains already exists throughout the soil profile in agricultural and nonagricultural soils. Examples include a leguminous tree, black locust (*Robina pseudoacacia* L.) (Batzli et al., 1992); *Rhizobium leguminosarum* in agricultural soil (e.g., Bottomley and Dughri, 1989; Brockman and Bezdicek, 1989); soils under acacia trees (*Acacia albida*) (Dupuy and Dreyfus, 1992); soils from an alfalfa field (Jenkins et al., 1987); the woody legume, mesquite (*Prospopis glandulosa*) in desert soils (Jenkins et al., 1989; Waldon et al., 1989) and in tropical soils of Hawaii (Woomer et al., 1988). Furthermore, this indigenous rhizobial population can be quite diverse; that is, it is comprised of a number of different strains. For example, Noel and Brill (1980) used sodium dodecyl sulfate (SDS) gel electrophoresis to evaluate nodule occupants from soybeans collected from two Wisconsin [USA] locations. At one location, 19 gel types were distinguished and classified into three groups.

Nodule occupancy by these indigenous rhizobia is not the result of superior nitrogen-fixing capacity, but it is derived to a large degree from a capacity to survive in the soil in population densities and habitats that allow contact with susceptible roots. The proportion of the total soil rhizobial population comprised by any individual strain is controlled by soil properties as well as plant growth traits. For example, the predominance of any particular rhizobial strain in the soils of the Noel and Brill (1980) study depended upon the time and depth of nodule formation on the host plant. Similarly, Mahler and Bezdicek (1980) detected that serogroup variability of *Rhizobium leguminosarum* strains was affected by soil temperature, moisture, and position on the slope in the catena. Both microclimate and macroclimate variation altered strain distribution of *Rhizobium leguminosarum* strains in peas (*Pisum sativum* L.) in a Palouse (Washington, USA) site. Subsequent studies of 192 *R. leguminosarum* strains from this site resulted in demonstration of 3 serogroups and 18 plasmid profile groups among the isolates. Similar variety and distribution has been observed for *R. meliloti* (Jenkins and Bottomley, 1985). A variety of protein profiles for *Rhizobium* isolates from alfalfa (*Medicago sativa* L.) were distinguishable on SDS–gel electrophoresis.

Thus, it can be concluded that in any given soil system supporting growth of legumes a variety of rhizobial strains are available for nodulation of root tissue. The adequacy of this indigenous population to form optimal numbers of nodules on legume populations generally depends on plant density. In native ecosystems where limited legume productivity is anticipated, indigenous rhizobial population densities are generally sufficient to support adequate nodulation of leguminous species. This generalized statement regarding a relative description of plant population can be placed in perspective by comparing it to intensively cultivated legumes. With high-yield, cultivated agricultural systems or with monoculture systems (e.g., alfalfa fields), plant biomass productivity is augmented by inoculation with exogenously grown rhizobia. Therefore, indigenous soil population densities can be concluded to be inadequate to support optimal nodulation. Hence, adding laboratory-produced or axenically grown field *Rhizobium* sp. isolates to soils and legumes seeds to stimulate crop productivity has long been a common agricultural practice. The increased nodulation of the legume roots and subsequent augmentation in inputs of biologically produced fixed nitrogen result in part from the higher probability of encounter between the host root and infective bacteria as well as the use of rhizobial isolates selected for an enhanced nitrogen-fixing capability.

12.3. RHIZOBIAL INOCULATION PROCEDURES

The primary concerns when analyzing the benefits of manipulating soil rhizobial populations have been a) development of an effective delivery system, b) determination of the interactions with the soil environment and soil microbial population affecting survival of the inoculant, and, more recently, c) elucidation of properties of the inoculant strain affecting its competition with indigenous rhizobial variants for nodule occupancy.

12.3.1. Inocula Delivery Systems

Axenically grown rhizobial strains with the capacity to fix nitrogen in a highly efficient manner are of little value if infective propagules cannot be delivered to the soil site where their activity is necessary. The first step in this process is the physical conveyance of viable bacterial cells to the soil. In a scientific experiment, this step is reasonably easy since a microbiologist can quite easily prepare an active bacterial culture that can be added to crop roots or seeds immediately. Difficulties are encountered in transferring this "technology" into a real-world situation where the viable bacterial cells may be abused tremendously before they are introduced to their new environment. A commercially prepared inoculum must be capable of surviving for long periods of time under somewhat adverse conditions between growth and delivery to the agriculturist and of enduring potential maltreatment on site before the legume seeds are inoculated.

A variety of delivery systems for rhizobial inocula have been evaluated. Examples include, ground peat or an oil carrier (Kremer and Peterson, 1983) with or without fertilizer (Kremer et al., 1982), irrigation water (Ciafardini et al., 1992), water (Crist et al., 1984), coal-based carriers (Crawford and Berryhill, 1983), soils (Chao and Alexander, 1984), and fluid gels (Jawson et al., 1989). A variety of commercial inoculants are available (e.g., Hume and Blair, 1992). Examples of some of the considerations in preparing rhizobial inocula include pretreatment of the support media (e.g., sterilization of the peat, Strijdon and Van Rensburg, 1981) and methods to prepare the inoculum for storage (e.g., drying of inocula and humidity for storage, Mary et al., 1985).

12.3.2. Survival of Rhizobial Inocula

A related aspect to optimization of fixed-nitrogen inputs from *Rhizobium*–legume symbioses is maximization of inoculum survival once they are amended to the soil system. Although the bacterial strain or variant selected for legume inoculation may have been derived from the soil–plant environment, cultivation under laboratory conditions necessarily induces changes in the phenotypic and genotypic properties of the cell. These alterations may or may not affect the capacity of the organism to adapt to the soil environment and to compete with indigenous populations. Axenically grown laboratory rhizobial strains added to legume seeds or amended directly to soil are stressed by the physical and chemical properties of the harsh soil environment as well as by interactions with the general soil microbial community. The capacity of the organism to overcome these barriers to survival determines its probability of existing in the soil site for sufficient duration to encounter a susceptible legume root and initiate infection.

Rhizobial strains amended to soils are capable of surviving for several years. For example, Crozat et al. (1982) found that their strains of *Rhizobium* (*Bradyrhizobium*) *japonicum* introduced into a variety of French soils survived at high rates ($\geq 10^4$ bacteria g^{-1} dry soil after 5 years). Similarly, Brunel et al. (1988), found that *Bradyrhizobium japonicum* USDA strains 125-Sp, USDA 138 and USDA 138-Sm survived for 8–13 years after their release into soil. Alien rhizobial strains (i.e., axenically grown and introduced or reintroduced into the soil system) persisted through passage though legume nodules produced in the soil. Survival of a portion of the introduced population may result from a capacity to adapt to the prevailing soil conditions and compete with indigenous microbes. Viteri and Schmidt (1987), using fluorescent antibody techniques to study variation in population densities of *Bradyrhizobium japonicum* seroclusters 123 and serogroups 110 and 128, found that amending soil samples with metabolizable sugars resulted in not only an increase in the general populations but also growth in the various rhizobial strains. This observation supports the conclusion that rhizobia can compete with the general soil flora. That is, these bacteria do not require the exclusive protection of the plant nodule to maintain soil populations.

Primary soil physical and chemical properties that impact survival and nodule formation and that are of concern to environmental microbiologists include soil pH, metal interactions, salinity, and moisture. Although these environmental system traits affect nodulation of agronomically important crops and that is the venue wherein they have been most studied, these considerations are particularly important in developing procedures to encourage fixed-nitrogen inputs into reclaimed or managed ecosystems. It is generally desirable to minimize anthropogenic intervention in the developing soil sites, that is, to develop site-resident biological mechanisms to control fixed-nitrogen inputs rather than to rely upon long-term fertilization plans. Thus, selection and control of an indigenous biological population capable of fixing nitrogen is essential.

Rhizobia and Acid Soils: Soil acidity affects the legume plant, the infecting rhizobia, as well as the product of their interaction, the symbiotic association. Generally host plant growth is less sensitivity to extremes of acidity than is the *Rhizobium* species. Infection of host roots and survival of the bacterial symbiant are generally limited to pH greater than 5.0, whereas the host plants may grow at values as low as 4.0. This pH sensitivity is important to the agricultural community because of the need for sustained growth of symbiotic associations in acid soils and to the environmental science community from the viewpoint of determining the potential for use of these highly effective fixed-nitrogen sources in acid-impacted sites (e.g., soils exemplified by those that are naturally acidic or receive acid mine slags or acid mine drainage). Examples of agricultural concerns include the difficulties associated with maintenance of alfalfa (Barber, 1980; Lowendorf and Alexander, 1983a), lucern (Munns and Fox, 1977; Pijnenborg et al., 1991); and clover (Richardson et al., 1988a) stands in acidic soils. Similar sensitivity to acidic soils has been observed for tropical legumes (18 legume species) as occurs with temperate legumes.

Both a direct and an indirect effect of acidic conditions on rhizobial activity are observed. Bacterial cells react directly to the elevated hydrogen ion concentration of acidic soils as well as to the alteration of metal solubility resulting from the reduction in the soil pH (indirect effect) (e.g., Franco and Munns, 1982; Keyser and Munns, 1979a). The latter interaction can be positive (e.g., inhibition of growth by augmented aluminum concentrations under acidic conditions) or negative (e.g., reduction of the essential mineral molybdenum by increased acidity).

A differential sensitivity of the rhizobial strains to the various changes in soil cation concentration resulting from increased soil acidity is observed. Keyser and Munns (1979b) in a study of cation effect on growth of *Rhizobium* (*Bradyrhizobium*) *japonicum* found that aluminum toxicity and acidity itself had a greater impact on rhizobial growth than did manganese toxicity and calcium deficiency. Note that aluminum toxicity and acid sensitivity independently affect rhizobial survival in soil (e.g., Hartel and Alexander, 1983; Hartel et al., 1983). It appears that survival of the cowpea, *Vigna unguiculata* (L.) Walp., rhizobia is pH but not aluminum sensitive. Hartel et al. (1983) concluded that aluminum content of soil was not a major factor controlling cowpea rhizobial survival but that it did significantly alter the nodulation process.

Although it has been proposed that survival of indigenous populations of *Rhizobium trifolii* in extremely acidic soils can occur through refuge in soil microsites of less extreme acidity (Richardson and Simpson, 1988), rhizobial strains with increased tolerance of acidic conditions can be isolated (e.g., Lowendorf and Alexander, 1983b; Keyser et al., 1979; Thornton and Davey, 1983; Tiwari et al., 1992) and constructed genetically (Chen et al., 1991). Increased tolerance of rhizobial strains growing under axenic condition does result in increased survival in sterile acid soils (Lowendorf et al., 1981). Thus, laboratory generation of acid-tolerant strains of rhizobia, which are also efficient nitrogen fixers, can increase fixed-nitrogen inputs into acid-impacted soil sites. It is reasonable to consider that the ability of the infecting bacterial strains to adjust or survive in the acid soils would also affect their probability of nodule occupancy (Dughri and Bottomley, 1983). Furthermore, interactions between the acidic conditions and cation concentrations can also affect nodule formation by rhizobia. For example, acidic conditions and levels of calcium and aluminum ions reduces induction of the nodulation genes in *Rhizobium leguminosarum* biovar *trifolii* (Richardson et al., 1988b). Thus, we can conclude that the concentrations of fixed nitrogen provided by the *Rhizobium*–legume symbioses in acidic and/or cation-impacted soils depend upon the capacity to maintain adequate population densities of infective bacteria in the soil to allow for frequent encounters between the bacterium and susceptible plant roots, the maintenance of a functional infection process in the acidic soils, and the production of effective nodules.

To develop rhizobial variants with an enhanced capacity to survive in acidic soils, we must understand the properties of the cell that can be varied to decrease sensitivity to soil acidity or aluminum toxicity. Analysis of the literature on this topic suggests that no single cellular property confers resistance or sensitivity to these adverse soil conditions. For example, DNA repair mechanisms (Johnson and Wood, 1990), extracellular polysaccharide production (Cunningham and Munns, 1984a), and ability to maintain intracellular pH (O'Hara et al., 1989) all appear to affect aluminum and/or acid tolerance of rhizobial strains. Variation among rhizobial strains is exemplified by studies of the role of extracellular polysaccharide production in protecting the bacterial cell. Cunningham and Munns (1984a) found that acid tolerance of rhizobial strains for six legume genera (*Cicer, Phaseolus, Leucaena, Lens, Melilotus,* and *Trifolum*) all correlated with extracellular production. The quantities of extracellular polysaccharide produced by 20 strains of *R. phaseoli* varied from 18 to 75 mg/100 ml culture and positively correlated with acid tolerance. Data from further studies (Cunningham and Munns, 1984b) indicated that the apparent protection did not relate to buffering capacity of the extracellular polysaccharide or its ability to chelate aluminum ions. In contrast, Kingsley and Bohlool (1992) evaluated the role of extracellular polysaccharide produced by *R. leguminosarum* bv. *phaseoli* (CIAT899) with mutant strains incapable of producing the polysaccharide. No difference in aluminum tolerance of the wild-type strain from the mutant were detected. Results from this study indicate that neither intracellular or extracellular products of the *R. leguminosarum* strain alleviated the aluminum toxicity of sensitive rhizobial strains. These studies demonstrate the potential to increase the use of the *Rhizobium*–legume symbioses for augmenting fixed-nitrogen inputs into acidic soil ecosystems as well as the complexity of the genetic control of this acid and aluminum sensitivity of the infective bacteria.

Although most interest in the effect of toxic metals on nitrogen fixation by *Rhizobium*–legume associations has involved growth of agriculturally important legumes in acidic soils, we must consider the impact of metals when evaluating the use of these associations in metal-impacted sites or in soils amended with materials that may contain environmentally significant levels of metals. For example, McGrath et al. (1988) determined the potential for reducing fixed-nitrogen yields from nodulated *Trifolium repens* in soils receiving metal-contaminated sewage sludge. Plants grown on the contaminated sludge-amended soil had reduced nitrogen concentrations and reduced yields compared to those produced on uncontaminated control soils. The root nodules on the impacted plants were numerous but ineffective in nitrogen fixation.

Soil Moisture Impact on Nodulation: The occurrence of rhizobial populations in desert soils and the effective nodulation of legumes growing therein (e.g., Jenkins et al., 1987, 1989; Waldon et al., 1989) attests to the fact that rhizobia can exist in soils with limiting moisture levels, but population densities tend to be lowest under the most desiccated situation with increasing densities as the moisture stress is relieved. The wide range of moisture levels characteristic of ecosystems where legumes have been shown to fix nitrogen suggests that rhizobial strains with varying sensitivity to soil moisture can be selected. Laboratory studies have shown that sensitivity to moisture stress varies with bacterial strain for a variety of rhizobial strains and species. Examples include with *Rhizobium leguminosarum* bv. *trifolii* (Boonkerd and Weaver, 1982; Fuhrmann et al., 1986), with *Rhizobium meliloti* (Busse and Bottomley, 1989), with cowpea rhizobia (Osa-Afiana and Alexander, 1982a), and with *Bradyrhizobium japonicum* (Mahler and Wollum, 1980). Thus, we can reasonably assume that rhizobial strains can be selected with moisture stress tolerance within the range of its legume host.

Because we can select strains with appropriate moisture tolerance for long-term survival of soil inocula, concerns regarding management of laboratory-developed and indigenous bacterial symbionts *in situ* in a native soil site relate to the interactive soil properties that modify or extend the microbe's moisture survival capacities. For example, survival of rhizobia in drying soils can be extended by increasing contents of soil organic matter and decreased by elevated soluble aluminum (Chao and Alexander, 1982) and proportion of clay in the soil (Osa-Afiana and Alexander, 1982b). Furthermore, organisms produced *in situ* tend to be better adapted to moisture stress than axenically grown organisms (Pena-Cabriales and Alexander, 1979). Optimization of soil moisture for growth of the host plant, which generally is more sensitive to moisture stress than is the bacteria, results in maximal development of *Rhizobium*–legume fixed-nitrogen inputs into the soil system.

Soil Salinity: An allied concern in regards to soil moisture in reclamation of environmentally impacted sites can be the effect of salinity on the survival of rhizobia in soil systems. For example, amendment of soils with products of industrial society can raise the soil salinity (e.g., sludge-borne salts, Madariaga and Angle, 1992). As was indicated in chapter 4, increasing salts may have a detrimental effect on soil microbial populations as a result of direct toxicity as well as through osmotic stresses.

Rhizobia vary in osmoregulation efficacy (e.g., Busse and Bottomley, 1989). This capacity of the microbial population to adapt was exhibited in a study of the survival of *Bradyrhizobium japonicum* in sludge-amended soils. The soluble salts of the sludge (not the heavy metals) were shown to be primarily responsible for a short-term reduction in bradyrhizobial populations following sludge application to soil. Singleton et al. (1982) found many *Rhizobium* strains with enhanced salinity tolerance compared to most agriculturally important varieties. In a study of the distribution of fast-growing and slow-growing mesquite-nodulating rhizobia in desert ecosystems, Jenkins et al. (1989) found that the distribution of the rhizobial strains within the rooting system was related to total soil salts.

This analysis of some soil properties affecting survival, growth, and nodulation capacities of soil rhizobial populations demonstrates the versatility and adaptability of these essential soil populations. The genotypic and phenotypic flexibility exists within these organisms to adapt to many of the stressed soil conditions normally encountered in our soil system. This observation does not imply efficient nitrogen-fixing capabilities under these conditions. The legume may be nodulated, but fixed-nitrogen production may be below the level required for optimal growth of the host plant. In these situations, symbiotic associations, such as the actinorhizal interaction, may be better adapted or more adaptable for use in site management or reclamation.

12.3.3. Biological Interactions in Legume Nodulation

Rhizobia can grow in native soil samples when the soils have been amended with fixed-carbon sources that support their metabolism (e.g., Pena-Cabriales and Alexander, 1983a; Viteri and Schmidt, 1987) or when they receive nutrients from growing roots (e.g., Pena-Cabriales and Alexander, 1983b). Success of this nutrient-based competition with indigenous populations depends on the capacity of the alien microbe to become established in an unoccupied niche. Rhizobia are best suited for competition for the newly created niche of the nitrogen-fixing nodule. Life as a member of the "free-living soil community" is more difficult. Rhizobia not only must compete for limited nutrients, but interactions with indigenous heterotrophic microbes and predators also reduce their capacity to maintain population densities at sufficient levels to ensure contact with susceptible legume roots (e.g., Li and Alexander, 1986; Pugashetti et al., 1982; Ramirez and Alexander, 1980). (See chapter 6 for further discussions regarding microbial competition.)

The outcome of competition between indigenous microbes can be manipulated. Aside from the brute-force method of overwhelming the system with massive population densities of the laboratory-grown rhizobial strain, use of fungicides and antibiotics can enhance the probability of rhizobial colonization (Hossain and Alexander, 1984; Li and Alexander, 1986). Hossain and Alexander (1984) augmented soybean rhizosphere colonization with *Bradyrhizobium japonicum* by using benomyl (a fungicide) and the antibiotics streptomycin and erythromycin. *Rhizobium* strains used with these procedures must be resistant to the inhibitors.

12.4. NODULE OCCUPANTS: INDIGENOUS VS. FOREIGN

Rhizobial inoculants are not only competitors with indigenous general microbial populations for nutrient resources but also participants in an intense race with indigenous rhizobial strains for occupancy of nodules. A study of the history of use of *Rhizobium* inoculants for cropping of legumes leaves the impression that it has been assumed that the soil-amended strain necessarily was the predominant nodule occupant. Perhaps a more pragmatic assessment of the situation would include the idea that the investigator and growers have not been as concerned with the identity of the nodule occupant as they were with augmentation of fixed-nitrogen resources and the concurrent improvement in crop yields. If crop yields increased, a benefit from inoculation with highly efficient rhizobial strains was assumed to have occurred. Unfortunately, detailed examination of legume nodules reveals that it is not unusual for the laboratory-cultivated rhizobial strains to loose the competition for a place in the nodule to the less efficient but better adapted indigenous soil populations.

Nodule occupancy frequencies for rhizobial strains inoculated into field situations ranging from nearly zero to essentially 100 percent have been reported. For example, van Rensburg and Strijdom (1985) compared *Rhizobium* nodule occupancy frequency of indigenous and axenically grown strains in *Trifolium* spp., *Medicago* spp., *Glycine max* and *Lotus pedunculatus*. They collected 420 nodules from crops growing in fields that had been inoculated 4–8 years previously with highly efficient rhizobial strains. The rate of nodule occupancy for the inoculated strain ranged from 17.7 percent in the *Glycine max* (soybean) to 100 percent with the *Lotus pedunculatus*. Klubek et al. (1988) in a study of *Bradyrhizobium japonicum* nodule occupancy in the Midwest (USA) detected frequency of occurrence of inoculated strains as low as 0.3 percent. Detection of highly variable capacity of laboratory-grown strains to compete with indigenous populations is commonly reported (e.g., Materon and Hagedorn, 1982; McDermot and Graham, 1989; Moawad and Bohlool, 1984).

The most obvious explanation for the variation in the frequency of nodule occupancy of soil-amended rhizobial strains involves the probability of encountering the added organism with susceptible portions of the legume root. In a study of seven legume species in Hawaii (Thies et al., 1991a), it was shown that the proportion of the nodules occupied by inoculated rhizobial strains is dependent upon the population density of the indigenous population. We can expect indigenous organisms to be better adapted to the soil system and perhaps even more appropriately distributed within the soil profile than alien organisms. The latter conclusion is based upon the observation that the physical location of indigenous organisms depends in part on the distribution of the root system in which they grew and the homogenization of the soil associated with intensive cultivation systems. Thies et al. (1991b) found that 59 percent of the observed variation in inoculation response could be attributed to the population density of the indigenous population.

Properties of rhizobia that enhance nodule occupancy rates are diverse and to a major extent poorly defined. Many of these properties are associated with the ability to maintain sufficient population densities in the soil to allow a reasonable probability of encounter with host roots. Therefore, when selecting inoculant strains, we must consider the properties of the soil in which they are to be used (e.g., moisture/salinity stresses, pH) as well as the capability to compete with indigenous microbes for available nutrients and to withstand incursions of predators and parasites. Further considerations relate the efficiency of the nodulation process. Initiation and development of effective nodules is a complex process. Many of the rhizobial cells attracted to the appropriate portion of the legume root fail to complete the infection process. To increase the probability of the inoculant with improved nitrogen-fixing efficiency contributing to the augmentation of ecosystem fixed-nitrogen resources, variants with increased capability of completing the nodule formation process are desirable.

12.5. ACTINORHIZAL ASSOCIATIONS

When we consider the potential to form nitrogen-fixing nodules on plant roots, the *Rhizobium*–legume association is frequently discussed nearly at the exclusion of

TABLE 12.1. Examples of Angiosperm Families Involved in Nitrogen-Fixing Actinorhizal Associations.

Family	Genera
Betulaceae	*Alnus*
Casiaromaceae	*Casuarina*
Coriariaceae	*Coriaria*
Elaeagnaceae	*Elaeagnus*
	Hypophae
	Shepherdia
Myricaceae	*Myrica*
	Comptonia
Rhamnaceae	*Ceanothus*
	Discaria
Rosaceae	*Cercocarpus*
	Dryas
	Pushia

any other interactions; yet, actinorrhizal interactions (nonleguminous nitrogen-fixing symbiotic associations between actinomycetes of the genus *Frankia* and a variety of woody angiosperms) are major contributors to nitrogen inputs in forests, wetlands, fields, and disturbed sites of temperate and tropical regions. Actinorhizal associations involve more than 160 species of angiosperms classified among six or seven orders. As shown in Table 12.1, a wide variety of genera of angiosperms can form actinorhizal associations. Not all genera in the various families cited are active in nitrogen fixation, and not all species of the cited genera form nitrogen-fixing associations.

Contributions of fixed nitrogen to native as well as managed ecosystems by these symbioses are comparable to those of the more studied *Rhizobium*–legume interactions. The nitrogen gains due to these interactions vary with soil, climate, and plant age. Typical contributions represented by *Alnus* associations are 12–2000 kg N ha^{-1} and by *Hippophae* associations are 27–179 kg N ha^{-1}. See Baker and Mullen (1992) for a review of the contributions of this association to global nitrogen cycling.

Furthermore, the energy requirements (i.e., photosynthetate contributions from host plants) and nitrogenase activity of actinorhizal symbioses are comparable to those of the *Rhizobium*–legume associations (Gauthier et al., 1981; Tjepkema and Winship, 1980). Note that, as with rhizobia, *Frankia* strains can fix nitrogen growing in axenic culture on defined media (Gauthier et al., 1981).

It has been recognized that the microbial symbiont is an actinomycete for more than 90 years. Early identification was based on microscopic detection of actinomycete-like mycelial structures in nodule tissue. (See Quispel, 1990, for a discussion of this topic.) The greatest limitation to the study of actinorhizal associations has been, until recently, the inability to culture the microbial partner exclusive of the plant nodule. Until the early 1980s, the microbial partner was generally regarded to be an unculturable actinomycete. Most hypotheses regarding participation of the microbial symbiont were developed therefore by observing the cells in the nodule or from biochemical studies of partially purified microbial biomass extracted from crushed nodules.

Currently, the *Frankia* sp. can be isolated and cultured (e.g., Baker et al., 1979; Benson, 1982). Procedures generally involve surface sterilization of the nodules with such substances as chlorox, hydrogen peroxide, or dilute silver nitrate. The sterilant is washed from the nodules and the nodules are crushed. The extraneous debris may be removed (Baker et al., 1979, used sucrose gradients) or laboratory culture media may be inoculated directly with the nodule extract. A variety of nitrogen-free media containing antibiotics have been used to culture the actinomycete. A key in culture of these organisms is the necessity of inhibiting the growth of fast-growing contaminants. Colonies of the slow-growing *Frankia* spp. are easily overgrown. Thus, antibiotics such as cycloheximide and nystatin are amended to the growth media (e.g., Benson, 1982). Once isolated, the proof that the axenically grown actinomycete participates in symbiotically produced fixed-nitrogen results from the successful nodulation of uninfected host plants with the cultured microbial strain (e.g., Diem et al., 1982; Mansour et al., 1990; Rogers and Wollum, 1974; Van Dijk and Sliumer-Stolk, 1990).

Polymerase chain reaction (PCR) amplification and related DNA analytical techniques portend to be valuable tools for elucidation of the diversity of *Frankia* strains involved in actinorhizal symbioses (e.g., Beayzova and Lechevalier, 1992; Sellstedt et al., 1992; Simonet et al., 1990, 1991) as well as the variability of these nitrogen-fixing actinomycetes in soil (e.g., Hahn et al., 1990a; Hilger and Myrold, 1991; Myrold et al., 1990) and in host nodules (Bloom et al., 1989; Hahn et al., 1990b).

Although at this time it is not recommended that the genus *Frankia* be divided into individual species (Lechevalier and Lechevalier, 1989), axenically grown *Frankia* strains are extremely diverse. For example, DNA hybridization studies (Bloom et al., 1989) revealed highly divergent populations of *Frankia* could be isolated from bayberry (*Myrica pensylvanica*) nodules. In fact, *Frankia* strains isolated from the same nodule were shown to be not only phenotypically different but also genotypically diverse. Similarly, restriction fragment length polymorphism analysis (RFLP) has been demonstrated to divide *Frankia* strains into groups that correlated with host specificity groups separated by cross-inoculation studies (Nittayarjarn et al., 1990) (As with rhizobia, *Frankia* strains can be separated into groups based on the specificity toward their host plant.)

12.6. CONCLUSIONS

Root-nodule-forming symbiotic associations are the predominant contributors of fixed nitrogen to terrestrial ecosystems. Both rhizobial and *Frankia* sp. strains form specific associations with legume and angiosperm hosts, respectively. Although adequate populations of infective diazotrophs commonly occur among soil microbes, the potential exists for manipulating these bacterial populations to increase their contribution of fixed nitrogen to the ecosystem. For rhizobial populations, it has been shown that sufficient genotypic variability for resistance to soil pH, moisture, salinity extremes, and nitrogen-fixing efficiency allows for the development of rhizobial strains that support increased fixed-nitrogen inputs compared to those provided by the activity of indigenous populations. Selection of such strains and their use in the field also require consideration of those phenotypically expressed traits that improve the microbes capacity to compete with soil heterotrophic populations as well as indigenous rhizobia. As our understanding of the ecological associations of *Frankia* spp. develops, similar manipulations to increase fixed-nitrogen production by actinorhizal symbioses in managed and wild-land ecosystems will be possible.

REFERENCES

Adebayo, A., I. Watanabe, and J. K. Ladha. 1989. Epiphytic occurrence of *Azorhizobium caulinodans* and other rhizobia on host and nonhost legumes. Appl. Environ. Microbiol. 55:2407–2409.

Aguilar, O. M., and G. Favelukes. 1982. Requirement for carbon dioxide for nonsymbiotic expression of *Rhizobium japonicum* nitrogenase activity. J. Bacteriol. 152:510–513.

Alexander, M. 1977. Introduction to Soil Microbiology. John Wiley & Sons. New York. 467 pp.

Ames, P., S. A. Schluederberg, and K. Bergman. 1980. Behavioral mutants of *Rhizobium meliloti*. J. Bacteriol. 141:722–727.

Baker, D. D., and B. C. Mullin. 1992. Actinorhizal symbiosis. *In* G. Stacey, R. H. Burris, and H. J. Evans (éds.), Biological Nitrogen Fixation, pp. 259–292. Chapman and Hall. New York.

Baker, D., J. G. Torrey, and G. H. Kidd. 1979. Isolation by sucrose-density fractionation and cultivation in vitro of actinomycetes from nitrogen-fixing root nodules. Nature (London) 281:76–78.

Barber, L. E. 1980. Enumeration, effectiveness, and pH resistance of *Rhizobium meliloti* populations in Oregon Soils. Soil Sci. Soc. Am. J. 44:537–539.

Barbour, W. M., D. R. Hattermann, and G. Stacey. 1991. Chemotaxis of *Bradyrhizobium japonicum* to soybean exudates. Appl. Environ. Microbiol. 57:2635–2639.

Battisti, L., J. C. Lara, and J. A. Leigh. 1992. Specific oligosaccharide form of the *Rhizobium meliloti* exopolysaccharide promotes nodule invasion in alfalfa. Proc. Natl. Acad. Sci. USA 89:5625–5629.

Batzli, J. M., W. R. Graves, and P. van Berkum. 1992. Diversity among rhizobia effective with *Robinia pseudoacacia* L. Appl. Environ. Microbiol. 58:2137–2143.

Beayzova, M., and M. P. Lechevalier. 1992. Low-frequency restriction fragment analysis of *Frankia* strains (*Actinomycetales*). Int. J. Syst. Bacteriol. 42:422–433.

Benson, D. R. 1982. Isolation of *Frankia* strains from alder actinorhizal root nodules. Appl. Environ. Microbiol. 44:461–465.

Bergman, K., M. Gulash-Hoffee, R. E. Hovestadt, R. C. Larosiliere, P. G. Ronco II, and L. Su. 1988. Physiology of behavioral mutants of *Rhizobium meliloti:* Evidence for a dual chemotaxis pathway. J. Bacteriol. 170:3249–3254.

Bhagwat, A. A., and J. Thomas. 1982. Legume-*Rhizobium* interactions: Cowpea root exudate elicits faster nodulation response by *Rhizobium* species. Appl. Environ. Microbiol. 43:800–805.

Bloom, R. A., B. C. Mullin, and R. L. Tate III. 1989. DNA restriction patterns and DNA–DNA solution hybridization studies of *Frankia* isolates from *Myrica pensylvanica* (Bayberry). Appl. Environ. Microbiol. 55:2155–2160.

Bohlool, B. B., and E. L. Schmidt. 1974. Lectins: A possible basis for specificity in the *Rhizobium*–legume root nodule symbiosis. Science 185:269–271.

Boonkerd, N., and R. W. Weaver. 1982. Survival of cowpea rhizobia in soil as affected by soil temperature and moisture. Appl. Environ. Microbiol. 43:585–589.

Bottomley, P. J., and M. H. Dughri. 1989. Population size and distribution of *Rhizobium leguminosarum* bv. *Trifolii* in relation to total soil bacteria and soil depth. Appl. Environ. Microbiol. 55:959–964.

Brewin, N. J. 1991. Development of the legume root nodule. Annu. Rev. Cell Biol. 7:191–226.

Brockman, F. J., and D. F. Bezdicek. 1989. Diversity within serogroups of *Rhizobium leguminosarum* biovar *Viceae* in the Palouse region of Eastern Washington as indicated by plasmid profiles, intrinsic antibiotic resistance, and topography. Appl. Environ. Microbiol. 55:109–115.

Brunel, B., J.-C. Cleyet-Marel, P. Normand, and R. Bardin. 1988. Stability of *Bradyrhizobium japonicum* inoculants after introduction into soil. Appl. Environ. Microbiol. 54:2636–2642.

Burns, R. C., and R. W. F. Hardy. 1975. Nitrogen Fixation in Bacteria and Higher Plants. Springer-Verlag. New York. 189 pp.

Busse, M. D., and P. J. Bottomley. 1989. Growth and nodulation responses of *Rhizobium meliloti* to water stress induced by permeating and nonpermeating solutes. Appl. Environ. Microbiol. 55:2431–2436.

Caetano-Anollés, G., D. K. Crist-Estes, and W. D. Bauer. 1988. Chemotaxis of *Rhizobium meliloti* to the plant flavone luteolin requires functional nodulation genes. J. Bacteriol. 170:3164–3169.

Caetano-Anollés, G., and P. M. Gresshoff. 1991a. Plant genetic control of nodulation. Annu. Rev. Microbiol. 45:345–382.

Caetano-Anollés, G., and P. M. Gresshoff. 1991b. Efficiency of nodule initiation and autoregulatory responses in a supernodulating soybean mutant. Appl. Environ. Microbiol. 57:2205–2210.

Chao, W.-L., and M. Alexander. 1982. Influence of soil characteristics on the survival of *Rhizobium* in soils undergoing drying. Soil Sci. Soc. Am. J. 46:949–952.

Chao, W.-L., and M. Alexander. 1984. Mineral soils as carriers for *Rhizobium* inoculants. Appl. Environ. Microbiol. 47:94–97.

Chen, H., A. E. Richardson, E. Gartner, M. A. Djordjevic, R. J. Roughley, and B. G. Rolfe. 1991. Construction of an acid-tolerant *Rhizobium leguminosarum* biovar *trifolii* strain with enhanced capacity for nitrogen fixation. Appl. Environ. Microbiol. 57:2005–2011.

Ciafardini, G., G. Marinelli, and R. Missich. 1992. Soil biomass of *Bradyrhizobium japonicum* inoculated via irrigation water. Can. J. Microbiol. 38:584–587.

Crawford, S. L., and D. L. Berryhill. 1983. Survival of *Rhizobium phaseoli* in coal-based legume inoculants applied to seeds. Appl. Environ. Microbiol. 45:703–705.

Crist, D. K., R. E. Wyza, K. K. Mills, W. D. Bauer, and W. R. Evans. 1984. Preservation of *Rhizobium* viability and symbiotic infectivity by suspensions in water. Appl. Environ. Microbiol. 47:895–900.

Crozat, Y., J. C. Cleyet-Marel, J. J. Giraud, and M. Obaton. 1982. Survival rates of *Rhizobium japonicum* populations introduced into different soils. Soil Biol. Biochem. 14:401–405.

Cunningham, S. D., and D. N. Munns. 1984a. The correlation between extracellular polysaccharide production and acid tolerance in *Rhizobium*. Soil Sci. Soc. Am. J. 48:1273–1276.

Cunningham, S. D., and D. N. Munns, 1984b. Effects of rhizobial extracellular polysaccharide on pH and aluminum activity. Soil Sci. Soc. Am. J. 48:1276–1279.

Currier, W. W., and G. A. Strobel. 1977. Chemotaxis of *Rhizobium* spp. to a glycoprotein produced by birdsfoot trefoil roots. Science 196:434–435.

Dazzo, F. B., G. L. Truchet, J. E. Sherwood, E. M. Hrabak, M. Abe, and S. H. Pankratz. 1984. Specific phases of root hair attachment in the *Rhizobium trifolii*–clover symbiosis. Appl. Environ. Microbiol. 48:1140–1150.

Dharmatilake, A. J., and W. D. Bauer. 1992. Chemotaxis of *Rhizobium meliloti* towards nodulation gene inducing compounds from alfalfa roots. Appl. Environ. Microbiol. 58:1153–1158.

Diem, H. G., D. Gauthier, and Y. Dommergues. 1982. Isolation and cultivation in vitro of an infective and effective strain of *Frankia* isolated from nodules of *Casuarina* sp. C. R. Seances Acad. Sci. Ser. III. Sci. Vie 295:759–763.

Dughri, M. H., and P. J. Bottomley. 1983. Effect of acidity on the composition of an indigenous soil population of *Rhizobium trifolii* found in nodules of *Trifolium subterraneum* L. Appl. Environ. Microbiol. 46:1207–1213.

Dupuy, N. C., and B. L. Dreyfus. 1992. *Bradyrhizobium* populations occur in deep soil under the leguminous tree *Acacia albida*. Appl. Environ. Microbiol. 58:2415–2419.

Elkan, G. H., and C. R. Bunn. 1992. The rhizobia. *In* A. Balows, H. G. Trüper, M. Dwarkin, W. Harder, and K.-H. Schleifer (eds.), The Prokaryotes, 2nd edition, pp. 2197–2213. Springer Verlag. New York.

Ervin, S. E., and D. H. Hubbell. 1985. Root hair deformations associated with fractionated extracts from *Rhizobium trifolii*. Appl. Environ. Microbiol. 49:61–68.

Franco, A. A., and D. N. Munns. 1982. Acidity and aluminum restraints on nodulation, nitrogen fixation, and growth of *Phaseolus vulgaris* in solution culture. Soil Sci. Soc. Am. J. 46:296–301.

Fuhrmann, J., C. B. Davey, and A. G. Wollum II. 1986. Desiccation tolerance of clover rhizobia in sterile soils. Soil Sci. Soc. Am. J. 50:639–644.

Gauthier, D., H. G. Diem, and Y. Dommergues. 1981. In vitro nitrogen fixation by two actinomycete strains isolated from *Causuarina* nodules. Appl. Environ. Microbiol. 41:306–308.

George, M. L. C., and F. M. Robert. 1991. Autoregulatory response of *Phaseolus vulgaris* L. to symbiotic mutants of *Rhizobium leguminosarum* bv. phaseoli. Appl. Environ. Microbiol. 57:2687–2692.

Gulash, M., P. Ames, R. C. Larosiliere, and K. Bergman. 1984. Rhizobia are attracted to localized sites on legume roots. Appl. Environ. Microbiol. 48:149–152.

Hahn, D., R. Kester, M. J. C. Starrenburg, and A. D. L. Akkermans. 1990a. Extraction of ribosomal RNA from soil for detection of *Frankia* with oligonucleotide probes. Arch. Microbiol. 154:329–335.

Hahn, D., M. J. C. Starrenburg, and A. D. L. Akkermans. 1990b. Oligonucleotide probes that hybridize with rRNA as a tool to study *Frankia* strains in root nodules. Appl. Environ. Microbiol. 56:1342–1346.

Hartel, P. G., and M. Alexander. 1983. Growth and survival of cowpea rhizobia in acid aluminum-rich soils. Soil Sci. Soc. Am. J. 47:502–506.

Hartel, P. G., A. M. Whelan, and M. Alexander. 1983. Nodulation of cowpeas and survival of cowpea rhizobia in acid, aluminum-rich soils. Soil Sci. Soc. Am. J. 47:514–517.

Hartmann, A., and N. Amarger. 1991. Genetic diversity of an indigenous *Rhizobium meliloti* field population assessed by plasmid profiles, DNA fingerprinting, and insertion sequence typing. Can. J. Microbiol. 37:600–608.

Higashi, S., and M. Abe. 1980. Promotion of infection thread formation by substances from *Rhizobium*. Appl. Environ. Microbiol. 39:297–301.

Hilger, A. B., and D. D. Myrold. 1991. Method for extraction of *Frankia* DNA from soil. Agric. Ecosys. Environ. 34:107–113.

Hossain, A. K. M., and M. Alexander. 1984. Enhancing soybean rhizosphere colonization by *Rhizobium japonicum*. Appl. Environ. Microbiol. 48:468–472.

Hume, D. J., and D. H. Blair. 1992. Effects of numbers of *Bradyrhizobium japonicum* applied in commercial inoculants on soybean seed yield in Ontario. Can. J. Microbiol. 38:588–593.

Hunter, W. J., and C. J. Fahring. 1980. Movement by *Rhizobium* and nodulation of legumes. Soil Biol. Biochem. 12:537–542.

Issa, S., L. P. Simmonds, and M. Wood. 1993a. Passive movement of chickpea and bean rhizobia through soils. Soil Biol. Biochem. 25:959–965.

Issa, S., M. Wood, and L. P. Simmonds. 1993b. Active movement of chickpea and bean rhizobia in dry soil. Soil Biol. Biochem. 25:951–958.

Jawson, M. D., A. J. Franzluebbers, and R. K. Berg. 1989. *Bradyrhizobium japonicum* survival in and soybean inoculation with fluid gels. Appl. Environ. Microbiol. 55:617–622.

Jenkins, M. B., and P. J. Bottomley. 1985. Evidence for a strain of *Rhizobium meliloti* dominating the nodules of alfalfa. Soil Sci. Soc. Am. J. 49:326–328.

Jenkins, M. B., R. A. Virginia, and W. M. Jarrell. 1987. Rhizobial ecology of the woody legume mesquite (*Prospis glandulosa*) in the Sonoran desert. Appl. Environ. Microbiol. 33:36–40.

Jenkins, M. B., R. A. Virginia, and W. M. Jarrell. 1989. Ecology of fast-growing and slow-growing mesquite-nodulating rhizobia in Chihuahuan and Sonoran desert ecosystems. Soil Sci. Soc. Am. J. 53:543–549.

Johnson, A. C., and M. Wood. 1990. DNA, a possible site for action of aluminum in *Rhizobium* spp. Appl. Environ. Microbiol. 56:3629–3633.

Kape, R., M. Parniske, and D. Werner. 1991. Chemotaxis and *nod* gene activity of *Bradyrhizobium japonicum* in response to hydroxycinnamic acids and isoflavonoids. Appl. Environ. Microbiol. 57:316–319.

Keyser, H. H., and D. N. Munns. 1979a. Tolerance of rhizobia to acidity, aluminum, and phosphate. Soil Sci. Soc. Am. J. 43:519–523.

Keyser, H. H., and D. N. Munns. 1979b. Effects of calcium, manganese, and aluminum on growth of rhizobia in acid media. Soil Sci. Soc. Am. J. 43:500–503.

Keyser, H. H., D. N. Munns, and J. S. Hohenberg. 1979. Acid tolerance of rhizobia in culture and in symbiosis with cowpea. Soil Sci. Soc. Am. J. 43:719–722.

Kiester, D. L., and W. R. Evans. 1976. Oxygen requirement for acetylene reduction by pure cultures of rhizobia. J. Bacteriol. 127:149–153.

Kingsley, M. T., and B. B. Bohlool. 1992. Extracellular polysaccharide is not responsible for aluminum tolerance of *Rhizobium leguminosarum* bv. Phaseoli (CIAT899). Appl. Environ. Microbiol. 58:1095–1101.

Klubek, B. P., L. L. Hendrickson, R. M. Zablotowicz, J. E. Skwara, E. C. Varsa, S. Smith, T. G. Islieb, J. Maya, M. Valdes, F. B. Dazzo, R. L. Todd, and D. D. Walgenback. 1988. Competitiveness of selected *Bradyrhizobium japonicum* strains in Midwestern USA soils. Soil Sci. Soc. Am. J. 52:662–666.

Kremer, R. J., and H. L. Peterson. 1983. Effects of carrier and temperature on survival of *Rhizobium* spp. in legume inocula: Development of an improved type of inoculant. Appl. Environ. Microbiol. 45:1790–1794.

Kremer, R. J., J. Polo, and H. L. Peterson. 1982. Effect of suspending agent and temperature on survival of *Rhizobium* in fertilizer. Soil Sci. Soc. Am. J. 46:539–542.

Kurz, W. G. W., and T. A. Larue. 1975. Nitrogenase activity in rhizobia in absence of plant tissue. Nature (London) 256:407–409.

Lechevalier, M. P., and H. A. Lechevalier. 1989. Genus *Frankia* Brunchorst 1885, 174[AL]. *In* S. T. Williams, M. E. Sharpe, and J. G. Holt (eds.), Bergy's Manual of Systematic Bacteriology, Vol. 4, pp. 2410–2417. The Williams & Wilkins Co. Baltimore.

Leps, W. T., G. P. Roberts, and W. J. Brill. 1980. Use of two-dimensional polyacrylamide electrophoresis to demonstrate that putative *Rhizobium* cross-inoculation mutants actually are contaminants. Appl. Environ. Microbiol. 39:460–462.

Li, D.-M., and M. Alexander. 1986. Bacterial growth rates and competition affect nodulation and root colonization by *Rhizobium meliloti*. Appl. Environ. Microbiol. 52:807–811.

Liu, R., V. M. Tran, and E. L. Schmidt. 1989. Nodulating competitiveness of a nonmotile Tn7 mutant of *Bradyrhizobium japonicum* in nonsterile soil. Appl. Environ. Microbiol. 55: 1895–1900.

Lowendorf, H. S., and M. Alexander. 1983a. Selecting *Rhizobium meliloti* for inoculation of alfalfa planted in acid soils. Soil Sci. Soc. Am. J. 47:935–938.

Lowendorf, H. S., and M. Alexander. 1983b. Identification of *Rhizobium phaseoli* strains that are tolerant or sensitive to soil acidity. Appl. Environ. Microbiol. 45:737–742.

Lowendorf, H. S., A. M. Baya, and M. Alexander. 1981. Survival of *Rhizobium* in acid soils. Appl. Environ. Microbiol. 42:951–957.

Madariaga, G. M., and J. S. Angle. 1992. Sludge-borne salt effect on survival of *Bradyrhizobium japonicum*. J. Environ. Qual. 21:276–280.

Mahler, R. L., and D. F. Bezdicek. 1980. Serogroup distribution of *Rhizobium leguminosarum* in peas in the Palouse of Eastern Washington. Soil Sci. Soc. Am. J. 44:292–295.

Mahler, R. L., and A. G. Wollum, II. 1980. Influence of water potential on the survival of rhizobia in Goldsboro loamy sand. Soil Sci. Soc. Am. J. 44:988–992.

Malek, W. 1992. The role of motility in the efficiency of nodulation by *Rhizobium meliloti*. Arch. Microbiol. 158:26–28.

Manian, S. S., R. Gumbleton, A. M. Buckley, and R. O'Gara. 1984. Nitrogen fixation and carbon dioxide assimilation in *Rhizobium japonicum*. Appl. Environ. Microbiol. 48:276–279.

Mansour, S. R., A. Dewedar, and J. G. Torrey. 1990. Isolation, culture, and behavior of *Frankia* strain HFPCgI4 from root nodules of *Casuarina glauca*. Bot. Gaz. 151:490–496.

Mary, P., D. Ocin, and R. Tailliez. 1985. Rates of drying and survival of *Rhizobium meliloti* strains during storage at different relative humidities. Appl. Environ. Microbiol. 50:207–211.

Materon, L. A., and C. Hagedorn. 1992. Competitiveness of *Rhizobium trifolii* strains associated with red clover (*Trifolium pratense* L.) in Mississippi soils. Appl. Environ. Microbiol. 44:1096–1101.

McDermott, T. R., and P. H. Graham. 1989. *Bradyrhizobium japonicum* inoculant mobility, nodule occupancy, and acetylene reduction in the soybean root system. Appl. Environ. Microbiol. 55:2493–2498.

McGrath, S. P., P. C. Brookes, and K. E. Giller. 1988. Effects of potentially toxic metals in soil derived from past applications of sewage sludge on nitrogen fixation by *Trifolium repens* L. Soil Biol. Biochem. 20:415–424.

Moawad, H., and B. B. Bohlool. 1984. Competition among *Rhizobium* spp. for nodulation of *Leucaena leucocephala* in two tropical soils. Appl. Environ. Microbiol. 48:5–9.

Munns, D. N., and R. L. Fox. 1977. Comparative lime requirements of tropical and temperate legumes. Plant Soil 46:533–548.

Myrold, D. D., A. B. Hilger, and S. H. Strauss. 1990. Detecting *Frankia* in soils using PCR. *In* P. M. Gresshoff, Le. E. Roth, G. Stacey, and W. E. Newton (eds.), Nitrogen Fixation: Achievements and Objectives, p. 429. Chapman and Hall. New York.

Napoli, C., and P. Albersheim. 1980. Infection and nodulation of clover by nonmotile *Rhizobium trifolii*. J. Bacteriol. 141:979–980.

Nittayarjarn, A., B. C. Mullin, and D. D. Baker. 1990. Screening of symbiotic frankiae for host specificity by restriction fragment length polymorphism analysis. Appl. Environ. Microbiol. 56:1172–1174.

Noel, K. D., and W. J. Brill. 1980. Diversity and dynamics of indigenous *Rhizobium japonicum* populations. Appl. Environ. Microbiol. 40:931–938.

O'Hara, G. W., T. J. Goss, M. J. Dilworth, and A. R. Glenn. 1989. Maintenance of intracellular pH and acid tolerance in *Rhizobium meliloti*. Appl. Environ. Microbiol. 55:1870–1876.

Osa-Afiana, L. O., and M. Alexander. 1982a. Differences among cowpea rhizobia in tolerance to high temperature and desiccation in soil. Appl. Environ. Microbiol. 43:435–439.

Osa-Afiana, L. O., and M. Alexander. 1982b. Clays and the survival of *Rhizobium* in soil during desiccation. Soil Sci. Soc. Am. J. 46:285–288.

Paau, A. S., C. B. Bloch, and W. J. Brill. 1980. Developmental fate of *Rhizobium meliloti* bacteroids in alfalfa nodules. J. Bacteriol. 143:1480–1490.

Pagan, J. D., J. J. Child, W. R. Scowcroft, and A. H. Gibson. 1975. Nitrogen fixation by *Rhizobium* culture on a defined medium. Nature (London) 256:406–407.

Pena-Cabriales, J. J., and M. Alexander. 1979. Survival of *Rhizobium* in soils undergoing drying. Soil Sci. Soc. Am. J. 43:962–966.

Pena-Cabriales, J. J., and M. Alexander. 1983a. Growth of *Rhizobium* in soil amended with organic matter. Soil Sci. Soc. Am. J. 47:241–245.

Pena-Cabriales, J. J., and M. Alexander. 1983b. Growth of *Rhizobium* in unamended soil. Soil Sci. Soc. Am. J. 47:81–84.

Pijnenborg, J. W. M., T. A. Lie, and A. J. B. Zehnder. 1991. Nodulation of lucerne (*Medicago sativa* L.) in an acid soil: Effects of inoculum size and lime-pelleting. Plant Soil 131:1–10.

Pugashetti, B. K., J. S. Angle, and G. H. Wagner. 1982. Soil microorganisms antagonistic towards *Rhizobium japonicum.* Soil Biol. Biochem. 14:45–49.

Quispel, A. 1990. Discoveries, discussions, and trends in research on actinorhizal root nodule symbioses before 1978. *In* C. R. Schwintzer and J. D. Tjepkema (eds.), The Biology of *Frankia* and Actinorhizal Plants, pp. 15–33. Academic Press. San Diego.

Ramirez, C., and M. Alexander. 1980. Evidence suggesting protozoan predation on *Rhizobium* associated with germinating seeds and in the rhizosphere of beans (*Phaseolus vulgaris* L.). Appl. Environ. Microbiol. 40:492–499.

Reuber, T. L., A. Urzainqui, J. Glazebrook, J. W. Reed, and G. C. Walker. 1991. *Rhizobium meliloti* exopolysaccharides. Structures, genetic analyses, and symbiotic roles. Ann. NY Acad. Sci. 646:61–68.

Richardson, A. E., A. P. Anderson, G. S. James, and R. J. Simpson. 1988a. Consequences of soil acidity and the effect of lime on the nodulation of *Trifolium subterraneum* L. growing in an acid soil. Soil Biol. Biochem. 20:439–445.

Richardson, A. E., and R. J. Simpson. 1988. Enumeration and distribution of *Rhizobium trifolii* under subterranean clover-based pasture growing in an acid soil. Soil Biol. Biochem. 20:431–438.

Richardson, A. E., R. J. Simpson, M. A. Djordjevic, and B. G. Rolfe. 1988b. Expression of nodulation genes in *Rhizobium leguminosarum* biovar *trifoli* is affected by low pH and by Ca and Al ions. Appl. Environ. Microbiol. 54:2541–2548.

Rogers, R. O., and A. G. Wollum II. 1974. Virulence if the *Alnus* endophyte after *in vitro* cultivation. Soil Sci. Soc. Am. J. 38:756–759.

Sellstedt, A., B. Wullings, U. Nyström, and P. Gustafsson. 1992. Identification of *Casuarina-Frankia* strains by use of polymerase chain reaction (PCR) with arbitrary primers. FEMS Microbiol. Lett. 93:1–6.

Sherwood, J. E., J. M. Vasse, F. B. Dazzo, and G. L. Truchet. 1984. Development and trifoliin A-binding ability of the capsule of *Rhizobium trifolii.* J. Bacteriol. 159:145–152.

Simonet, P., M.-C. Grosjean, A. K. Misra, S. Nazaret, B. Cournoyer, and P. Normand. 1991. *Frankia* genus-specific characterization by polymerase chain reaction. Appl. Environ. Microbiol. 57:3278–3286.

Simonet,P., P. Normand, A. Moiroud, and R. Bardin. 1990. Identification of *Frankia* strains in nodules by hybridization of polymerase chain reaction products with strain-specific oligonucleotide probes. Arch. Microbiol. 153:235–240.

Singleton, P. W., S. A. El Swaify, and B. B. Bohlool. 1982. Effect of salinity on *Rhizobium* growth and survival. Appl. Environ. Microbiol. 44:884–890.

Strijdon, B. W., and H. J. Van Rensburg. 1981. Effect of steam sterilization and gamma irradiation of peat on quality of *Rhizobium* inoculants. Appl. Environ. Microbiol. 41:1344–1347.

Thies, J. E., P. W. Singleton, and B. B. Bohlool. 1991a. Influence of the size of indigenous rhizobial populations on establishment and symbiotic performance of introduced rhizobia on field-grown legumes. Appl. Environ. Microbiol. 57:19–28.

Thies, J. E., P. W. Singleton, and B. B. Bohlool. 1991b. Modeling symbiotic performance of introduced rhizobia in the field by use of indices of indigenous population size and nitrogen status of the soil. Appl. Environ. Microbiol. 57:29–37.

Thornton, F. C., and C. B. Davey. 1983. Acid tolerance of *Rhizobium trifolii* in culture media. Soil Sci. Soc. Am. J. 47:496–501.

Tiwari, R. P., W. G. Reeve, and A. R. Glenn. 1992. Mutations conferring acid sensitivity in the acid-tolerant strains of *Rhizobium meliloti* WSM419 and *Rhizobium leguminoarum* biovar *vicae* WMS710. FEMS Microbiol. Lett. 100:107–112.

Tjepkema, J., and H. J. Evans. 1975. Nitrogen fixation by free-living *Rhizobium* in a defined liquid medium. Biochem. Biophys. Res. Commun. 65:625–628.

Tjepkema, J. D., and L. J. Winship. 1980. Energy requirement for nitrogen fixation in actinorhizal and legume root nodules. Science 209:279–280.

van Brussel, A. A. N., R. Bakhuizen, P. C. van Spronsen, H. P. Spaink, T. Tak, B. J. J. Lugtenberg, and J. W. Kijne. 1992. Induction of pre-infection thread structures in the leguminous host plant by mitogenic lipo-oligosaccharides of *Rhizobium*. Science 257:70–72.

Van Dijk, C., and A. Sliumer-Stolk. 1990. An infective strain type of *Frankia* in the soil of natural stands of *Alnus glutinosa* (L.) Gaertner. Plant Soil 127:107–121.

van Rensburg, H. J., and B. W. Strijdom. 1985. Effectiveness of *Rhizobium* strains used in inoculants after their introduction into soil. Appl. Environ. Microbiol. 49:127–131.

Viteri, S. E., and E. L. Schmidt. 1987. Ecology of indigenous soil rhizobia: Response of *Bradyrhizobium japonicum* to readily available substrates. Appl. Environ. Microbiol. 53:1872–1875.

Waldon, H. B., M. B. Jenkins, R. A. Virginia, and E. E. Harding. 1989. Characteristics of woodland rhizobial populations from surface and deep-soil environments of the Sonoran Desert. Appl. Environ. Microbiol. 55:3058–3064.

Woomer, P., P. W. Singleton, and B. B. Bohlool. 1988. Ecological indicators of native rhizobia in tropical soils. Appl. Environ. Microbiol. 54:1112–1116.

13 Denitrification

Some would use a term such as *pollutant* when considering the occurrence of nitrate in soil, whereas others would prefer to describe soil nitrate as an essential nutrient for ecosystem sustenance. One group would seek to minimize levels of the alleged environmentally offensive substance, whereas those interested in plant biomass production would seek to optimize nitrate production. Both groups would underscore the necessity of understanding processes leading to losses of soil nitrate to achieve their objectives.

To a great degree, the existence of such divergent opinions regarding the value of nitrate in soil results from the propensity of this compound to be in the wrong place at the wrong time at undesirable concentrations. Nitrate ions in the root zone of a food crop during maximal biomass production are inarguably beneficial, whereas migration of that same nitrate below the root zone to underlying aquifers transforms a beneficial soil chemical component into an undesirable water contaminant.

Societal implications associated with the timing and location of accumulating significant soil nitrate loadings reach far beyond this conflict. For example, traditional considerations have included concerns that nitrate contents of soil leachates or runoff waters may alter regional water quality (e.g., augmented lake eutrophication rates) and that there are potential public health difficulties (methemoglobenemia) associated with elevated nitrate-N loadings in drinking waters. Other concerns include management of fixed nitrogen amendment to soil systems to minimize societal costs and unwanted environmental impacts. Excessive losses of fertilizer nitrogen, whether added to agricultural, urban, or reclaimed soils, alter the economics of biomass production as well as increase the energy expenditures associated with industrial nitrogen fixation.

Furthermore, increased use of industrially fixed nitrogen may have an environmental impact far beyond the localized site. Fertilizer production requires inputs of large quantities of fossil fuels. Use of these resources to produce plant nutrients not only decreases the reserves of a nonrenewable commodity but also increases greenhouse gas production through its conversion to carbon dioxide and energy. Hence, whether the goal is to increase or to decrease the return of fixed nitrogen to atmospheric dinitrogen pools (denitrification), the necessity of understanding this essential portion of the nitrogen cycle is unquestioned.

Denitrification Defined: If nitrogen fixation is used as a reference for the initiation point of the nitrogen cycle, denitrification is its termination. **Denitrification** is the biological reduction of nitrogen oxides to nitrous oxide and/or dinitrogen. The process is catalyzed exclusive by a diverse group of bacteria that use nitrate or nitrite as terminal electron acceptors in their respiratory processes. Since the electrons are more readily transferred to molecular oxygen by the microbes involved, denitrification can occur only under anoxic conditions. From this basic definition of denitrification, the complexity of controlling its occurrence in heterogeneous soils becomes evident. For denitrification to occur, the microorganisms must have an energy supply (usually an organic carbon source) and reside in an anaerobic soil microsite; yet, most soil systems are predominantly aerobic, and the primary processes producing nitrate also have an obligatory requirement for molecular oxygen.

These limited examples of the role of denitrification in controlling levels of nitrate occurring in soil and the availability of this primary plant nutrient within soil ecosystems underscore the necessity of understanding the basic properties of the process and the factors in the soil environment that control the rate or reduction of nitrate to dinitrogen. Thus, this chapter elucidates the basic biological and biochemical properties of denitrification and determines the extent of its occurrence in terrestrial ecosystems.

13.1. PATHWAYS FOR BIOLOGICAL REDUCTION OF SOIL NITRATE

Soil fixed-nitrogen resources may be conserved through both assimilatory and dissimilatory nitrate reductive processes, or they may be reduced by dissimilatory nitrate reduction. Assimilatory and dissimilatory nitrate reduction both involve the transfer of electrons to nitrogen compounds, but they differ in the ultimate fate of the reduced nitrogen atom. With assimilatory reduction, nitrate is reduced to ammonium, which is incorporated into cellular biomass. Thus, with this process, the quantity of nitrogen reduced is proportional to cellular requirements for biomass production. In contrast, for dissimilatory nitrate reduction, the nitrogenous compounds accept electrons in support of cellular respiration. The final products, dinitrogen, nitrous oxide, or ammonium, are released from the cell and accumulate in the environment in concentrations far beyond those necessary for biomass synthesis. These two fates of the nitrate reductive processes are characterized by commonalties of some of the intermediates and products (e.g., ammonium, nitrite) but are also distinguished by differences in the specific enzymes catalyzing the reductive reactions.

Because assimilatory reduction of nitrate is closely linked to biomass production, it is catalyzed by all soil inhabitants capable of using nitrate as a nitrogen source. Ammonium, the terminal product of the process, is incorporated into plant and microbial biomass. Assimilatory reduction is distinguished from other nitrate-reducing processes in that, although a portion of the cellular-reducing power is transferred to the nitrogen, the cells rely upon *other* pathways as terminal electron acceptors (e.g., molecular oxygen, fermentative products.)

Three commonly evaluated microbial processes are classed under the title of dissimilatory nitrate reduction. These processes can be distinguished by their respective products: a) nitrite, b) ammonium, and c) nitrous oxide and dinitrogen (denitrification).

The first of these dissimilatory reductions of nitrate involves the simple reduction of nitrate to nitrite. The general occurrence of this process and the stability of the genetic trait were instrumental in the use of nitrate reduction to nitrite as a reliable taxonomic test for grouping bacterial isolates into genera and species. This reductive process is also significant in some *in situ* soil systems. In a soil site overlain by oxygen-bearing water, a nitrite and nitrate cycle may develop that provides energy for nitrite oxidizers (see Fig. 10.3). Under anoxic conditions, nitrate is reduced to nitrite by portions of the general soil bacterial population. This nitrite then diffuses from the anoxic sites where it is produced to aerobic microsites where nitrite oxidizers use it in their energy-generating processes. Evidence for the occurrence of such interactions in native sites is the occurrence of an imbalance between ammonium and nitrite oxidizer populations in some soil ecosystems. In other words nitrite oxidizer populations are greater in some soil systems than would be anticipated, especially where ammonium oxidizers are the sole provider of their energy resource, nitrite. This minicycle involving recirculation of nitrite and nitrate serves as an alternative energy source for the autotrophs.

Dissimilatory ammonium production and denitrification have several common characteristics. Both processes are coupled to organic matter decomposition (i.e., oxidation of fixed-carbon compounds provides cellular energy and reducing power), the nitrogen oxides that are reduced serve as terminal electron acceptors, and growth yields of the bacteria are increased as a result of the energy provided by passage of the electrons through an electron transport pathway, which includes cytochromes. (See section 13.2 for discussion of the role of cytochromes in denitrification.)

Although not previously considered to be a major soil process, ammonium production from nitrate reduction has been shown to occur in soils and sediments (e.g., Buresh and Patrick, 1978; Fazzolari et al., 1990; Koike and Hattori, 1978; Smith and Zimmerman, 1981). Dissimilatory ammonium production is catalyzed by a variety of enteric bacteria and spore forming *Bacillus* and *Clostridium* species in soil (e.g., Caskey and Tiedje, 1980; Prakash and Sadana, 1973; Rehr and Klemme, 1989; Smith and Zimmerman, 1981). As with denitrification, nitrous oxide has been found to be a product of metabolism of dissimilatory ammonium producers, but the nitrous oxide appears to be produced as a side reaction in these organisms (e.g., Smith, 1982).

13.2. BIOCHEMICAL PROPERTIES OF DENITRIFICATION

During biological denitrification, nitrate is transformed to dinitrogen by a series of reductive reactions as follows:

$$2HNO_3 \rightarrow 2HNO_2 \rightarrow [2NO] \rightarrow N_2O \rightarrow N_2$$

Nitric oxide is generally enclosed in brackets because it usually is not detected as a free intermediate. There is also some controversy whether nitric oxide is a true intermediate in the process or whether its presence represents a side reaction (e.g.,

Betlach and Tiedje, 1981; Garber and Hollocher, 1981, 1982; Hochstein and Tom-linson, 1988; Hollocher et al., 1980). In either case, the significant aspects of the reaction sequence from the soil microbiologist's viewpoint is that nitrate is reduced to nitrous oxide and dinitrogen with the transient accumulation of nitrite plus the occasionally detected transient accumulation of nitric oxide (e.g., Cady and Bar-tholomew, 1960; Cooper and Smith, 1963). As is indicated later, a variable ratio—depending upon the soil chemical properties—of nitrous oxide to dinitrogen is produced.

Carbon and Energy Sources for Denitrifiers: The electrons for nitrogen oxide reduction may be provided by oxidation of a variety of fixed-carbon compounds. Most commonly, investigators add glucose to soil samples as a carbon source in laboratory studies of the denitrification processes. Remember that a variety of organic and inorganic compounds are catalyzed during this process in native soil environments. This array of oxidizable substrates includes more complex organic compounds, some of which are of concern from the view of soil contamination (e.g., acetone, Platen and Schlink, 1989; anthranilic acid, Braun and Gibson, 1984; cresol, Bossert and Young, 1986; and toluene, Evans et al., 1992; also see a review of this topic prepared by Evans and Fuchs, 1988). Because the electrons for these oxidations are transferred through a chain of acceptors similar to that used with the reduction of molecular oxygen, the growth yields of microbes from oxidation of these fixed-carbon substances using nitrogen oxides as terminal electron acceptors are compara-ble with the two classes of acceptors (molecular oxygen and nitrogen oxides).

Using the traditional definition of anaerobic processes (i.e., energy-yielding cata-bolic reactions obligatorily occurring in the absence of molecular oxygen), deni-trification would be classified as meeting that definition. Yet, many soil microbiolo-gists do not consider denitrification to be a true anaerobic process, because the primary difference between nitrate respiration and oxygen-based respiration is that, in the former electron transport process, nitrate is merely replacing oxygen as the final electron acceptor. Denitrifiers are obligate aerobes; that is, they use molecular oxygen as a terminal electron acceptor and do not ferment organic substances in the absence of molecular oxygen. Denitrifiers also possess cytochromes that can trans-fer electrons to nitrogen oxides. For example, in the process catalyzed by *Paracoc-cus denitrificans* terminal electron transfer to nitrate, nitrite, and nitrous oxide are provided by cytochromes (e.g., Ballard and Ferguson, 1988; Bolgiano et al., 1989) (Fig. 13.1). If oxygen is present, different cytochromes are used for its reduction to water, and dissimilatory nitrate reduction is inhibited (e.g., Hernandez and Rowe, 1987).

Induction of Synthesis of Nitrogen Oxide Reductases: A further modification of cellular metabolism of the denitrifier growing in aerobic situations is that the requisite enzymes for denitrification are not synthesized or are produced in limited quantities. Nitrogen oxidase reductase synthesis is repressed by molecular oxygen. Once the oxygen is depleted in the microsite where the putative denitrifier resides, synthesis of these denitrifier enzymes is induced. Thus, development of a totally functional denitrification system in a previously aerobic ecosystem requires about 40 minutes to 3 hours. For some denitrifiers the time lapse necessary for synthesiz-

Fig. 13.1. Summary of the electron transport pathway leading from oxidation of fixed carbon compounds to nitrate, nitrite, or nitrous oxide as terminal electron acceptors.

ing nitrogen oxide reductases is reduced by an induction of synthesis of the first enzyme in the pathway, nitrate reductase, as the available molecular oxygen supply is being depleted. Full induction of the pathway occurs with complete exhaustion of molecular oxygen resources. Thus, the microbe can respond to the absence of molecular oxygen by reducing nitrate to nitrite but also avoid synthesis of the total denitrification enzyme array should anaerobiosis not be achieved. This differential synthesis of denitrification enzymes could result in a transient accumulation of nitrite.

13.3. ENVIRONMENTAL IMPLICATIONS OF NITROUS OXIDE FORMATION

Although at one time, the transient accumulation of nitrous oxide in reaction vessels wherein biological denitrification was occurring was ignored or considered to be a product of the closed reaction system, it is now known that in native terrestrial environments, nitrous oxide is a common product of denitrification. For example, fluxes of nitrous oxide from soils have been shown to range from 7 to 165 kg N ha^{-1} $year^{-1}$ in drained histosols in South Florida (Terry et al., 1981a). Nitrous oxide evolution from mineral-soil-based systems is usually less than reported from these drained organic soils. Annual losses from irrigated soils in California (USA) range from 19.6 to 41.8 kg N ha^{-1} (Ryden and Lund, 1980). Lesser yields have been detected from cropped soils in Colorado (Mosier and Hutchinson, 1981) and prairie soils (Mosier et al., 1981) as well as from agricultural soils in Wisconsin (USA) (Goodroad et al., 1984). Uncropped, native ecosystems also produce more limited quantities of nitrous oxide during plant growth seasons (Goodroad and Keeney, 1984). These data show that nitrous oxide production by soil processes are highly variable. Soil properties controlling the quantity of this product by soil microbes include soil redox potential (Letey et al., 1981; Smith et al., 1983), moisture tension (Davidson and Swank, 1986; Freney et al., 1979; Terry et al., 1981b), nitrate and oxygen concentrations (Blackmer and Bremner, 1978, 1979; Firestone et al., 1979, 1980; Terry and Tate, 1980) as well as the time of day that the measurements are taken (Blackmer et al., 1982). Generally, the proportion of gaseous nitrogen products composed of nitrous oxide increases with increasing soil acidity, reduced soil temperature, and augmented soil nitrate levels, but the exact ratio of nitrous oxide to

$$O_2 \xrightarrow{\text{Light}} 2O^{\cdot} + 2O_2 \longrightarrow 2O_3$$

$$N_2O + O^{\cdot} \longrightarrow 2NO$$

$$NO + O_3 \longrightarrow NO_2 + O_2$$

$$NO_2 + O^{\cdot} \longrightarrow NO + O_2$$

Fig. 13.2. Reaction sequence leading to ozone formation and the role of nitrous oxide in its destruction.

dinitrogen for a specific soil system varies with the combination of the soil chemical and physical properties existent therein and is therefore not reliably predicted.

The inclusion of nitrous oxide as a product of a major soil biological process is of general concern since nitrous oxide is an ozone-depleting gas. The reactions involved in this process are presented in Fig. 13.2. Note that not all the nitrous oxide evolving from the soil surface results from the activity of denitrifiers. A variety of soil heterotrophs as well as nitrifiers produce this gas as a metabolic product or by-product (e.g., Blackmer et al., 1980; Bleakley and Tiedje, 1982; Bollag and Tung, 1972; Davidson et al., 1986; Smith and Zimmerman, 1981).

13.4. MICROBIOLOGY OF DENITRIFICATION

Assessment of Total Soil Denitrifier Populations: The population density of denitrifiers in soil is commonly estimated by using most-probable-number procedures (e.g., Focht and Joseph, 1973; Volz, 1977). These procedures are generally based on detection of the conversion of nitrate or nitrite to gaseous end products in liquid culture media (e.g., Fig. 13.3). These methods are inherently imprecise (see chapter 3 for a discussion of the limitations of the most-probable-number procedures). More accurate and sensitive methods for enumerating these organisms will become available as the characteristics of the genome-encoding denitrification and the similarities in these gene sequences between the various denitrifiers present in soil samples are elucidated thereby enabling the use of PCR procedures for the quantification of denitrifiers in soil (e.g., Linne von Berg and Bothe, 1992; Ye et al., 1993).

Although denitrifier population density data do indicate the number of organisms present in a soil ecosystem that can catalyze the process, in most soils population densities of denitrifiers *do not* correlate with denitrification rates. Soil microbial communities commonly contain several million denitrifiers per gram of dry soil; yet, because of the fact that these organisms do not rely totally upon nitrate reduction for their growth, their presence in the system indicates only a potential for

Quantification of Denitrifiers

1. Prepare a serial dilution of a fresh soil sample [eg., 10^{-1}, 10^{-2}... 10^{-7}]
2. Inoculate 5 tubes containing 9 ml of nitrate broth with 1.0ml of dilution.
3. Incubate the inoculated media at 30°C for two weeks.
4. Quantify residual nitrate in tubes. Presence of any nitrate = +.
5. Calculate MPN using standard tables.

Fig. 13.3. Outline of a most-probable-number procedure for estimating denitrifier populations in soil (adapted from Focht and Joseph, 1973).

denitrification should conditions favor its occurrence. The denitrifier populations occurring in any particular soil with an oxygen-containing atmosphere may have developed through oxygen-based respiration. When the denitrification rate and denitrifier population have been shown to correlate, the cellular respiration has generally been based on nitrate reduction.

General Traits of Denitrifiers: Denitrifiers are a biochemically and taxonomically diverse group of bacteria. Although some denitrifiers are chemoautotrophs (e.g., using hydrogen or reduced sulfur compounds as energy sources) and others are photoautotrophs, most of these organisms generally derive their energy from oxidation of fixed-carbon substrates, including single-carbon compounds. The primary substrate and end product of the process are nitrate and dinitrogen, but some denitrifiers can reduce only nitrite (nitrite-dependent denitrifiers), whereas others lack nitrous oxide reductase, thereby producing nitrous oxide as the terminal product. The variety of metabolic types of bacteria capable of denitrifying defies any effort to group these organisms into one or a few bacterial genera.

Further difficulty with evaluating the occurrence of the "denitrifier trait" is caused by the fact that many organisms have been classified historically as being capable of denitrifying which by modern standards would *not* be grouped with denitrifiers. Care must be exercised in classing organisms as denitrifiers based strictly on the conversion of nitrate or nitrite to nitrous oxide or dinitrogen. Many bacterial strains can reduce nitrate and produce *limited* quantities of dinitrogen or nitrous oxide. Although some of these organisms have been termed denitrifiers, these marginal nitrate reducers are *not* true denitrifiers. To be a true denitrifier, a microbe must meet the following criteria:

- At least 80 percent of the nitrate or nitrite reduced by the bacterium must be converted to dinitrogen and nitrous oxide.

- There must be an increased growth yield due to the reduction of nitrate, nitrite, or nitrous oxide. This trait is the primary requirement for classing an organism as a denitrifier in that it shows that nitrogen oxide reduction is a dissimilatory process.

- The conversion of nitrate to nitrous oxide and dinitrogen must occur at a high rate. That is, the process must be central to cellular intermediary metabolism, not just a side reaction providing a minor pathway for electron transport.

- The presence of cytochrome cd or dissimilatory nitrite reductase should be demonstrable in the microbial cells.

Generic Identity of Denitrifiers: Most of the denitrifiers isolated from soils and waters are members of the genera *Pseudomonas* or *Alcaligenes,* but a large number of other bacterial genera contain strains of denitrifiers (e.g., Gamble et al., 1977). As with diazotrophs, denitrification is not a sufficiently definitive trait to allow separation of all denitrifiers into one or a limited number of genera. Not only are the individual bacterial strains of denitrifiers metabolically distinct, but the properties of the enzymes associated with denitrification process itself is also highly variable. For example, in a comparison of denitrification by *Pseudomonas stutzeri, Pseudomonas aeruginosa,* and *Paracoccus denitrificans,* it was found that the rates of anoxic growth of the organisms varied 1.5-fold, gas production varied over 8-fold, and cell yield differed by 3-fold (Carlson and Ingraham, 1983). The metabolic and taxonomic diversity of these organisms is exemplified by the following partial listing of some of the bacterial genera containing denitrifiers (for a more complete listing of denitrifiers, see Knowles, 1981, 1982):

- The genus *Achromobacter* is cited as containing a variety of denitrifiers, including methane oxidizers and nitrite-dependent denitrifiers. (These organisms have now been grouped among the species of the genus *Alcaligenes.*)
- *Alcaligenes eutrophus,* an organism originally classified as *Hydrogenomonas eutrophus,* is capable of autotrophic growth using hydrogen as an energy source, carbon dioxide for carbon, and nitrate as a terminal electron acceptor (e.g., Pfitzner and Schlegel, 1973). This bacterial species is a facultative autotroph, because it also uses organic carbon as an energy source. Other common denitrifiers in the genus *Alcaligenes* include *A. denitrificans* and *A. odorans.* Another facultative autotroph that oxidizes hydrogen for energy is *Paracoccus denitrificans.*
- Several nitrogen-fixing bacteria are also capable of denitrifying. These diazotrophs include some strains of *Azospirillum brasilense* (e.g., Chauret et al., 1992; Neyra et al., 1977) and some *Rhizobium* species (e.g., Chan et al., 1989; Coyne and Focht, 1987; Daniel et al., 1980; van Berkum and Keyser, 1985; Zablotowicz et al., 1978)
- Some denitrifiers are also thermophilic (e.g., Hollocher and Kristjánsson, 1992).
- Some strains of *Chromobacter* denitrify using nitrate or nitrite as terminal electron acceptors reducing them to nitrous oxide and dinitrogen whereas *Chromobacter violacium* can only reduce nitrogen oxides to nitrous oxide (Bazylinski et al., 1986).
- *Halobacterium marismortui* is a halophilic denitrifier that was isolated from the Dead Sea.
- *Hyphomicrobium* strains denitrify with methanol as the primary energy source. The practical significance of understanding the environmental limitations to activity of these strains relates to the fact that methanol has been used as an

additive to sewage effluent to provide the energy source necessary for maximizing denitrification rates for removing nitrate from the effluent.

- The photosynthetic bacterium *Rhodopseudomonas sphaeroids* forma sp. *denitrificans* is both a denitrifier and a diazotroph (Dunstan et al., 1982).

- *Thiobacillus denitrificans* is an example of a sulfur-oxidizing chemoautotroph that grows in the absence of oxygen with nitrate as a terminal electron acceptor. The energy sources for this organism include sulfide, elemental sulfur, and thiosulfate. *T. denitrificans* can also use molecular oxygen as a terminal electron acceptor.

13.5. QUANTIFICATION OF NITROGEN LOSSES FROM AN ECOSYSTEM VIA DENITRIFICATION

Historically, the most commonly used techniques for assessing nitrogen fixation have involved a) determination of nitrogen balances in the ecosystem and b) quantification of distribution and movement of ^{15}N-labeled fixed nitrogen through soil reservoirs. More specific assays of denitrification processes are a) assessment of disappearance of nitrogen oxides from soil slurries or b) use of acetylene inhibition of nitrous oxide reductase. The principles, advantages, and deficiencies of each of these techniques are assessed here.

13.5.1. Nitrogen Balance Studies

As discussed in section 11.1.3, assessing changes in nitrogen distribution among the various soil reservoirs as a means of quantifying process rates appears to be conceptually simple and logically attractive. In using nitrogen balance studies to quantify soil denitrification, denitrified nitrogen is considered to be the unaccountable or missing nitrogen; that is, denitrified nitrogen is calculated by subtracting total fixed nitrogen in the soil from nitrogen preexisting in the site plus fixed nitrogen added as fertilizer plus biological fixed nitrogen inputs. The procedure is also called the difference method, because all nitrogen occurring in the system is subtracted from the sum of that previously occurring therein plus any amendments.

Applying this procedure to the estimation of denitrification processes in soils is predicated upon three assumptions:

1. All the organic and fixed-nitrogen compounds contained in soil can be accurately quantified.

2. The assay procedures available for quantifying soil nitrogen compounds are sufficiently sensitive and soil sampling procedures are adequately reproducible and representative of *in situ* conditions that small changes in the various soil nitrogen compounds can be detected.

3. The sole means of gaseous nitrogen loss from soil is denitrification.

Unfortunately, none of these assumptions is valid. Reasonably sensitive assay procedures are available to assay soil organic and inorganic nitrogen pools. Their application to soil nitrogen cycle analysis is limited by the fact that changes in soil nitrogen reserves are commonly less than the precision of the assay method. Further problems with the direct quantification of nitrogen species in a particular soil site result from the fact that soil is an open system. In other words, water-soluble substances (e.g., nitrate, nitrite, and simple organic nitrogenous compounds) move freely into and out of the portion of the ecosystem of interest. Inability to accurately account for all fixed nitrogen inputs and losses from the soil site results in appreciable statistical variability in nitrogen balance data. This variation also results in invalidation of the second assumption.

Probably the biggest difficulty with applying nitrogen balance studies to assessment of denitrification is the third assumption (all nitrogen volatilization results from denitrification). There are at least three major routes of nitrogen volatilization from soil. Nitrogen may be lost through:

- Nonbiological losses of ammonia (ammonia volatilization)
- Chemical decomposition of nitrite
- Denitrification as dinitrogen and nitrous oxide

Ammonia volatilization can be a major route for loss of fixed nitrogen from soil. Up to 25 percent of the ammonium that is applied to soil as fertilizers or formed microbiologically may be lost as ammonia. Fortunately, for most soil systems, nitrogen volatilization is a minor problem, because ammonia losses are insignificant below pH 7.0. (Recall that at acidic pHs the equilibrium between ammonium and ammonia is shifted so that ammonium predominates.) Ammonia volatilization is particularly apparent in the use of farm yard manure and at times in association with fertilization with urea. Major environmental concern with this process is linked more to soil-fixed-nitrogen migration to nearby bodies of water than to its interference with the quantification of denitrification.

Nitrite decomposition is rarely a major route of exodus of fixed nitrogen from soils. The most commonly documented situation for accumulating meaningful nitrite concentrations is linked to alkaline soil conditions where nitrite accumulates due to ammonia toxicity to *Nitrobacter*. This phenomenon occurs in soils fertilized with anhydrous ammonia. In this process, nitrite reacts chemically with soil organic matter to produce dinitrogen and nitrous oxide plus some nitric oxide. In the absence of molecular oxygen, the nitric oxide may be evolved, but it is generally oxidized to nitrogen dioxide as soon as it is exposed to molecular oxygen. Although this reaction can occur in soil, nitrite will more commonly be denitrified in soil. Thus, the relative importance of the three pathways of nitrogen volatilization from soils may be ranked with denitrification being the major means of nitrogen loss, ammonia volatilization being ranked a distant second, and nitrite decomposition occurring only as a minor contributor.

The uncertainties associated with the nitrogen balance procedure in estimating soil nitrogen losses due to denitrification were exemplified by data from a study in which the results from this procedure were compared to that from a parallel quantification of nitrogen dynamics using ^{15}N to trace changes in soil nitrogen reservoirs (Rolston et al., 1979). In the more moist soils studied, actual denitrification losses were generally 0–30 kg N ha^{-1} less than amounts calculated from the nitrogen balance procedures. These differences ranged from 12 to 65 kg N ha^{-1} for drier soils. The authors concluded that this large variation in the nitrogen balance procedure resulted primarily from large sampling variability in nitrogen-leaching results (although variability of soil residual nitrate and organic nitrogen were also large). These results further emphasize the conclusion that even in soils where nitrogen volatilization due to chemical processes are minimal, the accuracy of the nitrogen balance procedure is limited as a result of difficulties in precisely accounting for all soil nitrogen forms present in and leaving the soil system.

13.5.2. Use of Nitrogen Isotopes to Trace Soil Nitrogen Movement

Although the short-lived radioactive nitrogen isotope (^{13}N) has been used to quantify denitrification processes (Firestone et al., 1979; Hollocher et al., 1980; Tiedje et al., 1979), due to the short half-life of the radioactive isotope, the heavy isotope of nitrogen (^{15}N) is more amenable to long-term evaluation of denitrification reactions in soils. For these studies, ^{15}N-labeled fixed-nitrogen sources (organic or inorganic) are amended to field soils or incubated soil samples (greenhouse or laboratory), and the ^{15}N-labeled dinitrogen or nitrous oxide is quantified.

As with the evaluation of biological nitrogen fixation (section 11.1.3), implementation of this procedure is predicated on a series of assumptions, the most significant being a) the denitrifier populations do not discriminate between nitrogen isotopes and b) the soil-amended ^{15}N-labeled fixed nitrogen is homogeneously mixed with native soil nitrogen resources and is, thus, of equivalent availability to the denitrifier population. As noted with biological nitrogen fixation these assumptions are questionable.

Discrimination of nitrogen isotopes by denitrifiers has been demonstrated (e.g., Blackmer and Bremner, 1977). These workers observed the clear microbial capacity to discriminate between ^{14}N and ^{15}N-labeled nitrate. Significant nitrogen isotope effects were detected in both reduction of nitrate to nitrite and nitrite to gaseous end products.

Furthermore, because of the heterogeneity of the soil system, uniform mixing of an external nitrogen supply with indigenously formed fixed nitrogen is difficult at best. Vanden Heuvel et al. (1988) found that the magnitude of this error increases with the range of ^{15}N enrichment between native soil N isotope distributions and amended nitrogen (for a constant mean ^{15}N enrichment), and it varies dependent upon the number of nitrate pools existent in the system. The error was not affected by overall nitrogen gas evolution rate or the ^{15}N enrichment of the amended fertilizer. The theoretical basis for this potential for overestimation and underestimation of nitrous oxide and dinitrogen evolution was documented through mathematical modeling by Boast et al. (1988).

Even though these cited studies demonstrate significant difficulties with the use of ^{15}N to quantify losses of fixed nitrogen from soil ecosystems due to denitrification, as with any laboratory procedure, its application to field, laboratory, and greenhouse studies is appropriate when the limitations are considered, especially during experimental design and data interpretation. Although the use of ^{15}N is appropriate for field experiments (especially for long-term studies as exemplified by Focht and Stolzy, 1978) when the limitations are considered during data interpretations, its use in greenhouse and laboratory incubation studies is particularly appropriate. In the latter situations extensive disruption of soil structure during sample preparation is common, and the conditions for denitrification are more controlled.

13.5.3. Disappearance of Endogenous or Added Nitrogen Oxides

Since nitrate, nitrite, and nitrous oxide are readily extracted from soil and quantified, it is reasonable to consider that the rate of change in concentrations (native or amended) of these chemicals can serve as an indication of potential denitrification rates in soils (e.g., Bremner and Shaw, 1958a). This procedure is based on the assumption that sole products of nitrate and nitrite reduction are gaseous. Except in systems where extensive dissimilatory ammonium production is occurring, alternative fates of the nitrogen oxide tends not to interfere with this analysis procedure. Greater limitation in the use of changes in nitrogen oxide levels as an assessment of denitrification kinetics in soil is derived from the fact that in most soil systems, nitrate concentrations are a few micrograms per gram of soil at best. Generally, nitrite is not detected in native soil samples. Thus, soils must generally be amended with exogenous nitrate or nitrite for a meaningful estimate of the capacity of the soil microbial community to denitrify nitrogen oxides to be attained. To ensure that the totality of the denitrification enzymes are measured (i.e., enzyme and not substrate is the limiting factor), enzyme-saturating concentrations of nitrate or nitrous oxide are generally added to soil with this technique. Therefore, because these nitrogen oxide concentrations are artificially elevated, denitrification rates assessed using this procedure must be understood to be potential rates rather than actual field values. The potential values are equivalent only to field denitrification rates in the unusual situations when soil nitrate concentrations are sufficient to saturate the denitrification enzymes present.

13.5.4. Acetylene Block Method for Assessing Denitrification Processes in Soil

Incubation of soil samples or amendment of the soil atmosphere *in situ* with 10^{-2} atmospheres of acetylene inhibits the reduction of nitrous oxide to dinitrogen (Yoshinari and Knowles, 1976; Yoshinari et al., 1977). Because commonly available gas chromatography procedures may be used to quantify nitrous oxide, this process provides a sensitive method for measuring denitrification enzymes. See Fig. 13.4 for an outline of the procedure. The validity of this procedure for determination of denitrification kinetics in soil slurries (e.g., Smith et al., 1978) and in field situations (e.g., soil cores, Parkin et al., 1985a, and direct field measurements, e.g., Aulakh et

Acetylene Block Method for Estimating
Soil Denitrification Potential

1. Place 10g soil sample into 60 ml serum bottle
2. Add water or KNO_3 solution to soil
3. Purge and evacuate bottles with Argon or Argon + 10% acetylene
4. Incubate for appropriate time interval with mixing
5. Assay headspace N_2O with gas chromatographic procedure

Fig. 13.4. Outline of a general procedure using the acetylene block method for quantifying denitrification potential in soil samples incubated in the laboratory. Samples receiving nitrate amendment provide estimates of maximum denitrification potential of the sample. Incubation times may vary from 1 hour to several days, depending upon the experimental objectives.

al., 1991; Mosier et al., 1986) has been confirmed by comparing the data derived from using the acetylene block methods with that from ^{15}N and ^{13}N enrichments. Acetylene inhibition of nitrous oxide reduction is a quick and reliable procedure for estimating *in situ* levels of denitrification enzymes in soil slurries, in soil cores, and in the field. The initially constant nitrous oxide evolution rate in soil slurries corresponds to preexisting denitrification enzymes in the soil sample (Smith and Tiedje, 1979a).

Complications associated with this procedure most commonly involve its application under field conditions and to some degree its use for soil column studies. The method is predicated on the assumption that sufficient acetylene is present in all microsites of the system to inhibit the totality of nitrous oxide reductase present. In both the field situation and with soil columns, care must be taken to ensure complete admixture of the acetylene gas with the soil atmosphere. This limitation is usually not observed with soil slurries, because the suspensions can be agitated at sufficient rates to guarantee distribution of the inhibitory gas throughout the mixture.

A further complication associated with long-term use of the acetylene inhibition procedure in field or column studies is that it is toxic to ammonium oxidizers (e.g., Berg et al., 1982), thereby inhibiting the natural source of nitrate for denitrification. Thus, implementing this procedure in the field or with long-term incubated soil columns requires sufficient duplication of sampling sites that recovery of the nitrifier population can occur between acetylene amendments. Additional difficulties may be encountered in the field should the acetylene amendment stimulate carbon mineralization, thereby augmenting nitrate reduction (Haider et al., 1983). Nitrate amendment to soil samples may stimulate both aerobic and anaerobic acetylene oxidation. Based on this observation, Haider et al. (1983) conclude that the response should be as was indicated by the inhibition of nitrifiers by acetylene, repeated application of acetylene to the same field plot should be avoided. Note that these complications are not universally associated with field implementation of the acetylene inhibition procedure. Ryden and Dawson (1982) found no difficulty with its repeated use on grassland soils. In conclusion, these data further underscore the necessity of conducting appropriate controls to ensure that the impact of acetylene on the soil system is direct (i.e., nitrous oxide changes because of nitrous oxide reductase inhibition) and not indirectly (i.e., changes occur as a result of microbial populations dissociated from nitrate reduction).

13.6. ENVIRONMENTAL FACTORS CONTROLLING DENITRIFICATION RATES

The primary factors that have been shown to control denitrification processes in the field are a) the nature and amount of organic matter available as energy sources to the denitrifiers, b) the soil nitrate concentration, c) the aeration/moisture status of the soil, d) soil pH, and e) soil temperature. Each of these factors and examples of their effect in native soil systems is evaluated here.

13.6.1. Nature and Amount of Organic Matter

Most members of the soil denitrifier population derive their energy for cell maintenance and growth from oxidation of photosynthetically fixed carbon. Although colloidal organic matter is a primary soil component, amendment of soil with metabolizable carbon generally stimulates denitrification (e.g., de Catanzaro and Beauchamp, 1975; Stanford et al., 1975b). For such studies, saturating concentrations of nitrate are usually amended to soil slurries that concurrently receive varying concentrations of a carbon source, such as glucose. The effect of carbon amendment on denitrification generally resembles Michaelis–Menten kinetics (e.g., Bowman and Focht, 1974; Reddy et al., 1982). Bowman and Focht (1974) found a Michaelis constant of 500 μg gluocse ml^{-1} when nitrate reduction was assessed in a Coachella fine sand. This stimulation of denitrification by carbon amendment suggests that the process is carbon limited. Further evidence for carbon limitations of denitrification is derived from the observation that denitrification rates generally correlate with quantities of total organic carbon in soil as well as with the available carbon, which is commonly estimated by measuring biological oxygen demand of soil water extracts (e.g., Beauchamp et al., 1980; Burford and Bremner, 1975; Katz et al., 1985; Stanford et al., 1975c). Exceptions to this generalization are associated with mineral soils receiving high organic matter inputs (e.g., sludge-amended soils).

This correlation of denitrification rates in soil with available organic matter leads to the prediction that treatment of soil to increase available native soil organic matter should result in an increase in the denitrification rate. Soil preparation procedures, such as freezing (e.g., McGarity, 1962) and air-drying (e.g., Patten et al., 1980) can result in an increase in the denitrification rate detected with the soil samples. Thus, even with the most careful handling of field soil samples, the possibility that the laboratory-derived results are an overestimation of field denitrification rates must be considered.

An interesting possibility based on the observed proportionality of denitrification rates to available fixed-carbon substrates is the potential for stimulating denitrification in the rhizosphere. Root exudates can be predicted to provide the energy source for the denitrifiers, and anoxic microsites can be anticipated to occur in the vicinity of the roots where these organisms can function. Addition of macerated roots to soils samples as well as invasion of soil by root tissue has been shown to stimulate denitrification (e.g., Garcia, 1975; Prade and Trolldenier, 1988). Others (Haider et al., 1985, 1987; Smith and Tiedje, 1979b) did not observe a rhizosphere effect on denitrification. Smith and Tiedje (1979) found that variation of soil nitrate concentrations did not alter their conclusion that denitrification was not stimulated in the rhizosphere. No effect was found with high nitrate concentrations, and a reduction of denitrification occurred in the presence of low nitrate. We can predict that, because field concentrations of nitrate are generally low, denitrification rates in the rhizosphere would be reduced over nonplant-impacted soil sites. These data exemplify the complexity of the interactions of physical, chemical, and biological factors in controlling denitrification in the rhizosphere. Although there is a clear impact of fixed-carbon concentration on the activity of denitrifiers, their function is also delimited by nitrate concentrations (for which they would be competing with plants as well as other bacterial populations) and anoxic microsites.

13.6.2. Nitrate Concentration

Because denitrification is an enzymically catalyzed process, the reaction rate of the process is anticipated to follow some form of Michaelis–Menten kinetics. That is, the rate of nitrate reduction to dinitrogen and nitrous oxide should increase until a saturating concentration is reached. (See section 5.4 for discussion of enzyme kinetics.) Once enzyme-saturating concentrations of the substrate have been reached, two alternatives are generally observed. There can be no effect of the augmented substrate levels or the substances may become inhibitory. The latter situation is generally detected when studying denitrification, where augmented nitrate concentrations inhibit nitrogen oxide reductases.

Nitrate reduction in soil is generally found to be either a zero- or first-order process. For example, McGarity (1961) found the denitrification rate to be independent of nitrate concentration (60–472 $\mu g\ g^{-1}$ in some South Australian soils). Bowman and Focht (1974) observed first-order kinetics with a Coachella fine sand. The Michaelis constant for the latter study was 170 μg nitrate ml^{-1}. Murray et al. (1989) measured Michaelis constants for mixed denitrifier populations from agricultural soils of 1.8–13.7 μM nitrate. Combinations of kinetic relationships may also be observed. Reddy et al. (1978) measured both first- and zero-order reaction kinetics with flooded soils. In their study, nitrate disappearance from the waters above the soil was zero order, whereas within the soil matrix, nitrate consumption followed first-order kinetics. In this situation, a nonbiological parameter entered into the reaction kinetics—diffusion of nitrate from the surface waters into the sediment soils. Diffusion limited the quantities of nitrate available to the soil microbes. Therefore, the available nitrate was reduced by the microbial community as quickly as it entered into their microsite. Myrold and Tiedje (1985) suggest diffusional limitation of nitrate to denitrifiers in large soil aggregates.

At high concentrations, nitrate inhibits nitrous oxide reductase. This inhibition varies with soil pH and history of the soil. The greater effect of high nitrate is detected at low pH, whereas extended flooding of the soil relieves the inhibitory effect of nitrate on the reductase.

13.6.3. Aeration/Moisture

As noted previously, when the biochemistry of denitrification was analyzed, oxygen represses the synthesis of the reductases associated with denitrification. As soil oxygen tensions are reduced, reduction of nitrate to nitrous oxide and dinitrogen increases (e.g., Allison et al., 1960; Cady and Bartholomew, 1961). Because of the slow rate of diffusion of molecular oxygen in water compared to air, this effect of aeration on denitrification rates in soils is highly related to soil moisture. In general, denitrification is not detected below 60 percent water-holding capacity. Above this value, the denitrification rate generally correlates with soil moisture.

The field implication of this observation is that we can anticipate that the majority of nitrate is denitrified in soil under flooded conditions. Generally low denitrification rates are observed continuously in a drained soil system, but maximal activity is associated with increases in moisture due to rainfall (Sexstone et al., 1985a). In the field, as soil moisture increases, losses of nitrate through denitrifier activity increases (e.g., Kroeckel and Stolp, 1988; Pilot and Patrick, 1972). This sensitivity of denitrifiers to increasing soil moisture explains a significant amount of the augmented loss of nitrate from no-till soils due to denitrifier activity compared to that occurring in conventionally tilled agricultural systems (Aulakh et al., 1984a, 1984b; Rice and Smith, 1982). This increase in denitrification results from augmentation of the number of anaerobic microsites in the soil. In other words, the low amount of denitrification occurring in drained soils occurs in anaerobic microsites and is *not* the result of activity of molecular oxygen resistant denitrifiers. Sexstone et al. (1985b) supported this conclusion by showing anaerobic centers within all soil aggregates wherein denitrification occurred. Anaerobiosis was not the sole factor controlling the denitrification rate in that this rate did not correlate with the size of the anaerobic zone.

13.6.4. pH

Although denitrification occurs within the pH range of approximately 3.9–9.0, maximum nitrogen oxide reduction occurs from pH 7.0–8.0. The rate decreases as the pH is lowered (e.g., Bremner and Shaw, 1958b; Focht, 1974; Waring and Gilliam, 1983). Acidophilic or acid-tolerant denitrifier populations are apparently selected in soils with histories of low pH. Parkin et al. (1985b) noted two distinctly different pH optima with soils of pH 3.9 and 6.3, which approximated the native pH values of the soils. Note that the acidic soil used in the latter study had a 20-year history of low pH, which would have provided more than adequate time for developing an acid-resistant soil microbial population.

Because of the occurrence of chemical denitrification in extremely acidic soils (e.g., Bollag et al., 1973), assessment of denitrification in such soils is difficult. Fortunately, the predominant products of the chemical and biological processes differ. Nitric oxide and nitrogen dioxide predominate from chemical reduction of nitrite, whereas nitrous oxide and dinitrogen are biologically produced (Bollag et al., 1973). Unfortunately, dinitrogen and nitrous oxide are also reported among the products of chemodenitrification (e.g., Bulla et al., 1970; Reuss and Smith, 1965). When the two processes are separated, significant biological reduction of nitrous oxides can be shown to occur. Muller et al. (1980) found that biological denitrification rates varied from 0.12 to 53.8 μg day^{-1} in a variety of low pH (pH $\geq$ 3.6) spodosols and peats collected from southern Finland. Complete inhibition of the reduction of nitrous oxide to dinitrogen was found in 99.3 percent of these soil samples. Similarly, Gilliam and Gambrell (1978) found that acidity (pH values as low as 4.5) was not a serious limitation to denitrification in Atlantic coastal plain soils (USA). As with the previous study, nitrous oxide was a major product of denitrification in these soils (Weier and Gilliam, 1986). Similarly, in a study of a silt loam soil with pH varying from 4.6 to 6.9, nitrous oxide comprised 83 percent of the gaseous nitrogen products of denitrification at pH 4.6 and 5.4. At pH 6.9, dinitrogen was the predominant product.

Frequently, several soil physical or chemical properties interact to retard denitrification under acidic conditions. For example, Dubey and Fox (1974) found that low pH combined with low soil organic matter precluded denitrification in humid tropical soils of Puerto Rico. Similarly, George and Antoine (1982) noted that the pH optimum varied with temperature in their studies of a salt marsh soil. In their study, nitrous oxide was produced only at low pH and nitrate concentrations.

13.6.5. Temperature

The minimum temperature for denitrification is generally associated with the occurrence of free water. Smid and Beauchamp (1976) from study of denitrification in the Ap horizon of a Huron clay loam suggested denitrification occurred at or near 0°C. Incubation of the same soil under anaerobic conditions resulted in a complete inhibition of denitrification at 5°C with a gradual increase in the rate to 30°C (Bailey and Beauchamp, 1973). Jacobson and Alexander in a study of two soils maintained in an anaerobic atmosphere (1980) found no nitrate-reducing activity at 1°C. Nitrate was slowly reduced at 7°C. As with the previous studies, the rate increased with increasing soil temperature. These data suggest a minimum temperature for denitrification approaching the freezing point for water with an apparent impact of the presence of molecular oxygen on the limiting temperature. The maximum temperature is approximately that value limiting biological activity in soil in general, 75°C. The optimum temperature (60–70°C) is well above the normally occurring soil temperature range. In the range of increasing temperature, the nitrate reduction rate increases at a rate approximating that of general soil biological processes. In other words, a Q_{10} of approximately 2 describes the relationship between increasing temperature and reaction rate (e.g., Stanford et al., 1975a).

These data indicate that significant denitrification can be anticipated to occur in any soil ecosystem at temperatures where general biological activity can be anticipated to occur. The Q_{10} value of about 2 indicates that the process is reasonably sensitive to slight fluctuations in soil temperature. Thus, not only should a seasonal variation in this process be anticipated, but a diurnal impact also should be anticipated. In other words, losses of fixed nitrogen as nitrous oxide and dinitrogen will necessarily vary with daily heating and cooling of the upper portions of the soil profile.

13.6.6. Interaction of Limitors to Denitrification in Soil Systems

Strictly considering each of the soil properties previously listed and their impact on denitrification processes provides a limited view of the variation of nitrate reduction in field sites. In reality, each of these variables is sufficiently different across the soil landscape to provide significant landscape-scale variation in denitrification rates. Pennock et al. (1992) conclude that landscape analysis should be incorporated into models of soil-fixed-nitrogen losses through denitrification because topography had a major impact on *in situ* denitrification rates. In their study, when the site was considered as a whole, variables that were most influential on nitrogen losses were volumetric water content and soil redox potential. In the level portions of the site, volumetric water content was most highly correlated with denitrification, whereas on the shoulder and foot slopes, respiration and bulk density were the most influential. Parkin (1987) found that "hot spots" of denitrification activity were associated with pockets of particulate organic matter. Parsons et al. (1991) found coefficients of variation between 74 and 268 percent for spatial variability of nitrogen gas loss from agricultural soils of central Kentucky (USA). Increases in nitrate reduction activity were related to increases in soil moisture and soil respiration. These data support the conclusion that, for a complete understanding of denitrification kinetics in a soil system to be developed, we must study the heterogeneity of the system, particularly with regard to soil moisture, organic matter, and most likely to some degree, nitrate distribution.

13.7. CONCLUDING COMMENTS

An understanding of the nuances of the impact of soil properties on biological denitrification and management of this process to regulate soil nitrate loadings is essential for optimizing agricultural production as well as for minimizing the potential for soil and water pollution. This terminal stage in the nitrogen cycle (return of fixed nitrogen to atmospheric reservoirs) is catalyzed by a diverse group of soil bacteria that are capable of using light or mineral (inorganic) or organic compounds as energy sources. The only readily discernible commonalties among denitrifiers are a) they are solely bacteria, b) they can reduce nitrogen oxides in the absence of free oxygen, and c) these nitrogen oxides serve as terminal electron acceptors for bacterial respiration. Thus, the primary determinants of denitrification rates in the field are the availability of energy sources and nitrate and the absence of molecular oxygen from the microenvironment of the bacteria. The diversity of the organisms involved leads to the prediction that if life is possible in the system, a denitrifier that can function therein should exist. This adaptability and ubiquity of denitrifiers supports the conclusion that denitrification is a process amenable to management for control of soil nitrate loadings.

REFERENCES

Allison, F. E., J. N. Carter, and L. D. Sterling. 1960. The effect of partial pressure of oxygen on denitrification in soil. Soil Sci. Soc. Am. Proc. 24:283–285.

Aulakh, M. S., J. W. Doran, and A. R. Mosier. 1991. Field evaluation of four methods for measuring denitrification. Soil Sci. Soc. Am. J. 55:1332–1338.

Aulakh, M. S., D. A. Rennie, and E. A. Paul. 1984a. Gaseous nitrogen losses from soils under zero-till as compared with conventional-till management systems. J. Environ. Qual. 13:130–136.

Aulakh, M. S., D. A. Rennie, and E. A. Paul. 1984b. The influence of plant residues on denitrification rates in conventional and zero tilled soils. Soil Sci. Soc. Am. J. 48:790–794.

Bailey, L. D., and E. G. Beauchamp. 1973. Effects of temperature on NO_3^- and NO_2^- reduction, nitrogenous gas production, and redox potential in a saturated soil. Can. J. Soil Sci. 53:213–218.

Ballard, A. L., and S. J. Ferguson. 1988. Respiratory nitrate reductase from *Paracoccus denitrificans*. Evidence for two b-type haems in they γ subunit and properties of a water-soluble active enzyme containing α and β subunits. Eur. J. Biochem. 174:207–212.

Bazylinski, D. A., E. Palome, N. A. Blakemore, and R. P. Blakemore. 1986. Denitrification by *Chromobacterium violacium*. Appl. Environ. Microbiol. 52:696–699.

Beauchamp, E. G., C. Gale, and J. C. Yoemans. 1980. Organic matter availability for denitrification in soils of different textures and drainage classes. Commun. Soil Sci. Plant Anal. 11:1221–1233.

Berg, P., L. Klemedtsson, and T. Rosswall. 1982. Inhibitory effect of low partial pressures of acetylene on nitrification. Soil Biol. Biochem. 14:301–303.

Betlach, M. R., and J. M. Tiedje. 1981. Kinetic explanation for accumulation of nitrite, nitric oxide, and nitrous oxide during bacterial denitrification. Appl. Environ. Microbiol. 42:1074–1084.

Blackmer, A. M., and J. M. Bremner. 1977. Nitrogen isotope discrimination in denitrification of nitrate in soils. Soil Biol. Bicohem. 9:73–77.

Blackmer, A. M., and J. M. Bremner. 1978. Inhibitory effect of nitrate on reduction of N_2O to N_2 by soil microorganisms. Soil Biol. Biochem. 10:187–191.

Blackmer, A. M., and J. M. Bremner. 1979. Stimulatory effect of nitrate on reduction of N_2O to N_2 by soil microorganisms. Soil Biol. Biochem. 11:313–315.

Blackmer, A. M., J. M. Bremner, and E. L. Schmidt. 1980. Production of nitrous oxide by ammonia-oxidizing chemoautotrophic microorganisms in soil. Appl. Environ. Microbiol. 40:1060–1066.

Blackmer, A. M., S. G. Robbins, and J. M. Bremner. 1982. Diurnal variability in rate of emission of nitrous oxide from soils. Soil Sci. Soc. Am. J. 46:937–942.

Bleakley, B. H., and J. M. Tiedje. 1982. Nitrous oxide production by organisms other than nitrifiers or denitrifiers. Appl. Environ. Microbiol. 44:1342–1348.

Boast, C. W., R. L. Mulvaney, and P. Baveye. 1988. Evaluation of nitrogen-15 tracer techniques for direct measurement of denitrification in soil: I. Theory. Soil Sci. Soc. Am. J. 52:1317–1322.

Bolgiano, B., L. Smith, and H. C. Davies. 1989. Electron transport reactions in a cytochrome *c*-deficient mutant of *Paracoccus denitrificans*. Biochim. Biophys. Acta 973:227–234.

Bollag, J.-M., S. Drzymala, and L. T. Kardos. 1973. Biological versus chemical nitrite decomposition in soil. Soil Sci. 116:44–50.

Bollag, J.-M., and G. Tung. 1972. Nitrous oxide release by soil fungi. Soil Biol. Biochem. 4:271–276.

Bossert, I. D., and L. Y. Young. 1986. Anaerobic oxidation of *p*-cresol by a denitrifying bacterium. Appl. Environ. Microbiol. 52:1117–1122.

Bowman, R. A., and D. D. Focht. 1974. The influence of glucose and nitrate concentrations upon denitrification rates in sandy soils. Soil Biol. Biochem. 6:297–301.

Braun, K., and D. T. Gibson. 1984. Anaerobic degradation of 2-aminobenzoate (anthranilic acid) by denitrifying bacteria. Appl. Environ. Microbiol. 48:102–107.

Bremner, J. M., and K. Shaw. 1958a. Denitrification in soil. I. Methods of investigation. J. Agr. Sci. 51:22–39.

Bremner, J. M., and K. Shaw. 1958b. Denitrification in soil. II. Factors affecting denitrification. J. Agr. Sci. 51:40–52.

Bulla, L. A., Jr., C. M. Gilmour, and W. B. Bollen. 1970. Non-biological reduction of nitrite in soil. Nature (London) 225:664.

Buresh, R. J., and W. H. Patrick, Jr. 1978. Nitrate reduction to ammonium in anaerobic soil. Soil Sci. Soc. Am. J. 42:913–918.

Burford, J. R., and J. M. Bremner. 1975. Relationships between the denitrification capacities of soil and total, water-soluble and readily decomposible soil organic matter. Soil Biol. Biochem. 7:389–394.

Cady, F. B., and W. V. Bartholomew. 1960. Sequential products of anaerobic denitrification in Norfolk soil material. Soil Sci. Soc. Am. Proc. 24:477–482.

Cady, F. B., and W. V. Bartholomew. 1961. Influences of low pO_2 on denitrification processes and products. Soil Sci. Soc. Am. Proc. 25:362–365.

Carlson, C. A., and J. L. Ingraham. 1983. Comparison of denitrification by *Pseudomonas stutzeri, Pseudomonas aeruginosa,* and *Paracoccus denitrificans*. Appl. Environ. Microbiol. 45:1247–1253.

Caskey, W. H., and J. M. Tiedje. 1980. The reduction of nitrate to ammonium by a *Clostridium* sp. isolated from soil. J. Gen. Microbiol. 119:217–223.

Chan, Y.-K., L. R. Barran, and E. S. P. Bromfield. 1989. Denitrification activity of phage types representative of two populations of indigenous *Rhizobium meliloti*. Can. J. Microbiol. 35:737–740.

Chauret, C., W. L. Barraquio, and R. Knowles. 1992. Molybdenum incorporation in denitrifying *Azospirillum brasilense* Sp7. Can. J. Microbiol. 38:1042–1047.

Cooper, G. S., and R. L. Smith. 1963. Sequence of products formed during denitrification in some diverse western soils. Soil Sci. Soc. Am. Proc. 27:659–662.

Coyne, M. S., and D. D. Focht. 1987. Nitrous oxide reduction in nodules: Denitification or N_2 fixation? Appl. Environ. Microbiol. 53:1168–1170.

Daniel, R. M., I. M. Smith, J. A. D. Phillip, H. D. Ratcliffe, J. W. Drozd, and T. A. Bull. 1980. Anaerobic growth and denitrification by *Rhizobium japonicum* and other rhizobia. J. Gen. Microbiol. 120:517–521.

Davidson, E. A., and W. T. Swank. 1986. Environmental parameters regulating gaseous nitrogen losses from two forested ecosystems via nitrification and denitrification. Appl. Environ. Microbiol. 52:1287–1292.

Davidson, E. A., W. T. Swank, and T. O. Perry. 1986. Distinguishing between nitrification and denitrification as sources of gaseous nitrogen production in soil. Appl. Environ. Microbiol. 52:1280–1286.

de Catanzaro, J. B., and E. G. Beauchamp. 1985. The effect of some carbon substrates on denitrification rates and carbon utilization in soil. Biol. Fert. Soil 1:183–187.

Dubey, H. D., and R. H. Fox. 1974. Denitrification losses from humid tropical soils of Puerto Rico. Soil Sci. Soc. Am. Proc. 38:917–920.

Dunstan, R. H., B. C. Kelley, and D. J. D. Nicholas. 1982. Fixation of dinitrogen derived from denitrification of nitrate in a photosynthetic bacterium, *Rhodopseudomonas sphaeroides* forma sp. *denitrificans*. J. Bacteriol. 150:100–104.

Evans, P. J., W. Ling, N. J. Palleroni, and L. Y. Young. 1992. Quantification of identification by strain T1 during anaerobic degradation of toluene. Appl. Microbiol. Biotechnol. 37:136–140.

Evans, W. C., and G. Fuchs. 1988. Anaerobic degradation of aromatic compounds. Ann. Rev. Microbiol. 42:289–317.

Fazzolari, E., A. Mariotti, and J. C. Germon. 1990. Dissimilatory ammonia reduction vs denitrification in vitro and in inoculated agricultural soil samples. Can. J. Microbiol. 36:786–793.

Firestone, M. K., R. B. Firestone, and J. M. Tiedje. 1980. Nitrous oxide from soil denitrification: Factors controlling its biological production. Science 208:749–751.

Firestone, M. K., M. S. Smith, R. B. Firestone, and J. M. Tiedje. 1979. The influence of nitrate, nitrite, and oxygen on the composition of the gaseous products of denitrification in soil. Soil Sci. Soc. Am. J. 43:1140–1144.

Focht, D. D. 1974. The effect of temperature, pH, and aeration on the production of nitrous oxide and gaseous nitrogen—a zero order kinetic model. Soil Sci. 118:173–179.

Focht, D. D., and H. Joseph. 1973. An improved method for the enumeration of denitrifying bacteria. Soil Sci. Soc. Am. Proc. 37:698–699.

Focht, D. D., and L. H. Stolzy. 1978. Long-term denitrification studies in soils fertilized with $(^{15}NH_4)_2SO_4$. Soil Sci. Soc. Am. J. 42:894–898.

Freney, J. R., O. T. Denmead, and J. R. Simpson. 1979. Nitrous oxide emission from soils at low moisture contents. Soil Biol. Biochem. 11:167–173.

Gamble, T. M., M. R. Betlach, and J. M. Tiedje. 1977. Numerically dominant denitrifying bacteria from world soils. Appl. Environ. Microbiol. 33:926–939.

Garber, E. A. E., and T. C. Hollocher. 1981. ^{15}N tracer studies on the role of NO in denitrification. J. Biol. Chem. 256:5459–5465.

Garber, E. A. E., and T. C. Hollocher. 1982. ^{15}N, ^{18}O tracer studies on the activation of nitrite by denitrifying bacteria. J. Biol. Chem. 257:8091–8097.

Garcia, J.-L. 1975. Effet rhizosphère du riz sur la denitrification. Soil Biol. Biochem. 7:139–141.

George, U. S., and A. D. Antoine. 1982. Denitrification potential of a salt marsh soil: Effect of temperature, pH and substrate concentration. Soil Biol. Biochem. 14:117–125.

Gilliam, J. W., and R. P. Gambrell. 1978. Temperature and pH as limiting factors in loss of nitrate from saturated Atlantic coastal plain soils. J. Environ. Qual. 7:526–532.

Goodroad, L. L., and D. R. Kenney. 1984. Nitrous oxide emission from forest, marsh, and prairie ecosystems. J. Environ. Qual. 13:448–452.

Goodroad, L. L., D. R. Keeney, and L. A. Peterson. 1984. Nitrous oxide emissions from agricultural soils in Wisconsin. J. Environ. Qual. 13:557–561.

Haider, K., A. R. Mosier, and O. Heinemeyer. 1983. Side effects of acetylene on the conversion of nitrate in soil. Z. Pflanzenernaehr. Bodenk. 146:623–633.

Haider, K., A. Mosier, and O. Heinemeyer. 1985. Phytotron experiments to evaluate the effect of growing plants on denitrification. Soil Sci. Soc. Am. J. 49:636–641.

Haider, K., A. Mosier, and O. Heinemeyer. 1987. The effect of growing plants on denitrification at high soil nitrate concentrations. Soil Sci. Soc. Am. J. 51:97–102.

Hernandez, D., and J. J. Rowe. 1987. Oxygen regulation of nitrate uptake in denitrifying *Pseudomonas aeruginosa*. Appl. Environ. Microbiol. 53:745–750.

Hochstein, L. I., and G. A. Tomlinson. 1988. The enzymes associated with denitrification. Ann. Rev. Microbiol. 42:231–261.

Hollocher, T. C., E. Garber, A. J. L. Cooper, and R. E. Reiman. 1980. [13]N, ^{15}N isotope and kinetic evidence against hyponitrite as an intermediate in denitrification. J. Biol. Chem. 255:5027–5030.

Hollocher, T. C., and J. K. Kristjánsson, 1992. Thermophilic denitrifying bacteria: A survey of hot springs of Southwestern Iceland. FEMS Microbiol. Ecol. 101:113–119.

Jacobson, S. N., and M. Alexander. 1980. Nitrate loss from soil in relation to temperature, carbon source and denitrifier populations. Soil Biol. Biochem. 12:501–505.

Katz, R., J. Hagin, and L. T. Kurtz. 1985. Participation of soluble and oxidizable soil organic compounds in denitrification. Biol. Fert. Soils 1:209–213.

Knowles, R. 1981. Denitrification. *In* E. A. Paul and J. N. Ladd (eds.), Soil Biochemistry, Vol. 5, pp. 323–369. Dekker. New York.

Knowles, R. 1982. Denitrification. Microbiol. Rev. 46:43–70.

Koike, I., and A. Hattori. 1978. Denitrification and ammonia formation in anaerobic coastal sediments. Appl. Environ. Microbiol 35:278–282.

Koskinen, W. C., and D. R. Keeney. 1982. Effect of pH on the rate of gaseous products of denitrification in a silt loam soil. Soil Sci. Soc. Am. J. 46:1165–1167.

Kroeckel, L., and H. Stolp. 1988. Influence of oxygen on denitrification and aerobic respiration in soil. Biol. Fert. Soils 1:189–193.

Letey, J., N. Valoras, D. D. Focht, and J. C. Ryden. 1981. Nitrous oxide production and reduction during denitrification as affected by redox potential. Soil Sci. Soc. Am. J. 45:727–730.

Linne von Berg, K.-H., and H. Bothe. 1992. The distribution of denitrifying bacteria in soils monitored by DNA-probing. FEMS Microbiol. Ecol. 86:331–340.

McGarity, J. W. 1961. Denitrification studies on some South Australian soils. Plant Soil 14:1–21.

McGarity, J. W. 1962. Effect of freezing of soil on denitrification. Nature (London) 196:1342–1343.

Mosier, A. R., W. D. Guenzi, and E. Z. Schweizer. 1986. Field denitrification estimates by nitrogen-15 and acetylene inhibition techniques. Soil Sci. Soc. Am. J. 50:831–833.

Mosier, A. R., and G. L. Hutchinson. 1981. Nitrous oxide emissions from cropped fields. J. Environ. Qual. 10:169–173.

Mosier, A. R., M. Stillwell, W. J. Parton, and R. G. Woodmansee. 1981. Nitrous oxide emissions from a native shortgrass prairie. Soil Sci. Soc. Am. J. 45:617–619.

Muller, M. M., V. Sundman, and J. Skujins. 1980. Denitrificatoin in low pH spodosols and peats determined with the acetylene inhibition method. Appl. Environ. Microbiol. 40:235–239.

Murray, R. E., L. L. Parsons, and M. S. Smith. 1989. Kinetics of nitrate utilization by mixed populations of denitrifying bacteria. Appl. Environ. Microbiol. 55:717–721.

Myrold, D. D., and J. M. Tiedje. 1985. Diffusional constraints on denitrification in soil. Soil Sci. Soc. Am. J. 49:651–657.

Neyra, C. A., J. Döbereiner, R. Lalande, and R. Knowles. 1977. Denitrification by N_2-fixing *Spirillum lipoferum*. Can. J. Microbiol. 23:300.

Parkin, T. B. 1987. Soil microsites as a source of denitrification variability. Soil Sci. Soc. Am. J. 51:1194–1199.

Parkin, T. B., A. J. Sextone, and J. M. Tiedje. 1985a. Comparison of field denitrification rates determined by acetylene-based soil core and nitrogen-15 methods. Soil Sci. Soc. Am. J. 49:94–99.

Parkin, T. B., A. J. Sexstone, and J. M. Tiedje. 1985b. Adaptation of denitrifying populations to low soil pH. Appl. Environ. Microbiol. 49:1053–1056.

Parsons, L. L., R. E. Murray, and M. S. Smith. 1991. Soil denitrification dynamics: Spatial and temporal variations of enzyme activity, populations, and nitrogen gas loss. Soil Sci. Soc. Am. J. 55:90–95.

Patten, D. K., J. M. Bremner, and A. M. Blackmer. 1980. Effects of drying and air-dry storage of soils on their capacity for denitrification of nitrate. Soil Sci. Soc. Am. J. 44:67–70.

Pennock, D. J., C. van Kessell, R. E. Farrell, and R. A. Sutherland. 1992. Landscape-scale variations in denitrification. Soil Sci. Soc. Am. J. 56:770–776.

Pfitzner, J., and H. G. Schlegel. 1973. Denitrification in *Hydrogenomonas eutropha* strain H16. Arch. Mikrobiol. 90:199–211.

Pilot, L., and W. H. Patrick, Jr. 1972. Nitrate reduction in soils: Effect of soil moisture tension. Soil Sci. 114:312–316.

Platen, H., and B. Schink. 1989. Anaerobic degradation of acetone and higher ketones via carboxylation by newly isolated denitrifying bacteria. J. Gen. Microbiol. 135:883–891.

Prade, K., and G. Trolldenier. 1988. Effect of wheat roots on denitrification at varying soil air-filled porosity and organic-carbon content. Biol. Fert. Soil 7:1–6.

Prakash, O., and J. C. Sadana. 1973. Metabolism of nitrate in *Achromobacter fischeri*. Can. J. Microbiol. 19:15–25.

Reddy, K. R., W. H. Patrick Jr., and R. E. Phillips. 1978. The role of nitrate diffusion in determining the order and rate of denitrification in flooded soil: I. Experimental results. Soil Sci. Soc. Am. J. 42:268–272.

Reddy, K. R., P. S. C. Rao, and R. E. Jessup. 1982. The effect of carbon mineralization on denitrification kinetics in mineral and organic soils. Soil Sci. Soc. Am. J. 46:62–68.

Rehr, B., and J.-H. Klemme. 1989. Formate dependent nitrate and nitrite reduction to ammonia by *Citrobacter freundii* and competition with denitrifying bacteria. Antonie Van Leeuwenhoek 56:311–321.

Reuss, J. O., and R. L. Smith. 1965. Chemical reactions of nitrites in acid soils. Soil Sci. Soc. Am. Proc. 29:267–270.

Rice, C. W., and M. S. Smith. 1982. Denitrification in no-till and plowed soils. Soil Sci. Soc. Am. J. 46:1168–1173.

Rolston, D. E., F. E. Broadbent, and D. A. Goldhamer. 1979. Field measurement of denitrification: II. Mass balance and sampling uncertainty. Soil Sci. Soc. Am. J. 43:703–708.

Ryden, J. C., and K. P. Dawson. 1982. Evaluation of the acetylene-inhibition technique for the measurement of denitrification in grassland soils. J. Sci. Food Agric. 33:1197–1296.

Ryden, J. C., and L. J. Lund. 1980. Nitrous oxide evolution from irrigated land. J. Environ. Qual. 9:387–393.

Sexstone, A. J., T. B. Parkin, and J. M. Tiedje. 1985a. Temporal response of soil denitrification rates to rainfall and irrigation. Soil Sci. Soc. Am. J. 49:99–103.

Sexstone, A. J., N. P. Revsbech, T. B. Parkin, and J. M. Tiedje. 1985b. Direct measurement of oxygen profiles and denitrification rates in soil aggregates. Soil Sci. Soc. Am. J. 49:645–651.

Smid, A. E., and E. G. Beauchamp. 1976. Effects of temperature and organic matter on denitrification in soil. Can. J. Soil Sci. 56:385–391.

Smith, C. J., M. F. Wright, and W. H. Patrick. Jr. 1983. The effect of soil redox potential and pH on the reduction and production of nitrous oxide. J. Environ. Qual. 12:186–188.

Smith, M. S. 1982. Dissimilatory reduction of NO_2- to NH_4^+ and N_2O by a soil *Citrobacter* sp. Appl. Environ. Microbiol. 43:854–860.

Smith, M. S., M. K. Firestone, and J. M. Tiedje. 1978. The acetylene inhibition method for short-term measurement of soil denitrification and its evaluation using nitrogen-13. Soil Sci. Soc. Am. J. 42:611–615.

Smith, M. S., and J. M. Tiedje. 1979a. Phases of denitrification following oxygen depletion in soil. Soil Biol. Biochem. 11:261–267.

Smith, M. S., and J. M. Tiedje. 1979b. The effect of roots on soil denitrification. Soil Sci. Soc. Am. J. 43:951–955.

Smith, M. S., and K. Zimmerman. 1981. Nitrous oxide production by nondenitrifying soil nitrate reducers. Soil Sci. Soc. Am. J. 45:865–871.

Stanford, G., S. Dzienia, and R. A. Vander Pol. 1975a. Effect of temperature on denitrification rate in soils. Soil Sci. Soc. Am. Proc. 39:867–870.

Stanford, G., J. O. Legg, S. Dzienia, and E. C. Simpson. Jr., 1975b. Denitrification and associated nitrogen transformations in soils. Soil Sci. 120:147–152.

Stanford, G., R. A. Vander Pol, and D. Dzienia. 1975c. Denitrification rates in relation to total and extractable soil carbon. Soil Sci. Soc. Am. Proc. 39:284–289.

Terry, R. E., and R. L. Tate III. 1980. The effect of nitrate on nitrous oxide reduction in organic soils and sediments. Soil Sci. Soc. Am. J. 44:744–746.

Terry, R. E., R. L. Tate III, and J. M. Duxbury. 1981a. Nitrous oxide emissions from drained, cultivated organic soils of South Florida. Air Pollut. Contr. Assoc. J. 31:1173–1176.

Terry, R. E., R. L. Tate III, and J. M. Duxbury. 1981b. The effect of flooding on nitrous oxide emissions from an organic soil. Soil Sci. 132:228–232.

Tiedje, J. M., R. B. Firestone, M. K. Firestone, M. R. Betlach, M. S. Smith, and W. H. Caskey. 1979. Methods for the production and use of nitrogen-13 in studies of denitrification. Soil Sci. Soc. Am. J. 43:709–715.

van Berkum, P., and H. H. Keyser. 1985. Anaerobic growth and denitrification among different serogroups of soybean rhizobia. Appl. Environ. Microbiol. 49:772–777.

Vanden Heuvel, R. M., R. L. Mulvaney, and R. G. Hoeft. 1988. Evaluation of nitrogen-15 tracer techniques for direct measurement of denitrification in soil: II. Simulation studies. Soil Sci. Soc. Am. J. 52:1322–1326.

Volz, M. G. 1977. Denitrifying bacteria can be enumerated in nitrite both. Soil Sci. Soc. Am. J. 41:549–551.

Waring, S. A., and J. W. Gilliam. 1983. The effect of acidity on nitrate reduction and denitrification in lower coastal plain soils. Soil Sci. Soc. Am. J. 47:246–251.

Weier, K. L., and J. W. Gilliam. 1986. Effect of acidity on denitrification and nitrous oxide evolution from Atlantic coastal plain soils. Soil Sci. Soc. Am. J. 50:1202–1205.

Ye, R. W., M. R. Fries, S. G. Bezborodnikov, B. A. Averill, and J. M. Tiedje. 1993. Characterization of the structural gene encoding a copper-containing nitrite reductase and homology of this gene to DNA of other denitrifiers. Appl. Environ. Microbiol. 59:250–254.

Yoshinari, T., R. Hynes, and R. Knowles. 1977. Acetylene inhibition of nitrous oxide reduction and measurement of denitrification and nitrogen fixation in soil. Soil Biol. Biochem. 9:177–183.

Yoshinari, T., and R. Knowles. 1976. Acetylene inhibition of nitrous oxide reduction by denitrifying bacteria. Biochem. Biophys. Res. Commun. 69:705–710.

Zablotowicz, R. M., D. L. Eskew, and D. D. Focht. 1978. Denitrification in *Rhizobium*. Can. J. Microbiol. 24:757–760.

14 The Sulfur and Related Biogeochemical Cycles

Biogeochemical cycles can easily be grouped into two categories based on the significance of soil minerals to nutrient availability. With the carbon and nitrogen cycles, the soil-based portions of the cycles are strongly biologically based. Once carbon dioxide and dinitrogen are incorporated into biomass, the bulk of the cycling processes involve interconversions of organic and water-soluble substances. Nearly all the soil carbon and nitrogen are contained within the organic components contained within the ecosystem. In contrast, with the cycling of sulfur, phosphorus, and metals, major reservoirs of the nutrients include the soil mineral fractions (e.g., pyrite and phosphatic rocks). Availability of these latter plant nutrients for biomass synthesis is determined by their rates of dissolution and formation.

Thus, this chapter elucidates the major biological processes associated with these nutrient cycles functioning at the intersection of soil minerals and the living community. Emphasis is directed toward the biological aspects of the processes, but those reactions involved in converting various minerals to organic forms are elucidated.

14.1. SULFUR IN THE SOIL ECOSYSTEM

Sulfur atoms are widely dispersed in soil being found in a wide variety of organic and inorganic compounds. These atoms are an integral part of soil humus, biomass, and minerals. Volatile sulfurous substances are among the gases of the soil atmosphere, and ionic forms contribute to the salt content of soil interstitial water. The primary water-soluble ionic sulfur form in soils is sulfate. As a gas, sulfur atoms are detected in hydrogen sulfide, which may occur at significant concentrations in the atmosphere of waterlogged soils. Sulfur contributes to the soil particulate fraction in both soil organic matter and soil minerals. Furthermore, sulfur-bearing minerals include the original source of sulfur, pyrite (FeS_2), which occurs in igneous rocks; gypsum ($CaSO_4 \cdot 2H_2O$); and epsomite ($MgSO_4 \cdot 7H_2O$).

Quantities of organic and mineral sulfur forms in soil are extremely variable, ranging from a value as low as 0.002 percent in highly leached, weathered soils of humid regions to concentrations as high as 5 percent, which are detected in calcareous and saline soils of arid and semiarid regions (Stevenson, 1986). Much of this soil sulfur is derived from internal sources, but external inputs (e.g., fertilizers, air pollutants [sulfur dioxide], acid mine drainage [sulfuric acid], and a variety of xenobiotic compounds [pesticides]) may also be significant.

Fig. 14.1. Examples of organic sulfur compounds found in soil.

The importance of the sulfur atom transformations to life-supporting processes is documented by the fact that not only are sulfur-containing organic compounds (e.g., biotin, coenzyme A, thiamine, glutathione, cysteine, cystine, methionine —Fig. 14.1) essential cellular components, but inorganic sulfur substances also serve as energy sources (e.g., sulfide, S^{2-}, and elemental sulfur, S^0), and electron acceptors (e.g., sulfate, SO_4^{2-}) for soil microbes.

A primary property of the sulfur atom contributing to its use and versatility in biological systems is its wide range of oxidation states. Common valence states of sulfur relating to biologically active molecules are -2 (e.g., in organic compounds

TABLE 14.1. Oxidation State of Major Sulfur Compounds in Soil.

Compound	Oxidation State
Organic S (R-SH)	−2
Sulfide (S^{2-})	−2
Elemental S (S^0)	0
Thiosulfate ($S_2O_3^{2-}$)	+2 (Average/S)
Tetrathionate ($S_4O_6^{2-}$)	+2 (Average/S)
Sulfur Dioxide (SO_2)	+4
Sulfite (SO_3^{2-})	+4
Sulfate (SO_4^{2-})	+6

as sulfhydrals and in mineral sulfides), 0 (elemental sulfur), and +6 (e.g., organosulfates and mineral sulfates). See Table 14.1 for further documentation of the variation in sulfur atom valence in commonly detected soil sulfur compounds. Microbes are the primary mediators for cycling the sulfur atom among these various oxidation states, although spontaneous chemical processes contribute to these transformations to a limited degree.

14.2. BIOGEOCHEMCIAL CYCLING OF SULFUR IN SOIL

A generalized schematic of the sulfur cycle is depicted in Fig. 14.2. As with those biogeochemical cycles previously evaluated, sulfur atoms are transferred between a variety of organic and inorganic reservoirs. Primary soil-resident sulfur pools are organic residues (biomass plus the decomposition products and humic substances derived from them) as well as soil minerals. In contrast to the nitrogen cycle, except in limited situations (e.g., anthropogenically generated sulfur dioxide, which is produced primarily from the consumption of fossil fuels), atmospheric contributions of sulfur compounds to this biogeochemical cycle are meager. This diversity in sulfur reservoirs and the intermixing of biological and mineral sulfur pools in soil suggests a complexity in sulfur cycling in soil not previously encountered in our study of carbon and nitrogen cycles. This situation is exemplified by the observation that three minicycles can be highlighted in this conceptual model of the soil sulfur cycle:

- **Cycling of sulfur between sulfide (and elemental sulfur) reservoirs and sulfate:** This process involves biological reduction of sulfate under anoxic conditions as well as oxidation of sulfide and elemental sulfur. The latter energy-generating process is catalyzed by soil microbes that require molecular oxygen or nitrate as their terminal electron acceptor.
- **Cycling of sulfur between soil biomass reservoirs and soluble sulfate (assimilation and mineralization):** Sulfate incorporated into biomass may be

Atmosphere

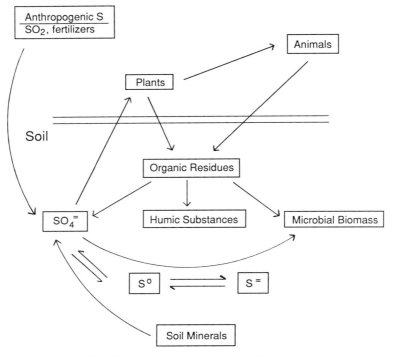

Fig. 14.2. Major features of the sulfur cycle.

derived from mineralization of native soil organic matter and plant and animal residues. Organic sulfur of biomass may be mineralized prior to its incorporation into newly synthesized cells. The term *may* is used in this situation, because microbial cells may incorporate plant biomass substituents directly into their cellular biomass.

· **Complexing of sulfide and sulfate into soil minerals and their solubilization to biologically available substances:** Some mineral sulfur may be lost from the foregoing mineralization–assimilation cycle through transfer to water-insoluble reservoirs. Sulfide accumulation in soil minerals has been defined as **sulfidization,** whereas the exposure of sulfide-bearing minerals to oxidizing conditions resulting in production of sulfate is **sulfurization.** These processes may involve dissolution or precipitation of the minerals through chemical processes (e.g., precipitation of sulfide by iron or solubilization of minerals under acidic conditions) as well as biological oxidation or reduction of the primary soil mineral components.

This conceptual model of the sulfur cycle emphasizes the primary sulfur resources in soil ecosystems. Examples of other sulfur-containing reservoirs that may be of significant environmental importance, at least on a localized basis, and that are not depicted in this specific model include various waste products of society, which may be incorporated into soil (e.g., sludges, composts, garbage, fly ash), agricultural chemicals (including pesticides), and air pollutants (e.g., products of fossil fuel combustion in electrical power generation and in automobile exhausts). These substances enter the soil sulfur cycle through reactions similar to those already depicted. For example, sulfur dioxide from fossil fuel consumption enters the interstitial water where it is oxidized to sulfate, whereas all biodegradable organic sulfur compounds are mineralized as is shown for native biomass of the ecosystem.

The most obvious atmospheric contribution to the sulfur cycle is the previously mentioned sulfur dioxide, but we must recognize that particulate (dry deposition) sulfur inputs may also be significant on a localized basis. As with the sulfur dioxide, particulate sulfur inputs (primarily as elemental sulfur, but also as various sulfates) contribute to soil acidity by increasing resident sulfate concentrations. An extreme example of the impact of deposition of elemental sulfur on an ecosystem is provided by studies of forest ecosystems in Alberta, Canada, which received large amounts of industrially produced sulfur. Kennedy et al. (1988) and Maynard et al. (1986) both report dramatic declines in aboveground vegetation quantity and diversity due to increasing soil sulfur loadings. Sulfur inputs reported by Maynard et al. (1986) ranging from 4100 to 51,400 $\mu g\, g^{-1}$ soil resulted in decreasing soil pH (4.4 to 2.4) and augmentation of the *Thiobacillus thiooxidans* populations. Visser and Parkinson (1989) in their study of forest soils receiving comparable elemental sulfur loadings found a decline in microbial biomass that correlated with soil pH and a parallel reduction in the ability of the soil community to mineralize glucose. The environmental impact of the perturbation of the sulfur cycle resulting from deposition of unusually large quantities of elemental sulfur are a logical outgrowth of the capacity of the native soil microbial population to use this sulfur substance as an energy source.

14.3. BIOLOGICAL SULFUR OXIDATION

Chemoorganotrophic and chemolithotrophic bacteria derive their energy from oxidation of organic and mineral substrates. The large range of oxidation states of the sulfur atom (-2 to $+6$) make it a nearly ideal candidate as a primary energy source for resourceful soil microbes. In soil ecosystems, a variety of soil microbes can derive their energy from oxidation of inorganic sulfur compounds. This exploitation of sulfur oxidation is a minor biogeochemical process in that terrestrial ecosystems are predominantly driven by heterotrophic processes, but sulfur oxidation is an important energy source for some portions of the terrestrial ecosystem, ranging from the exotic deep sea hydrothermal vents to more mundane agricultural ecosystems receiving elemental sulfur amendments. Both sulfide and elemental sulfur are substrates in this energy-yielding process.

The diversity of the oxidation status of the sulfur atom allows for the concurrent oxidation and reduction of the same compound by the same microbe. In this disproportionation process, the sulfur atom serves both as an electron donor and an electron acceptor for the microorganism. Examples of the process are demonstrated by the disproportionation of thiosulfate and sulfite by *Desulfovibrio sulfodismutans*. Sulfate and sulfide are formed from the sulfur compounds by the following reactions:

$$S_2O_3^{2-} + H_2O \rightarrow SO_4^{2-} + H_2S$$

$$4\ SO_3^{2-} + 2H^+ \rightarrow 2SO_4^{2-} + H_2S$$

14.3.1. Microbiology of Sulfur Oxidation

Oxidation of sulfur-containing compounds, especially elemental sulfur, has a major impact on overall ecosystem function, primarily due to the fact that the terminal product of the reaction, sulfate, causes acidification of the soil. Elemental sulfur has long been used as a soil amendment when reduction of soil pH is desired. Environmental degradation has been linked with excessive oxidation of sulfur as can result from the mining of sulfide-bearing ores or atmospheric deposition of elemental sulfur due to anthropogenic activities. Hence, an understanding of the biological interactions involved with the oxidation of sulfur compounds and the nuances of environmental management of the process is essential for proper stewardship of soil resources as well as for reclamation of mismanaged systems.

The Thiobacilli: Although a diverse group of soil microbes can oxidize sulfur and sulfide for energy, members of the bacterial genus *Thiobacillus* are the primary mediators of this process in soil ecosystems. As indicated later, some studies suggest that heterotrophic bacteria and fungi may be of some importance in this biogeochemical process, but the significance of their contribution has yet to be fully documented.

The thiobacilli are chemoautotrophic or facultatively chemoautotrophic bacteria that are widely distributed in soil ecosystems. These generally motile, gram-negative bacteria are distinguished from other members of the soil community by their capacity to catalyze at least some portion of the oxidative continuum of sulfide, elemental sulfur, thiosulfate, and/or tetrathionate to sulfate. Among the energy substrates of the five most studied members of this genus (*T. thiooxidans, T. ferroxidans, T. thioparus, T. novalis,* and *T. denitrificans*), elemental sulfur, thiosulfate, tetrathionate, and sulfide are most commonly listed. Note that each species of the genus *Thiobacillus* can oxidize a specific portion of the spectrum of oxidizable sulfur compounds occurring in soil. For example, not all species or strains within a given species can oxidize elemental sulfur. Furthermore, not all *Tihobacillus* species are limited to deriving their energy from oxidation of sulfur compounds, fixed carbon and ferrous iron may also be oxidized. The 18 described species can be divided into three groups based on their sensitivity to acidity and capacity to oxidize fixed carbon, neutrophilic chemoautotrophs, neutrophilic facultative autotrophs, and acidophiles (Kuenen et al., 1992). Of the most studied members of this genus, *T. novalis* is a facultative chemoautotroph, and *T. ferrooxidans* recovers some energy from ferrous iron oxidation (Table 14.2).

TABLE 14.2. Metabolic Properties of Commonly Studied *Thiobacillus* species.

Thiobacillus species	Optimum pH	Energy Substrates	Electron Acceptors
T. thiooxidans	2.0–3.5	S^0, $S_2O_3^{2-}$, $S_4O_6^{2-}$	O_2
T. ferrooxidans	2.0–3.5	S^0, $S_2O_3^{2-}$, Fe^{2+}	O_2
T. thioparus	approx. 7.0	S^0, S^{2-}, $S_2O_3^{2-}$, $S_4O_6^{2-}$	O_2
T. novalis	approx. 7.0	$S_2O_3^{2-}$, organic compounds	O_2
T. denitrificans	approx. 7.0	S^0, S^{2-}, $S_2O_3^{2-}$, $S_4O_6^{2-}$	O_2, NO_3^-

These sulfur-oxidizing bacteria are further distinguished by the nature of their terminal electron acceptor and their variable resistance to acidic conditions. The latter trait is particularly valuable for the survival of members of the genus in that extremely acidic soils (pH 1–3 are commonly reported) may result from their formation of sulfuric acid. Most members of the genus are obligate aerobes, although *Thiobacillus denitrificans* uses oxygen or nitrate as its terminal electron acceptor, and *T. ferrooxidans* transfers electrons to ferric ion under anoxic conditions. A wide range of tolerance to acidity is characteristic of the genus. For example, *T. thiooxidans* and *T. ferrooxidans* have pH optima between 2.0 and 3.5. In contrast, *T. thioparus, T. denitrificans,* and *T. novellus* grow optimally at near neutral pH values (Table 14.2). For a more general discussion of the properties and function of *Thiobacillus* species, see Alexander (1977), Harrison (1984), Kuenen et al. (1992), Stevenson (1986), or Visniac and Santer (1957).

The kinetics of formation of sulfate by these microbes can be exemplified by a listing of the reactions involved in elemental sulfur, thiosulfate, and tetrathionate by *T. thiooxidans:*

$$2S + 3O_2 + 2H_2O \rightarrow 2H_2SO_4$$

$$Na_2S_2O_3 + 2O_2 + H_2O \rightarrow H_2SO_4 + Na_2SO_4$$

$$2Na_2S_4O_6 + 7O_2 + 6H_2O \rightarrow 6H_2SO_4 + 2Na_2SO_4$$

T. ferrooxidans oxidizes elemental sulfur and thiosulfate by reactions similar to those listed previously for *T. thiooxidans,* but this oxidation is combined with the oxidation of ferrous to ferric iron as follows:

$$12FeSO_4 + 3O_2 + 6H_2O \rightarrow 4Fe_2(SO_4)_3 + 4Fe(OH)_3$$

This species also contains enzymes capable of reducing ferric iron (Fe^{3+}) (Sugio et al., 1987), tetravalent manganese (Mn^{4+}) (Sugio et al., 1988a), Cupric (Cu^{2+}) ion (Sugio et al. 1990), and molybdic (Mo^{6+}) ion (Sugio et al., 1988b). The environmental significance of the latter processes has not been elucidated.

The oxidative process for *T. denitrificans* is coupled to the reduction of nitrate as follows:

$$5S^0 + 6KNO_3 + 2H_2O \rightarrow K_2SO_4 + 4KHSO_4 + 3N_2$$

The environmental significance of the latter process is that it provides a mechanism for sulfur oxidation in anoxic environments. Alternatively, ferric ion may also serve as a terminal electron acceptor for sulfur-oxidizing *T. ferrooxidans* growing in the absence of molecular oxygen. Pronk et al. (1992) have shown a linear relationship between ferrous iron accumulation and cell density in *T. ferroxidans* reducing ferric iron and oxidizing elemental sulfur. Suzuki et al. (1990) have also shown that several strains of *T. ferrooxidans* can oxidize elemental sulfur under anoxic conditions when using ferric ion as a terminal electron acceptor. Some of the strains oxidized the sulfur with ferric ion as the electron acceptor at rates comparable to those of the aerobic process, but with others, sulfur oxidation occurred at a rate 50 percent or more less than that occurring in the presence of oxygen.

A variety of other metal-mobilizing, sulfur-oxidiaing *Thiobacillus* species have been evaluated (e.g., *T. acidophilus,* Meulenbert et al., 1992; *T. prosperus,* Huber and Stetter, 1989; *T. versutus,* Beffa et al., 1991; and *T. cuprinus,* Huber and Stetter, 1990). Other autotrophic bacteria that catalyze sulfur oxidation under aquatic or more specialized conditions have been described. These autotrophs include phototrophic bacteria (Brune, 1989) and *Sulfolobus* species that function in hot springs (Alexander, 1977).

Heterotrophic Sulfur Oxidizers: Several heterotrophic microbes also catalyze sulfur oxidation. *Aspergillus niger, Mucor flavus,* and *Trichoderma harzianum,* for example, oxidize elemental sulfur (Grayston et al., 1986). These fungi produced substantial amounts of sulfate from sulfur oxidation with thiosulfate (*Aspergillus niger*) or thiosulfate and tetrathionate (*Mucor flavus*) accumulating as intermediates. *Aureobasidium pullulans* (de Bary) Arnaud, isolated from sycamore phylloplane that had been exposed to atmospheric pollution, oxidized elemental sulfur to dithionite, tetrathionate, and sulfate (Killham et al., 1981). Similar products are produced by yeast isolates of the genus *Rhodotorula* (Kurek, 1979, 1985) as well as bacteria (Pepper and Miller, 1978) and actinomycete (Yagi et al., 1971) isolates. Because these heterotrophic microorganisms use carbonaceous substrates for their carbon and energy source and generally accumulate oxidized sulfur compounds after the period of active growth, their role in sulfur oxidation in native soil ecosystems is questioned.

Indirect evidence suggests that there is at least some role for the heterotrophs in elemental sulfur oxidation in soil. Lawrence et al. (1988) evaluated the effect of elemental sulfur fertilization on a variety of measures of sulfur oxidation in two Grey Luvisolic soils from Canada. In one soil, increases in *Thiobacillus* populations were induced by elemental sulfur amendment, whereas in the other soil no change in these populations was detected. Heterotrophic sulfur oxidations were the most abundant sulfur oxidizers in both soils. In a related study (Lawrence and Germida, 1988), elemental sulfur oxidation in 28 Canadian soils correlated linearly with soil microbial biomass C ($r = 0.68$, $p < 0.01$) and soil respiration ($r = 0.88$, $p < 0.01$). Further evidence implicating a heterotrophic contribution to the sulfur oxidation included a) augmentation of heterotrophic biomass through glucose amendment increased sulfur oxidation rates and b) inhibition of general heterotrophic microbial activity reduced sulfur oxidation rates.

In contrast, Lee et al. (1987), in their study of 48 New Zealand soils, found that elemental sulfur amendment resulted in increases of thiobacilli populations with no change or decline in heterotrophic sulfur-oxidizing populations. These reports suggest that environmental selection is likely to be of great significance in determining the dominant populations involved in sulfur oxidation in soil ecosystems.

14.3.2. Environmental Conditions Affecting Sulfur Oxidation

Although sulfur-oxidizing microbes are considered to be ubiquitous, populations of autotrophic thiobacilli can be essentially undetectable in some soils. These populations are inducible by sulfur amendment. For example, thiobacilli population augmentation by sulfur amendment has been demonstrated with several soils from Kansas (Attoe and Olson, 1966), Canada (Janzen and Bettany, 1987a), and New Zealand (Lee et al. 1988a). The increased sulfur oxidizer population resulting from prior sulfur amendments can apparently cause increased sulfur oxidation rates in subsequent seasons (Lee et al., 1988b). These observations relate specifically to *Thiobacillus* populations. Because, as indicated previously, a variety of heterotrophic microbes also can oxidize sulfur in soil, it is reasonable to conclude that sulfur oxidation is not precluded in soils due to nonexistence of requisite microbes. Significant autotrophic or heterotrophic populations can be anticipated to occur in any biologically active soil. The rate may be limited, but at least the numbers of autotrophic microbes should increase following amendment with sulfur substrates. Therefore, it can be concluded that, because the requisite populations can be assumed to be present in any given soil ecosystem, oxidation of sulfur is dependent primarily upon occurrence of physical and chemical conditions conducive to the function of these microbes.

Physiologically, microbes require an energy source (an oxidizable substrate) and an electron acceptor. For most living organisms, these are a reduced carbon supply (e.g., glucose) and oxygen. For elemental sulfur oxidation, the sulfur itself serves as the electron source (i.e., the oxidizable substrate), and oxygen is the electron acceptor. One species of elemental sulfur oxidizer, *Thiobacillus denitrificans,* which can oxidize sulfur under anoxic conditions, has been isolated. This process is considered to be aerobic in that cytochromes are still used in the electron transport system, and the energy yield approaches that gained when oxygen serves as the electron acceptor. Although some stains of *Thiobacillus denitrificans* have been demonstrated to be unable to oxidize extracellularly offered elemental sulfur anaerobically (Schedel and Trueper, 1980), others function quite efficiently with nitrate as an electron acceptor and amended elemental sulfur as the energy source. For example, the system is sufficiently efficient to be proposed as a means of removing excess nitrate from waste water. Elemental sulfur serves as a cheap source of energy for the denitrification process (Batchelor and Lawrence, 1978). High percentages of denitrification (over 95 percent) have been demonstrated using this process (Hashimoto et al., 1987).

Elemental sulfur as an energy source in soil is limited by the capacity of the microorganism to interact with the substrate. Physical attachment of thiobacilli to sulfur particles is required for enzymatic oxidation of sulfur and is frequently the controlling factor in sulfur oxidation rate. This was shown clearly 50 years ago in a study of *Thiobacillus oxidans* by Vogler and Umbreit (1941). Espejo and Romero

(1987) demonstrated that for *Thiobacillus ferrooxidans* only bacteria attached to the sulfur particle were capable of growth. No replication occurred once the bacteria were detached from the sulfur particle. Growth of the microbes was not uniform on the sulfur particle surface but rather it proceeded at distortions present on the surface. Staining of sulfur prills with crystal violet revealed that even after more than 4 months of incubation, intact regions remained on the substrate surface, whereas destruction of the surface had occurred quite deeply at other locations on the prill.

A unique feature of *Thiobacillus ferrooxidans* was that desorbed microbes could survive through the oxidation of ferrous ion to ferric ion. Sulfur oxidation rate was not affected by the presence of ferrous ion in the growth medium (Espejo et al., 1988). For *Thiobacillus albertis* specific cellular adhesions, bacterial glycocalyx are produced to facilitate the interaction of the thiobacilli cells with the sulfur granule (Bryant et al., 1984). Elemental sulfur oxidation by the latter *Thiobacillus* species is proportional to particle size, surface area per unit weight, and the crystal microstructure of the elemental sulfur (Laishley et al., 1986).

Because the elemental sulfur particles do not dissolve in water and direct contact of the microbial cell with the sulfur particle is necessary for oxidation to proceed, it can be concluded that surface area of the sulfur particles controls their oxidation rate in soils. This has been demonstrated repeatedly in a wide variety of soils over the last several decades (e.g., see Bertramson et al., 1950; Frederick and Starkey, 1948; Janzen and Bettany, 1987b; Laishley et al., 1986; Lee et al., 1988a, 1988b; McCaskill and Blair, 1987; Nor and Tabatabai, 1977; Watkinson, 1989). For each of these studies, a size distribution of sulfur particles was added to soil and sulfate concentration or pH changes due to sulfur oxidation were quantified as a function of particle size. In each case, rapid oxidation was detected with small particles, and limited reaction of the larger particles occurred. For example, Lee et al. (1988a) found that after 340 days, 80–90 percent of elemental sulfur particles less than 0.15 mm diameter were oxidized as opposed to only 24–55 percent of the particles size ranges greater than 0.15 mm diameter. McCaskill and Blair (1987) found that with mixtures of particle sizes a biphasic oxidation curve could be detected. A reduction in conversion rate occurred once the 0.1 and 0.2 mm diameter particles were exhausted. Frederick and Starkey (1948) evaluated particles of sulfur in pipe-sealing mixtures. They found that over a 3-week incubation period sulfur oxidation was essentially nondetectable with particle sizes greater than about 3 mm mean diameter.

Along with availability of sulfur-oxidizing microbes in soil and their growth substrates, sulfur oxidation is controlled by soil moisture, pH, and temperature. Because of the physical stability of elemental sulfur particles in soil, the impact of these factors on sulfur oxidation relates to their effect on the microbes catalyzing the reactions.

In soil, the rates of oxidation of sulfur-containing compounds and elemental sulfur tend to increase with increasing soil moisture, reach an optimum level, and then rapidly declines. (e.g., Attoe and Olson, 1966; Janzen and Bettany, 1987b; Lettl et al., 1981; Moser and Olson, 1953). With slight variation dependent upon soil texture, optimum soil moisture levels are near field capacity.

Complex interactions between soil moisture, soil texture, and microbial populations alter elemental sulfur oxidation rates (e.g., see Janzen and Bettany, 1987a, 1987b;

Watkinson, 1989). In one study of soils with limiting oxidizer population densities (as indicted by responses to inoculation), elemental S conversion rate was not affected by clay contents ranging from 9 to 52 percent (McCaskill and Blair, 1987). This contrasts with data from Janzen and Bettany (1987b) who found that with 40 diverse Canadian soils, elemental sulfur oxidation rate was negatively correlated with soil clay contents. Watkinson (1989) found that the improved soil porosity resulting from amendment of phosphate rock to mineral soils augmented elemental sulfur oxidation. It can be concluded that the prevailing controller of elemental sulfur oxidation in soils of varying moisture and texture is the availability of the obligatory electron acceptor, molecular oxygen.

Although elemental sulfur is used to reduce soil acidity, the elemental sulfur oxidation rate generally increases with soil pH. Both Attoe and Olson (1966) and Janzen and Bettany (1987b) found a positive correlation of elemental sulfur oxidation rate and soil pH in a wide variety of soils. Furthermore, Lettl et al. (1981) noted that oxidation of elemental sulfur was stimulated by amendment of soil with calcium carbonate.

Elemental sulfur oxidation takes place between 4 and 55°C (Stevenson, 1986), a range well within that normally observed in soil systems, and the sulfur oxidation rate has been shown to be directly proportional to the temperature in a variety of soil systems (Ahonen and Tuovinen, 1990; Attoe and Olson, 1966; Chapman 1989; Janzen and Bettany, 1987c; Nor and Tabatabai, 1977). A wide range of Q_{10} values was reported in these studies; 1.9–3.1 (Chapman, 1989), 2.1 (Ahonen and Tuovinen, 1990), and 3.2–4.3 (Janzen and Bettany, 1987c).

14.4. BIOLOGICAL SULFUR REDUCTION PROCESSES

As is the situation with reduction of nitrate in the nitrogen cycle, sulfate can be reduced by both assimilatory and dissimilatory paths. Assimilatory sulfate reduction, which is catalyzed by a variety of organisms, including higher plants, algae, fungi, and bacteria, provides the sulfide necessary for cellular biomass production. Dissimilatory sulfate reduction, sulfidogenesis, is catalyzed by a specialized group of obligately anaerobic bacteria. In soil the organisms involved with the latter reaction are generally found in the genera *Desulfovibrio* and *Desulfotomaculum*. With the latter organisms, sulfate is serving as an electron acceptor for intermediary metabolism. As with dissimilatory nitrate reduction, the electrons are transferred through a cytochrome-based electron transport process to the terminal electron acceptor.

With dissimilatory sulfate reduction, sulfate is reduced through the transfer of eight electrons produced by the oxidation of a fixed carbon substrate as described by the following reaction:

$$SO_4^{2-} + 8e^- + 8H^+ \rightarrow H_2S + 2H_2O + 2OH^-$$

The sulfate atom is activated in this process through the consumption of adenosine triphosphate (ATP). A variety of fixed-carbon compounds may be oxidized to provide the electrons for this process. The most commonly studied energy sources are organic acids (e.g., lactate, acetate, fatty acids), but an increasing number of more complex substances (e.g., various aromatic compounds) are being added to the list of electron donors.

Dissimilatory sulfate reduction in soil systems is distinguished by its occurrence only in specialized ecosystems. Aside from the requirement for an oxidizable substrate and sulfate, the primary environmental property delimiting function of these anaerobic bacteria is anoxic conditions. Sulfidogenesis is detected only in soil ecosystems exhibiting extremely reducing conditions. A reducing potential of about -0.25 volts (E_0') is required before dissimilatory sulfate reduction occurs. Thus, this sulfate reduction process is limited to sites similar to those where carbon dioxide reduction (methanogenesis) occurs (i.e., swampy or chronically flooded soils).

Although sulfate is a common component of soil interstitial water, its presence in concentrations supportive of extensive dissimilatory sulfate reduction is limited. A typical example of a land-based ecosystem that would allow extensive sulfide generation is a marsh site inundated with sea or brackish water. The best indication of current or previously existent sulfidogenic conditions is detection of metal sulfide deposits in the soil profile or detection of hydrogen sulfide fluxes from the site.

Note that, although the primary substrate considered for sulfidogenesis is sulfate, elemental sulfur can also be reduced by a variety of microorganisms. For example, *Desulfurolobus ambivalens* has the capacity to oxidize or to reduce elemental sulfur (Zillig et al., 1986). *Desulfotomaculum termoacetoxidans* (Min and Zinder, 1990), *Desulfuromonas acetoxidans* (Gebhardt et al., 1985), and *Spirillum* 5175 N (Zoephel et al., 1988) all reduce elemental sulfur to hydrogen sulfide.

Sulfidogenesis may be manipulated to ameliorate the impact of a variety of environmental pollutants, both organic and inorganic. For example, this process can be managed to reduce organic contamination of chronically flooded soils, reduce drainage water acidity, and ameliorate the difficulties associated with heavy metal contamination.

Anaerobic Biodegradation: A variety of xenobiotic compounds containing halogenated aromatic rings may enter soil ecosystems in undesirable concentrations. Under aerobic conditions, the halogenated substances are reasonably stable, but under anoxic conditions, bacteria have been shown to be capable of removing the halogen atom. Because at least some of these substances are mineralized by the sulfidogenic bacteria, the combination of dehalogenating and sulfate-reducing populations may be used to remove this troublesome class of compounds from sites that can be easily managed under anoxic conditions. These techniques may be particularly operative in sediments polluted with xenobiotic compounds.

Reduce Acidity of Acid Mine Drainage: Acid mine drainage results from exposure of previously buried sulfide-bearing soil minerals to oxidative conditions in the presence of an active sulfur-oxidizing population. The environmental impact of the highly acidified waters is further accentuated by their high metal loadings resulting from the dissolution of the soil and mine slag minerals. Thus, soils and waters receiving these drainage waters are adversely impacted not only by the acidic conditions but also by toxic levels of heavy metals. A means of reducing the environmental impact in these situations is to pass the drainage waters through a swampy area where dissimilatory sulfate reducers can convert the sulfate to sulfide. The metal concentrations of the waters are reduced by their precipitation as metal sulfides. (See Mills, 1985, for a more detailed discussion of this process.)

Reduction of Complications of Metal Contamination in Soil: Sulfidogenesis can also be exploited in situations where soils have been contaminated by heavy metals. As with the acid mine drainage problem outlined previously, metal loadings of the leachate waters from the contaminated site may be removed by precipitation as metal sulfides in a sulfate-reducing system. Mobility of the metals may also be reduced in the site through creation of appropriate conditions for sulfide genesis within the soil profile. The primary difficulty with both the use of sulfidogenesis to remove metal contaminants from acid mine drainage and to reclaim metal-contaminated sites is that the anoxic conditions of the purification site must be maintained. Should the soil become aerobic, sulfur-oxidizing bacteria will use the sulfide for an energy source thereby mobilizing the previously stabilized heavy metals.

14.5. MINERALIZATION AND ASSIMILATION OF SULFUROUS SUBSTANCES

Mineralization of organic sulfur compounds is a strictly microbially catalyzed process, which is analogous to carbon and nitrogen mineralization processes. Although the relative rates of nitrogen, carbon, and sulfur mineralization in a specific soil system may differ (e.g., see Bettany et al., 1980, for an example of a field study and Tabatabai and Al-Khafaji, 1980, for a laboratory-based study demonstrating this difference), each is dependent upon activity of a variety of heterotrophic bacteria and fungi. This commonalty of mediators results in a similarity in impact of variation of environmental conditions on process rates. In other words, sulfur mineralization is anticipated to occur in any ecosystem wherein heterotrophic microbes can function. Under aerobic conditions, complete mineralization of an organic substrate yields carbon dioxide, ammonium, and sulfate, whereas under anoxic conditions carbon dioxide, methane, sulfide, and ammonium are primary products. Optimal conditions for mineralization of sulfur-containing organic compounds include a near neutral pH, mesophilic temperatures, and approximately field capacity soil moisture. It is easily concluded a priori that because sulfur mineralization occurs over essentially the totality of the physical and chemical conditions supportive of microbial life, the process is catalyzed by an extremely diverse group of common soil bacteria and fungal species.

The mineral sulfur produced through the catalytic activity of the soil decomposer community may be assimilated by the plant community or immobilized (assimilated) by the soil microbes. Because ecosystem conditions generally favor assimilation of mineral sulfate into plant or microbial biomass, the bulk of the sulfur contained in soil systems is in the organic fractions (e.g., see David et al., 1982; Swank et al., 1984). Because these assimilatory processes are catalyzed by the plant community and the heterotrophic soil bacteria and fungi, the physical and chemical limitations to the activities would be similar to those previously described for overall soil microbial activity.

Nonbiomass fates of mineral sulfur include the reductive and oxidative processes described previously. In other words, sulfate may be reduced through sulfidogenic reactions, and should aerobic conditions develop, the sulfide could be oxidized for energy by the sulfur oxidizers. For this latter transformation to occur, the hydrogen sulfide would have to diffuse to aerobic soil microsites, or the site where it was produced would have to oxygenated.

As has been observed with carbon and nitrogen mineralization, the rate of mineralization of sulfur compounds is proportional to the biodegradation susceptibility of the organic compound itself as well as its physical availability to the degrader population. That is, the decomposition rate is directly dependent upon the capability of the microbial population to synthesize the requisite enzymes necessary for the catalysis and the proximity of the organic substance to the active microbial population. Both biodegradation-sensitive (common cellular substituents) and more biodegradation-resistant sulfurous compounds (humified substances) are found in soil (e.g., see Anderson et al., 1981, or Bettany et al., 1979). The conversion of organic sulfur to sulfate in an aerobic soil system may or may not be proportional to the total soil organic matter content. This relationship is totally dependent upon the relative contribution of biodegradable and accessible organic sulfur to biologically stabilized (e.g., humic acid) sulfur.

14.6. THE PHOSPHOROUS CYCLE

The processes associated with transformations of phosphorus occurring in soil are described by the principles previously delineated for other soil-based biogeochemical cycles. For example, mineralization and assimilation are catalyzed by the general soil biological community. The phosphorus cycle, like the sulfur cycle, involves the interconversion of mineral and organic nutrient forms with the quantity of water-soluble phosphate being controlled not only by biological mineralization processes but also by dissolution rates from soil mineral fractions. Unique properties of the phosphorous cycle include the observation that phosphorus does not undergo any valence changes in the cycle and that there is no gaseous component to the cycle. Based on the similarities of the biological and chemical properties of the phosphorus cycle to those of other nutrient cycles, only the salient features of the phosphorus cycle are described here. For a more detailed analysis of the nuances of the phosphorus cycle, see Stevenson (1986).

In a depiction of the phosphorus cycle with biomass accentuated, the pivotal process is mineralization of organic matter into plant and microbially available phosphate (Fig. 14.3). As with the mineralization processes central to the carbon, nitrogen, and sulfur cycles, phosphorus mineralization is catalyzed by the general soil heterotrophic microbial population (e.g., Molla et al., 1984). Hence, as in the previously described process, mineralization rate is controlled by the physical and chemical factors delimiting overall microbial heterotrophic activity. Phosphate mineralization and its assimilation into biomass is generally optimal under those soil conditions suitable for aboveground plant community development. As with the previous cycles, mineraliza-

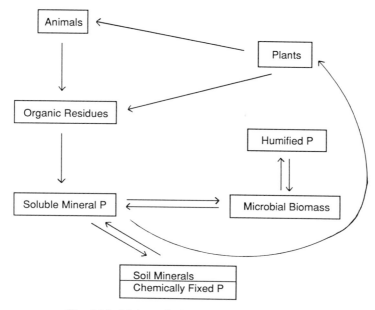

Fig. 14.3. Major soil phosphorus cycle reservoirs.

tion occurs under both aerobic and anaerobic conditions, but maximal activity is detected in the presence of molecular oxygen.

A variety of relatively easily decomposed phosphorus containing compounds are found in biomass. The most abundant phosphatic substances are inositol phosphates, which make up 10–50 percent of the organic phosphate fraction. Other organic phosphates are phospholipids (1–5 percent) and nucleic acids (0.2–2.5 percent) (Stevenson, 1986). Lesser concentrations of phosphopyridines (e.g., nicotinamide–adeninedinudeotide, NAD^+) and nucleotides (e.g., ATP) are detectable in soil organic phosphorus fractions.

In common with the sulfur cycle is the occurrence of significant mineral phosphorus pools in most soil systems. Native phosphate in soils is derived from rock phosphate, apatite, which has the empirical formula $3[Ca_3(PO_4)_2]\cdot CaX_2$, where X equals Cl^-, F^-, OH^-, or CO_3^{2-}. Secondary minerals, which are formed from phosphate in soil, are wavellite $[Al_3(PO_4)_2(OH)_3\cdot 5H_2O]$, variscite $[Al(PO_4)\cdot 2H_2O]$, dufrenite $[FePO_4\cdot Fe(OH)_3]$, strengite $(Fe(PO_4)\cdot H_2O)$, and vivianite $[Fe_3(PO_4)_2\cdot 8H_2O]$. Along with the effect of phosphate mineralization rate, the quantities of plant-available phosphate present are determined by the solubilization rate of phophatic minerals. Because the quantities of phosphate in organic forms range from 15 to 80 percent in soil ecosystems (Stevenson, 1986), in some ecosystems the solubilization rate of mineral phosphate is a major delimiter of aboveground biomass productivity.

Soil microbes are not only mediators of phosphate mineralization and immobilization processes, but they may also be instrumental in solubilization of rock phosphates. Directly synthesized products of microbial activity that solubilizes rock phosphate are organic acids and carbon dioxide. The organic acids liberate phosphate through chelation of the associated metal ions. Carbon dioxide dissolved in soil interstitial water forms carbonic acid as follows:

$$CO_2 + H_2O \rightarrow H_2CO_3$$

As the pH of an acidic soil increases, organic acid production becomes increasingly important in determining dissolution rate of rock phosphates (e.g., Traina et al., 1986).

14.7. MICROBIALLY CATALYZED SOIL METAL CYCLING

Soil-resident reservoirs of biomass–nutrient cations (as well as toxic cations) are easily grouped into three categories: biomass-available, water-soluble, and mineral-insoluble precipitates (Fig. 14.4). Biomass productivity may be determined by the existence of excessive quantities as well as the occurrence of minimal levels of nutrient and other cations in soil interstitial waters. The soil biological community is instrumental in catalyzing a variety of processes that directly or indirectly result in the transfer of cations between these three soil reservoirs.

14.7.1. Interactions of Metals with Living Systems

Benefits from maintaining appropriate levels of cations in interstitial waters result from their essential roles in cellular biochemistry. Plant essential trace metals include iron, zinc, manganese, calcium, boron, molybdenum, and, most likely, nickel. Animals also require cobalt, chromium, and tin. In contrast, cations of major interest from the view of negative interactions with biomass production include cadmium, mercury, and lead.

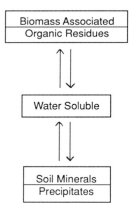

Fig. 14.4. Primary metal reservoirs in soil.

In most ecosystems, the balance between aboveground productivity and trace metal availability is maintained by the movement of soil-resident cations between the various reservoirs (Fig. 14.4). Limitations of plant nutrients are generally restricted to ecosystems where augmentation of biomass production above that essentially dictated by native soil mineral cycling rates occurs (i.e., intensive agricultural production) or in cases where metals are removed from the system through harvesting of aboveground biomass. In these situations, aboveground biomass needs are met through external amendments and/or modification of soil pH to optimize available soil resources.

In contrast to the situation with higher plant and animal nutrition, metal limitations for microbial productivity in soil are considered to be a rarity. Soil microbes possess highly efficient mechanisms for transport of cations into the cell. Furthermore, as is discussed later, many members of the soil microbial community produce chelators, which increase water-soluble cation resources.

In contrast to these situations where generally positive benefits are accrued to the ecosystem through solubilization of biomass-essential cations, an increasing array of environmental problems are arising where management of the soil for stimulation of ecosystem sustenance and/or development requires enhancement of metal precipitation (i.e., decreasing cation availability to the microbial community). Historically, the most commonly cited toxicity problem involving soil cations has been associated with aluminum concentrations in acidic soils. Aluminum is toxic to biological processes. As the soil pH declines, the solubility of aluminum increases. Thus, best management practices of acidic soils for crop production has included adding lime to the soil to raise the pH as well as amendment the soil with organic matter to complex the water-soluble ions (see Stevenson, 1982, for a more complete analysis of aluminum solubility in soil). More recent considerations involving toxic metals in soil have been associated with environmental contamination (e.g., smelter and foundry operations) and from disposal of societal wastes (e.g., metal contamination of sludges and composts).

The implications of increased soils loadings of toxic cations on ecosystem function are best exemplified by the succession of aboveground and belowground communities associated with contamination from smelters. The availability of the toxic cation may become the controlling factor in population development with the impact of those environmental properties normally considered to determine community diversity (e.g., moisture and organic matter content) being minimized, as was observed by Nordgren et al. (1983) in soils near a brass mill in South Sweden. In these soils sampled along a copper and zinc gradient (with maximum levels of 20,000 μg copper and zinc g^{-1} soil), the influence of soil organic matter content and moisture on fungal community composition was minimal compared to the effect of heavy metal pollution on these populations.

A more general impact on ecosystem communities has been noted for heavy-metal-contaminated soils near a zinc smelter at Palmerton, PA (USA). Soil heavy metal loadings (litter horizon within 1 km of the smelter) in the 1970s included about 26,000 ppm zinc, 10,000 ppm iron, 2300 ppm cadmium, and 340 ppm copper (Strojan, 1978) with as much as 90 percent of the metals being retained within the top 15 cm of the soil profile (Baucher, 1973). These heavy metal loadings declined to approximately one-tenth these values at a control area about 40 km east of the smelter. In this 1978 study, consideration of the average amount of organic matter in the more highly polluted site and comparison of decomposition rates of leaf litter in litter bags led to the conclusion that organic matter decomposition rates were reduced by the heavy metal contamination. In a 1975 report (Jordan, 1975), reduced germination of tree seeds due to high zinc concentrations was proposed as a mechanism preventing the establishment of invader tree species in regions where indigenous trees, shrubs, and herb populations were in decline from the metal contamination. Additionally, in this site, the prevalent metal contaminant concentrations exceed those necessary to result in major reductions of total bacterial, actinomycete, and fungal populations (Jordan and Lechevalier, 1975). Major alterations of the soil microbial as well as aboveground populations were occurring.

Currently, the aboveground status of the site at Palmerton, PA, can be defined as an "establishing" grass community due to overlaying of the soil with a sludge-fly ash–limestone–grass seed mixture or an essentially barren system in untreated sites (Fig. 14.5). In other words, the imbalance in ecosystem function due to the heavy metal pollution has resulted in total destruction of the forest ecosystem. The once flourishing forest ecosystem has been reduced to a bare mountainside. Reclamation of the site requires redevelopment of a functional soil microbial community accompanied by the establishment of an aboveground plant community consisting of a grass community.

With both of these examples, extremely high metal loadings were detected in the environmental samples. Remember that the toxicity of the metals is proportional to their availability to the microbial community. Thus, inhibitory concentrations of metals, derived from a pure culture study where all the toxicant is water-soluble, frequently do not correspond to field observations where total soil metal concentration can be distributed between a variety of chemical forms that vary in availability

Fig. 14.5. Metal contaminate site in Palmerton, PA (USA), showing the effect of organic matter, lime, and fly ash amendments on grass growth.

to the soil microbial community (e.g., Lighthart et al., 1983). Therefore, when evaluating metal-loading toxicity in native ecosystems, soil properties that alter distribution between water-soluble and water-insoluble species must be considered. For example, in a study of planktonic, sedimentary, and epilithic bacterial community adjustment to heavy metal concentrations (Dean-Ross and Mills, 1989), changes in microbial populations did not reflect the known toxicity of the contaminating metals. It was concluded that the occurrence of a high pH in the samples that reduced the solubility of the toxic metals apparently resulted in a lack of correlation between metal concentrations and response of the bacterial communities to the toxicants.

14.7.2. Microbial Response to Elevated Metal Loadings

The most obvious effect of high metal loadings to the soil microbial community is toxicity. Death of sensitive microbes and plants may result directly from dissolved cations as well as indirectly from the release of cations following catabolism of the organic matter to which they are bound. As implied by these data, alteration of soil microbial community dynamics and functions can be anticipated to result from soil contamination with heavy metals. A primary impact of metal contamination is the inhibition of catabolism of organic substances by the toxic metals. This was reflected in the studies of metal-contaminated forest sites cited previously by the augmented accumulation of plant debris in regions of highest metal concentrations.

Microbial populations may be inhibited by release of organic-matter-associated cations through microbial catabolism of the organic matter. The microbial recovery of energy from the fixed carbon substances results in the accumulation of toxic concentrations of the metal. This phenomenon is exemplified by studies of the decomposition of iron, aluminum, zinc, and copper salts complexed to polysaccharides (Martin et al., 1966), metal–nitrilotriacetate (calcium, manganese, magnesium, copper, zinc, cadmium, iron, and sodium chelates were examined) complexes (Firestone and Tiedje, 1975) as well as copper, mercury, zinc and cadmium salts of 2,4-dichloro-phenoxyacetic acid methyl ester (Said and Lewis, 1991). As the concentration of the metal liberated from the organic complex increased, microbial activity decreased. The degree of inhibition varied with toxicity of the metal to the decomposer populations.

Overall reduction of microbial respiration by increasing concentrations of a variety of heavy metals is exemplified by the study of Lighthart et al. (1983) cited previously. Additionally, examples of the capacity of heavy metal contamination to reduce or inhibit soil biological processes include a variety of enzymatic activities (e.g., Cole, 1977; Doelman and Haanstra, 1979, 1989; Stott et al., 1985) as well as various soil nitrogen transformations (immobilization, mineralization, and nitrification, Chang and Broadbent, 1982; anaerobic nitrogen transformations, Blais et al., 1988; and nitrogen fixation, Wickliff et al., 1980).

The diversity of the soil microbial genome suggests that a variety of organisms must exist in soil with augmented resistance to heavy metal pollutants. For example, a variety of soil bacteria have been isolated with augmented resistance to metal contaminants, including cadmium (Boop et al., 1983), copper (Dressler et al., 1991; Yang et al., 1993), mercury (Kelley and Reanney, 1984), nickel (Schmidt et al., 1991), and silver (Haefeli et al., 1984). Selective enirchment of metal-tolerant bacterial populations in polluted soils has been demonstrated with classical microbiological procedures (e..g, Duxbury and Bicknell, 1983) as well as DNA probe-mediated methods (e.g., Diels and Mergeay, 1990).

14.7.3. Microbial Modification of Metal Mobility in Soils

The soil properties controlling the availability of both essential nutrient cations and toxicants are primarily soil pH, redox potential (E_0') and organic matter concentrations, but soil-buffering capacity, chlorinity, inorganic cations and anions, clay mineralogy, and moisture (e.g., Babich and Stotzky, 1983) also contribute to metal solubility in these ecosystems. The microbial community is instrumental in modifying each of these soil factors in a manner that enhances or reduces soil interstitial water metal loadings. Microbial interactions affecting metal mobility in soil can be grouped as indirect or direct. Indirect interactions include pH alteration, siderophore production, and modification of the physical environment (e.g., changing the redox potential), whereas direct interactions include valance change, substitution, methylation, transalkylation, incorporation into cell substituents, and release from organic association through mineralization.

Acidification and Chelation: Soil microbes produce a variety of inorganic and organic acids that contribute to the solubilization of metals. The inorganic acids could be considered to be by-products of energy metabolism. Two of the more commonly encountered products are carbonic and sulfuric acid. Aerobic microbes yield carbon dioxide as their terminal oxidation product. This gaseous substance exists in equilibrium in soil between carbonic acid, bicarbonate, and carbonate by the following relationship:

$$CO_2 \rightleftharpoons HCO_3^- \rightleftharpoons CO_3^{2-}$$

Carbonic acid is a weak acid capable of dissolving soil minerals.

A more dramatic example of mineral acid production is the sulfuric acid produced by oxidation of sulfides and elemental sulfur by soil thiobacilli. This process has been exploited by recovery of mineral from ores (e.g., Carlson et al., 1992) and has been proposed as a means of removing mineral sulfides from municipal sludges (e.g., Jain and Tyagi, 1992). Ecosystem destruction results from this process when oxidation of mineral sulfides and associated dissolution of heavy metals (typically iron) results in the production of the orange-colored leached known as acid mine drainage (e.g., Mills, 1985).

Organic acids, which may solubolize soil minerals either directly through acidification of the microsite or through chelation reactions, are produced by aerobic and anaerobic soil microbial processes. Under anaerobic conditions a variety of organic acids accumulate due to incomplete oxidation of fixed-carbon substrates. The most common substance produced under aerobic conditions are the siderophores, usually studied for their role in iron nutrition of bacteria as well as higher plants. Bacterially synthesized siderophores may be instrumental in solubolizing significant quantities of iron for higher plant biomass synthesis. Such substances produced by rhizosphere bacteria have been shown to stimulate the associated plant productivity (e.g., Kloepper et al., 1980). Similarly, rhizobial siderophores have been demonstrated to stimulate clover productivity (Derylo and Skorupska, 1992). A large number of members of the general soil community also produce these metal chelators, for example *Pseudomonas* species (e.g., Ankenbauer et al., 1988; Bar-Ness et al., 1991; Cody and Gross, 1987; Hoefte et al., 1991). The fact that these substances are synthesized within the soil ecosystem is supported by the capacity to isolate a variety of such substances from soil samples (e.g., Akers, 1983a, 1983b; Powell et al., 1982).

Direct and Indirect Oxidation and Reduction of Soil Minerals: Oxidation of metals by soil bacteria are exemplified by the oxidation of ferrous ion as an energy source by *Thiobacillus ferrooxidans* as discussed previously and by mercury oxidation by a variety of bacteria (e.g., Holm and Cox, 1975) as well as by the oxidation of metal oxides, such as conversion of arsenite to aresenate by *Alcaligenes faecalis* (Phillips and Taylor, 1976). Similarly, a number of bacteria can reduce a number of metallic cations and metallic oxides, including iron (e.g,. Lovley and Phillips, 1986), manganese and iron (e.g., Nealson and Myers, 1992), uranium (Lovley et al., 1991; Lovley and Phillips, 1992), and chromate (Eary and Rai, 1991).

Indirect modification of metal mobility by changing its valence state can result from microbial modification of the redox potential of the soil microsite. This process is most easily envisioned in sites where microbial energy metabolism results in the exhaustion of molecular oxygen, nitrate, and other electron acceptors until a highly reducing condition is created. Then the spontaneous chemical-conversion-oxidized metals (e.g., ferric ion) to a more reduced state (ferrous ion) is enhanced. With the metal exemplified, should sulfide be produced through the dissimilatory transfer of electrons to sulfate, then iron sulfides will accumulate in the soil, in some situations resulting in the formation of gley soils (see Tate, 1987).

An interesting indirect modification of oxidation state of metal compounds may occur in the rhizosphere of plants that actively oxygenate their root zone. These plants, such as rice, transport molecular oxygen from the aboveground tissue into the root zone. Oxidation of technectium in the rice (*Oryza sativa* L.) was shown to result from this mechanism (Sheppard and Evenden, 1991).

Bioaccumulation and Solubilization of Metal Ions Through Mineralization: As is true of all living cells, soil microbial biomass incorporates essential metals into cell substituents, but accumulation of metallic cations on or within living cells can greatly exceed these metabolic requirements. Bioaccumulation is delineated by the achievement of cellular loadings of metals far beyond any levels that might be associated with cellular metabolism. This process is exemplified among the bacterial populations with strontium accumulation by *Micrococcus luteus* (Faison et al., 1990) and precipitation of cadmium by *Clostridium thermoaceticum* at the cell surface as well as in the surrounding medium (Cunningham and Lundie, 1993). Similarly, cesium (Tomioka et al., 1992), zinc (Sakurai et al., 1990) are accumulated by bacteria. Note that the cells need not be living for augmentation of cell-associated metals to occur. Kurek et al. (1982) found that dead microorganisms can absorb more cadmium than did the live cells used in their studies.

The return of biomass-associated metals to soil-water-soluble or mineral forms occurs through mineralization of the microbial biomass following the death of the cell. As opposed to the cycling of carbon, nitrogen, and sulfur discussed previously, the return of the metal to a mineral state likely provides little benefit to the decomposer population. This metal cycling could be classified as a gratuitous product of microbial cycling of carbon.

Other Soil Metal Transformations: As indicated previously, soil minerals may also be methylated and transalkylated by soil microbial populations. The attachment of alkyl groups to the metal may alter the metal mobility within the ecosystem through enhanced volatility and water solubility. These processes are significant in instances of metal toxicity (as exemplified by the effect of methylation of mercury on mobility and toxicity) and thus may be significant in situations of soil pollution with metals that may be methylated (see Tate, 1987, for further discussion of metals in soil ecosystems).

14.7.4. Managing Soils Contaminated with Metals

Metal-contaminated soils present a unique challenge to the soil microbiologist charged with the responsibility for contributing to the development of renovation plans. Conceptually, most microbial reclamation procedures involve managing the availability of the contaminant to the soil community of degraders and encouraging mineralization of the offending substance. In contrast, reducing elevated metal loadings involves either removal of the contaminant from the site (leaching or excavation of the soil itself) or modification of the properties of the polluted soil in a manner that encourages sequestering of the toxic metal into biologically unavailable forms. Because the three major soil properties controlling metal mobility are pH, redox potential, and organic matter content, traditional soil reclamation involving metal management has included manipulation of these parameters.

Some shifting of soil metals from water soluble to the particulate fractions can be achieved through soil amendment with lime and/or organic matter. Increasing soil pH to approximate neutrality by adding lime encourages a change in the equilibrium between soluble and insoluble metal forms. With the addition of organic matter, the soluble metal loadings are reduced through complexing of the metal with the organic material itself. (See Tate, 1987, for a more detailed discussion of control of soil metals through these mechanisms.) Either of these nonbiological management procedures provide only a temporary solution to the metal toxicity; that is, organic matter amendment may be biologically decomposed. Because the natural tendency of soil systems is to become increasingly acidic, continued monitoring and adjustment of soil acidity may be necessary in particularly sensitive ecosystems.

Neither of these procedures presents any difficulty to the development or continued function of the soil microbial community. Organic matter encourages the stability of soil decomposer communities through the provision of carbon and energy.

Furthermore, because the soil pH is raised to values approximating neutrality which is within the range of optimal function of the majority of the members of the soil biological community, negative effects of this adjustment of soil properties are not anticipated.

A more intensive stewardship of metal-contaminated sites may actually involve alteration of the biological properties of the system. Activity of both the aerobic and the anaerobic microbial populations can be manipulated to encourage sequestering of metal contaminants. A logical extension of an understanding of the dynamics of sulfidogenesis and chemical interactions of the sulfide produced with soil metals is management of the cycle to purify soil interstitial waters. Provision of reducing conditions in the presence of sulfate ions may allow production of sufficient sulfide to reduce soluble metal loadings to acceptable levels. The difficulty with this procedure is that the system must be maintained in an anoxic conditions. Once oxygen is introduced, sulfur oxidizers use the sulfide as an energy source, thereby liberating the associated metal ion.

14.8. CONCLUSIONS

The foregoing analysis of the sulfur, phosphorus, and mineral cycles in soils underscores not only their essentiality to total ecosystem biomass productivity but also a primary role in the reversal of adverse situations resulting from anthropogenic activities. In other words, excessively acidic soils may result from elevated influxes of reduced sulfur compounds, heavy metal contamination may preclude sustenance of soil microbial communities, and mobility of soil phosphate may accelerate eutrophication of regional surface waters. Ameliorating these unfavorable soil or water conditions, while optimizing biomass productivity in soil ecosystems containing restrictive concentrations of the essential components of these cycles, requires a firm appreciation for not only the biological aspects of the biogeochemical cycles but also the chemistry of formation and dissolution of soil minerals. These biogeochemical cycles are highlighted by the major role of soil primary and secondary minerals in maintaining concentrations of bioavailable water-soluble nutrients. This intersection of the mineral with the biological world and the implications of societal mismanagement of soil ecosystems involving soil phosphate, sulfur and metal loadings can be managed to encourage the development of sustainable ecosystems.

REFERENCES

Ahonen, L., and O. H. Tuovinen. 1990. Kinetics of sulfur oxidation at suboptimal temperatures. Appl. Microbiol. 56:560–562.

Akers, H. A. 1983a. Isolation of the siderophore schizokinen from soil of rice fields. Appl. Environ. Microbiol. 45:1704–1706.

Akers, H. A. 1983b. Multiple hydroxamic acid microbial iron chelators (siderophores) in soils. Soil Sci. 135:156–159.

Alexander, M. 1977. Soil Microbiology. John Wiley, & Sons. New York. 467 pp.

Anderson, D. W., S. Saggar, J. R. Bettany, and J. W. B. Stewart. 1981. Particle size fractions and their use in studies of soil organic matter: I. The nature and distribution of forms of carbon, nitrogen, and sulfur. Soil Sci. Soc. Am. J. 45:767–772.

Ankenbauer, R. G., T. Toyokuni, A. Staley, K. L. Rinehart Jr., and C. D. Cox. 1988. Synthesis and biological activity of pyrochelin, a siderophore of *Pseudomonas aeruginosa*. J. Bacteriol. 170:5344–5351.

Attoe, O. J., and R. A. Olson. 1966. Factors affecting rate of oxidation in soils of elemental sulfur and that added in rock phosphate–sulfur fusions. Soil Sci. 101:317–325.

Babich, H., and G. Stotzky. 1983. Physicochemical factors of natural reservoirs affect the transformation and exchange of heavy metals toxic to microbes. *In* R. Hallberg (ed.), Environmental Biogeochemistry, Ecol. Bull. (Stockholm), Vol. 35, pp. 315–323. Libertryck. Stockholm.

Bar-Ness, E., Y. Chen, Y. Hadar, H. Marschner, and V. Roemheld. 1991. Siderophores of *Pseudomonas putida* as an iron source for dicot and monocot plants. Plant Soil 130:231–241.

Batchelor, B., and A. W. Lawrence. 1978. A kinetic model for autotrophic denitrification using elemental sulfur. Water Res. 12:1075–1084.

Baucher, M. J. 1973. Contamination of soil and vegetation near a zinc smelter by zinc, cadmium, copper, and lead. Environ. Sci. Technol. 7:131–135.

Beffa, T., M. Berczy, and M. Aragno. 1991. Chemolithoautotrophic growth on elemental sulfur (S) and respiratory oxidation of S by *Thiobacillus versutus* and another sulfur-oxidizing bacterium. FEMS Microbiol. Lett. 84:285–290.

Bertramson, B. R., M. Fried, and S. L. Tisdale. 1950. Sulfur studies of Indiana soils and crops. Soil Sci. 70:27–41.

Bettany, J. R., S. Saggar, and J. W. B. Stewart. 1980. Comparison of the amounts and forms of sulfur in soil organic matter fractions after 65 years of cultivation. Soil Sci. Soc. Am. J. 44:70–74.

Bettany, J. R., J. W. B. Stewart, and S. Saggar. 1979. The nature and forms of sulfur in organic matter fractions of soils selected along an environmental gradient. Soil Sci. Soc. Am. J. 43:981–985.

Blais, J. S., W. D. Marshall, and R. Knowles. 1988. Toxicity of alkyl lead salts to anaerobic nitrogen transformations in sediment. J Environ. Qual. 17:457–462.

Bopp, L. H., A. M. Chakrabarty, and H. L. Ehrlich. 1983. Chromate resistance plasmid in *Pseudomonas fluorescens*. J. Bacteriol. 155:1105–1109.

Brune, D. C. 1989. Sulfur oxidation by phototrophic bacteria. Biochim. Biophys. Acta 975:189–221.

Bryant, R. D., J. W. Costerton, and E. J. Laishley. 1984. The role of *Thiobacillus albertis* glycocalyx in the adhesion of cells to elemental sulfur. Can. J. Microbiol. 30:81–90.

Carlson, L., E. B. Lindström, K. B. Halalberg, and O. H. Tuovinen. 1992. Solid-phase products of bacterial oxidation of arsenical pyrite. Appl. Environ. Microbiol. 58:1046–1049.

Chang, F.-H., and F. E. Broadbent. 1982. Influence of trace metals on some soil nitrogen transformations. J. Environ. Qual. 11:1–4.

Chapman, S. J. 1989. Oxidation of micronized elemental sulfur in soil. Plant Soil 116:69–76.

Cody, Y. A., and D. C. Gross. 1987. Characterization of pyoverdin$_{pss}$, the fluorescent siderophore produced by *Pseudomonas syringae* pv. *syringae.* Appl. Environ. Microbiol. 53:928–934.

Cole, M. A. 1977. Lead inhibition of enzyme synthesis in soil. Appl. Environ. Microbiol. 33:262–268.

Cunningham, D. P., and L. L. Lundie, Jr. 1993. Precipitation of cadmium by *Clostridium thermoaceticum.* Appl. Environ. Microbiol. 59:7–14.

David, M. B., J. M. Mitchell, and J. P. Nakas. 1982. Organic and inorganic sulfur constituents of a forest soil and their relationship to microbial activity. Soil Sci. Soc. Am. J. 46:847–852.

Dean-Ross, D., and A. L. Mills. 1989. Bacterial community structure and function along a heavy metal gradient. Appl. Environ. Microbiol. 55:2002–2009.

Derylo, M., and A. Skorupska. 1992. Rhizobial siderophore as an iron source for clover. Physiol. Plant 85:549–553.

Diels, L., and M. Mergeay. 1990. DNA probe-mediated detection of resistant bacteria from soils highly polluted by heavy metals. Appl. Environ. Microbiol. 56:1485–1491.

Doelman, P., and L. Haanstra. 1979. Effect of lead on soil respiration and dehydrogenase activity. Soil Biol. Biochem. 11:475–479.

Doelman, P., and L. Haanstra. 1989. Short- and long-term effects of heavy metals on phosphatase activity in soils: An ecological dose-response model approach. Biol. Fert. Soils 8:235–241.

Dressler, C., U. Kües, D. H. Nies, and B. Friedrich. 1991. Determinants encoding resistance to several heavy metals in newly isolated copper-resistant bacteria. Appl. Environ. Microbiol. 57:3079–3085.

Duxbury, T., and B. Bicknell. 1983. Metal-tolerant bacterial populations from natural and metal-polluted soils. Soil Biol. Biochem. 15:243–250.

Eary, L. E., and D. Rai. 1991. Chromate reduction by subsurface soils under acidic conditions. Soil Sci. Soc. Am. J. 55:676–683.

Espejo, R. T., B. Escobar, E. Jedlicki, P. Uribe, and R. Badilla-Ohlbaum. 1988. Oxidation of ferrous iron and elemental sulfur by *Thiobacillus ferrooxidans.* Appl. Environ. Microbiol. 54:1694–1699.

Espejo, R. T., and P. Romero. 1987. Growth of *Thiobacillus ferroxidans* on elemental sulfur. Appl. Environ. Microbiol. 53:1907–1912.

Faison, B. D., C. A. Cancel, S. N. Lewis, and H. I. Adler. 1990. Binding of dissolved strontium by *Micrococcus luteus.* Appl. Environ. Microbiol. 56:3649–3656.

Firestone, M. K., and J. M. Tiedje. 1975. Biodegradation of metal–nitrilotriacetate complexes by a *Pseudomonas* species: Mechanism of reaction. Appl. Microbiol. 29:758–764.

Frederick, L. R., and R. L. Starkey. 1948. Bacterial oxidation of sulfur in pipe sealing mixtures. J. Am. Water Works Assoc. 40:729–736.

Gebhardt, N. A., R. K. Thauer, D. Linder, P.-M. Kaulfers, and N. Pfennig. 1985. Mechanism of acetate oxidation to carbon dioxide with elemental sulfur in *Desulfuromonas acetoxidans.* Arch. Microbiol. 141:392–398.

Grayston, S. J., W. Nevell, and M. Wainwright. 1986. Sulfur oxidation by fungi. Trans. Br. Mycol. Soc. 87:193–198.

Haefeli, C., C. Franklin, and K. Hardy. 1984. Plasmid-determined resistance in *Pseudomonas stutzeri* isolated from a silver mine. J. Bacteriol. 158:389–392.

Hashimoto, S., K. Furukawa, and M. Shioyama. 1987. Autotrophic denitrification using elemental sulfur. J. Ferment. Technol. 65:683–692.

Harrison, A. P., Jr. 1984. The acidophilic thiobacilli and other acidophilic bacteria that share their habitat. Annu. Rev. Microbiol. 38:265–292.

Hoefte, M., K. Y. Seong, E. Jurkevitch, and W. Verstraete. 1991. Pyroverdin production by the plant growth beneficial *Pseudomonas* strain 7NSK2: Ecological significance in soil. Plant Soil 130:249–257.

Holm, H. W., and M. F. Cox. 1975. Transformation of elemental mercury by bacteria. Appl. Microbiol. 29:491–494.

Huber, H., and K. O. Stetter. 1989. *Thiobacillus prosperus* sp. nov., represents a new group of halotolerant metal-mobilizing bacteria isolated from a marine geothermal field. Arch. Microbiol. 151:479–485.

Huber, H., and K. O. Stetter. 1990. *Thiobacillus cuprinus* new species, A novel facultatively organotrophic metal-mobilizing bacterium. Appl. Environ. Microbiol. 56:315–322.

Jain, D. K., and R. D. Tyagi. 1992. Leaching of heavy metals from anaerobic sewage sludge by sulfur-oxidizing bacteria. Enzyme Microb. Technol. 14:376–383.

Janzen, H. H., and J. R. Bettany. 1987a. Oxidation of elemental sulfur under field conditions in central Saskatchewan Canada. Can. J. Soil Sci. 67:609–618.

Janzen, H. H., and J. R. Bettany. 1987b. Measurement of sulfur oxidation in soils. Soil Sci. 143:444–452.

Janzen, H. H., and J. R. Bettany. 1987c. The effect of temperature and water potential on sulfur oxidation in soils. Soil Sci. 144:81–89.

Jordan, M. J. 1975. Effects of zinc smelter emissions and fire on a chestnut–oak woodland. Ecology. 56:78–91.

Jordan, J. J., and M. P. Lechevalier. 1975. Effects of zinc-smelter emissions on forest soil microflora. Can. J. Microbiol. 21:1855–1865.

Kelley, W. J., and D. C. Reanney. 1984. Mercury resistance among soil bacteria: Ecology and transferability of genes encoding resistance. Soil Biol. Biochem. 16:1–18.

Kennedy, K. A., P. A. Addison, and D. G. Maynard. 1988. Effect of elemental sulfur on the vegetation of a lodgepole pine stand. Environ. Pollut. 51:121–130.

Killham, K., N. D. Lindley, and M. Wainwright. 1981. Inorganic sulfur oxidation by *Aureobasidium pullans*. Appl. Environ. Microbiol. 42:629–631.

Kloepper, J. W., J. Leong, M. Teintze, and M. N. Schroth. 1980. Enhanced plant growth by siderophores produced by plant growth-promoting rhizobacteria. Nature (London) 286:885–886.

Kuenen, J. G., L. A. Robertson, and O. H. Tuovinen. 1992. The genera *Thiobacillus, thiomicrospira,* and *Thiosphaera. In* A. Balows, H. G. Trüper, M. Dwarkin, W. Harder, and K.-H. Schleifer (eds.) The Prokaryotes, pp. 2197–2213. Springer-Verlag. New York.

Kurek, E. 1979. Oxidation of inorganic sulfur compounds by yeast. Acta Microbiol. Pol. 28:169–172.

Kurek, E. 1985. Elemental sulfur and thiosulfate oxidation by *Rhodotorula* sp. Zentrabl. Microbiol. 140:497–500.

Kurek, E., J. Czaban, and J.-M. Bollag. 1982. Sorption of cadmium by microorganisms in competition with other soil constituents. Appl. Environ. Microbiol. 43:1011–1015.

Laishley, E. J., R. D. Bryant, B. W. Kobryn, and J. B. Hyne. 1986. Microcrystalline structure and surface area of elemental sulfur as factors influencing its oxidation by *Thiobacillus albertis*. Can. J. Microbiol. 32:237–242.

Lawrence, J. R., and J. J. Germida. 1988. Relationship between microbial biomass and elemental sulfur oxidation in agricultural soils. Soil Sci. Soc. Am. J. 52:672–677.

Lawrence, J. R., V. V. S. R. Gupta, and J. J. Germida. 1988. Impact of elemental sulfur fertilization on agricultural soils. II. Effects on sulfur-oxidizing populations and oxidation rates. Can. J. Soil Sci. 68:475–484.

Lee, A., C. C. Boswell, and J. H. Watkinson. 1988a. Effect of particle size on the oxidation of elemental sulfur: Thiobacilli numbers, soil sulfate, and its availability to pasture. N. Z. J. Agric. Res. 30:373–386.

Lee, A., J. H. Watkinson, and D. R. Lauren. 1988b. Factors affecting oxidation rates of elemental sulfur in a soil under a ryegrass dominant sward. Soil Biol. Biochem. 20:809–816.

Lee, A., J. H. Watkinson, G. Orbell, J. J. Bagyara, and D. R. Lauren. 1987. Factors influencing dissolution of phosphate rock and oxidation of elemental sulfur in some New Zealand soils. N. Z. J. Agric. Res. 30:373–386.

Lettl, A., O. Langkramer, and V. Lochman. 1981. Some factors influencing production of sulfate by oxidation of elemental sulfur and thiosulfate in upper horizons of spruce forest soils. Folia Microbiol. 26:158–163.

Lighthart, B., J. Baham, and V. V. Volk. 1983. Microbial respiration and chemical speciation in metal-amended soils. J. Environ. Qual. 12:543–548.

Lovley, D. R., and E. J. P. Phillips. 1986. Organic matter mineralization with reduction of ferric iron in anaerobic sediments. Appl. Environ. Microbiol. 51:683–689.

Lovley, D. R., and E. J. P. Phillips. 1992. Reduction of uranium by *Desulfovibrio desulfuricans*. Appl. Environ. Microbiol. 58:850–856.

Lovley, D. R., E. J. P. Phillips, Y. A. Gorby, and E. R. Landa. 1991. Microbial reduction of uranium. Nature (London) 350:413–416.

Martin, J. P., J. O. Ervin, and R. A. Shepherd. 1966. Decomposition of the iron, aluminum, zinc, and copper salts or complexes of some microbial and plant polysaccharides in soil. Soil Sci. Soc. Am. Proc. 30:196–200.

Maynard, D. G., J. J. Germida, and P. A. Addison. 1986. The effect of elemental sulfur on certain chemical and biological properties of surface organic horizons of a forest soil. Can. J. For. Res. 16:1050–1054.

McCaskill, M. R., and G. J. Blair. 1987. Particle size and soil texture effects of elemental sulfur oxidation. Agron. J. 79:1079–1083.

Meulenbert, R., J. T. Pronk, W. Hazeu, P. Bos, and J. G. Kuenen. 1992. Oxidation of reduced sulphur compounds by intact cells of *Thiobacillus acidophilus*. Arch. Microbiol. 157:161–168.

Mills, A. L. 1985. Acid mine waste drainage: Microbial impact on the recovery of soil and water ecosystems. *In* R. L. Tate III and D. A. Klein (eds.), Soil Reclamation Processes, Microbiological Analyses and Applications, pp. 35–81. Dekker. New York.

Min, H., and S. H. Zinder. 1990. Isolation and characterization of a thermophilic sulfate-reducing bacterium *Desulfotomaculum thermoacetoxidans* new species. Arch. Microbiol. 153:399–404.

Molla, M. A. Z., A. A. Chowdhury, A. Islam, and S. Hoque. 1984. Microbial mineralization of organic phosphate in soil. Plant Soil 78:393–394.

Moser, U. S., and R. V. Olson. 1953. Sulfur oxidation in four soils as influenced by soil moisture tension and sulfur bacteria. Soil Sci. 76:251-257.

Nealson, K. H., and C. R. Myers. 1992. Microbial reduction of manganese and iron: New approaches to carbon cycling. Appl. Environ. Microbiol. 58:439-443.

Nor, Y. M., and M. A. Tabatabai. 1977. Oxidation of elemental sulfur in soils. Soil Sci. Soc. Am. J. 41:736-741.

Nordgren, A., E. Bååth, and B. Söderström. 1983. Microfungi and microbial activity along a heavy metal gradient. Appl. Environ. Microbiol. 45:1829-1837.

Pepper, I. L., and R. H. Miller. 1978. Comparison of the oxidation of thiosulfate and elemental sulfur by two heterotrophic bacteria and *Thiobacillus thiooxidans*. Soil Sci. 126:9-14.

Phillips, S. E., and M. L. Taylor. 1976. Oxidation of arsenite to arsenate by *Alcaligenes faecalis*. Appl. Environ. Microbiol. 32:392-399.

Powell, P. E., G. R. Cline, C. P. P. Reid. and P. J. Szaniszlo. 1982. Occurrence of hydroxamate siderophore iron chelators in soils. Nature (London) 287:833-834.

Pronk, J. T., J. C. De Bruyn, P. Bos, and J. G. Kuenen. 1992. Anaerobic growth of *Thiobacillus ferrooxidans*. Appl. Environ. Microbiol. 58:2227-2230.

Said, W. A., and D. L. Lewis. 1991. Quantitative assessment of the effects of metals on microbial degradation of organic chemicals. Appl. Environ. Microbiol. 57:1498-1503.

Sakurai, I., Y. Kawamura, H. Koike, Y. Inoue, Y. Kosako, T. Nakase, Y. Knodou, and S. Sakurai 1990. Bacterial accumulation of metallic compounds. Appl. Environ. Microbiol. 56:2580-2583.

Schedel, M., and H. G. Trueper. 1980. Anaerobic oxidation of thiosulfate and elemental sulfur in *Thiobacillus denitrificans*. Arch. Microbiol. 124:205-210.

Schmidt, T., R.-D. Stoppel, and H. G. Schlegel. 1991. High-level nickel resistance in *Alcaligenes xylosoxidans* 31A and *Alcaligenes eutrophus* KTO2. Appl. Environ. Microbiol. 57:3301-3309.

Sheppard, S. C., and W. G. Evenden. 1991. Can aquatic macrophytes mobilize technetium by oxidizing their rhizosphere? J. Environ. Qual. 20:738-744.

Stevenson, F. J. 1982. Humus Chemistry. Genesis, Composition, Reactions. John Wiley, and Sons. New York. 443 pp.

Stevenson, F. J. 1986. Cycles of Soil: Carbon, Nitrogen, Phosphorus, Sulfur, Micronutrients. John Wiley & Sons. New York. 380 pp.

Stott, D. E., W. A. Dick, and M. A. Tabatabai. 1985. Inhibition of pyrophosphatase activity in soils by trace elements. Soil Sci. 139:112-117.

Strojan, C. J. 1978. Forest leaf litter decomposition in the vicinity of a zinc smelter. Oecologia (Berlin) 32:203-212.

Sugio, T., Y. Tsujita, K. Hirayama, K. Inagaka, and T. Tano. 1988a. Mechanism of tetravalent manganese reduction with elemental sulfur by *Thiobacillus ferrooxidans*. Agric. Biol. Chem. 52:185-190.

Sugio, T., Y. Tsujita, K. Inagaki, and T. Tano. 1990. Reduction of cupric ions with elemental sulfur by *Thiobacillus ferrooxidans*. Appl. Environ. Microbiol. 56:693-696.

Sugio, T., Y. Tsujita, T. Katagiri, K. Inagaki, and T. Tano. 1988b. Reduction of molybdic ions with elemental sulfur by *Thiobacillus ferrooxidans*. J. Bacteriol. 170:5956-5959.

Sugio, T., W. Mizunashi, K. Inagaki, and T. Tano. 1987. Purification and some properties of sulfur ferric ion oxidoreductase from *Thiobacillus ferrooxidans*. J. Bacteriol. 169:4916-4922.

Suzuki, I., T. L. Takeuchi, T. D. Yuthasastrakosol, and J. K. Oh. 1990. Ferrous iron and sulfur oxidation and ferric iron reduction activities of *Thiobacillus ferrooxidans* are affected by growth on ferrous iron and sulfur or a sulfide ore. Appl. Environ. Microbiol. 56:1620–1626.

Swank, W. T., J. W. Fiktzgerald, and J. T. Ash. 1984. Microbial transformation of sulfate in forest soils. Science 223:182–184.

Tabatabai, M. A., and A. A. Al-Khafaji. 1980. Comparison of nitrogen and sulfur mineralization in soils. Soil Sci. Soc. Am. J. 44:1000–1006.

Tate, R. L., III. 1987. Soil Organic Matter: Biological and Ecological Effects. John Wiley & Sons. New York. 291 pp.

Tomioka, N., H. Uchiyama, and O. Yagi. 1992. Isolation and characterization of cesium-accumulating bacteria. Appl. Environ. Microbiol. 58:1019–1023.

Traina, S. J., G. Sposito, D. Hesterberg, and U. Kafkafi. 1986. Effect of pH and organic acids on orthophosphate solubility in an acidic, montmorillonitic soil. Soil Sci. Soc. Am. J. 50:45–52.

Visniac, W. V., and M. Santer. 1957. The thiobacilli. Bacteriol. Rev. 21:195–213.

Visser, S., and D. Parkinson, 1989. Microbial respiration and biomass in soil of a lodgepole pine stand acidified with elemental sulfur. Can. J. For. Res. 19:955–961.

Vogler, K. G., and W. W. Umbreit. 1941. The necessity for direct contact in sulfur oxidation by *Thiobacillus thiooxidans*. Soil Sci. 51:331–337.

Watkinson, J. H. 1989. Measurement of the oxidation rate of elemental sulfur in soil. Aust. J. Soil Res. 27:365–376.

Wickliff, C., H. J. Evans, K. R. Carter, and S. A. Russell. 1980. Cadmium effects on the nitrogen fixation system of red alder. J. Environ. Qual. 9:180–184.

Yagi, S., S. Katai, and T. Kimura. 1971. Oxidation of elemental sulfur to thiosulfate by *Streptomyces*. Appl. Microbiol. 22:157–159.

Yang, C.-H., J. A. Menge, and D. A. Cooksey. 1993. Role of copper resistance in competitive survival of *pseudomonas fluroescens* in soil. Appl. Environ. Microbiol. 59:580–584.

Zillig, W., S. Yeats, I. Holz, A. Boeck, M. Rettenberger, F. Gropp, and G. Simon. 1986. *Desulfurolobus ambivalens* new genus, new species, An autotrophic archaebacterium facultatively oxidizing or reducing sulfur. Syst. Appl. Microbiol. 8:197–203.

Zoephel, A., M. C. Kennedy, H. Beinert, and P. M. H. Kroneck. 1988. Investigations on microbial sulfur respiration. 1. Activation and reduction of elemental sulfur in several strains of eubacteria. Arch. Microbiol. 150:72–77.

Concluding Challenge

In the preceding pages, the multifaceted world of the soil microbes was revealed. Indeed, the individual habitats of these organisms are each nearly unique. Soil microsites are comprised of a diversity of physical components assembled in varying degrees of structure ranging from an apparently structureless array of sand grains to complex, ever-changing combinations of mineral matter into soil aggregates. Furthermore, the microbes encounter an infinite combination of soil organic compounds ranging from simple organic acids to highly complex humic acids. This assemblage of mineral and physical components of the microbial world is further varied by the presence of a variety of inorganic substances, many of which provide the energetic basis for growth and development of a portion of the soil microbial community known as the chemolithotrophs.

The complexity of the soil biological community is only partially revealed by this abbreviated illumination of the complexity of the soil environment. Growth and development of individual microbial cells and the associated alteration of soil components through cellular energy metabolism and the by-products of this activity appear to be nearly chaotic. Microbial growth in test tubes is highly ordered and easily described by a variety of somewhat simple mathematical relationships. But, the world of the test tube is rather idealized. In soil, a highly productive cell may be allowed only one to a few cellular divisions before stress of space, nutrients, or biological interactions reduce the proliferation rate. In reality, it is difficult to envision a situation in soil where the maximal division rate observed in culture would ever occur. Furthermore, quiescent organisms may suddenly be spurred into a few rounds of seemingly unimpeded division by the influx of an energy source or temporary relief from growth-inhibiting conditions, but this time of luxury can just as quickly be disrupted by exhaustion of the newly found growth sources or imposition of new impediments to proliferation.

The preceding chapters provided an overall description of the basic principles of the discipline of soil microbiology, which provides some order for describing our highly disordered world and for predicting the behavior of the soil microbial community as the properties of their habitats are altered. With a continually growing appreciation of the complexity of the soil microbial community and the interactions between its members, it becomes readily apparent that the information revealed in these pages serves only as guideposts along the path to a more complete understanding of the home of the soil microbes and the genetic potential contained in that population. This pathway could be likened to a divided highway leading to the ultimate in understanding and wisdom in this discipline in that the seemingly contradictory endeavors of delving into the minutia of the system must be combined with just as diligent a drive to conceptualize the role of the soil microbes in the total ecosystem. This latter system must extend far beyond the simple soil of a meadow or mineland reclamation site to the totality of terrestrial systems and perhaps beyond.

This treatise was introduced with the objective of presenting the basic principles of soil microbiology as well as examples from the primary literature accumulated over the short history of this discipline. Any soil microbiology textbook, no matter how complete, can provide only a snapshot of the current status of our studies. A viable scientific endeavor must be continually growing. Great gains in describing the soil environment have been made, but the activities of society have created environmental challenges that may baffle even the most learned soil microbiologist. The challenges of reclaiming severely damaged soil sites, the goals of properly managing our terrestrial home, and the necessity of manipulating the soil system to provide the needs to an ever-growing, ever-changing society demand that we refine our knowledge of the capacity of the soil microbes to adapt to their continually evolving physical home and our capability to manage this minute soil community to meet our needs for survival. Thus, those who have mastered the introduction to soil microbiology provided within these pages are now challenged to use the information as a form of intellectual energy source to advance our knowledge base into the realms necessary to meet the demands of the twenty-first century. As we entered the twentieth century, soil microbiology, if it could really be said to exist then, would have been viewed as a diversion into explaining the intriguing "wee beasties" of the soil upon which we walk. In the second century of its existence, soil microbiology must play an integral role in the development of ecosystem-sustaining stewardship practices.

Index